Challenges in Physics Education

Series Editors

Marisa Michelini, Dipartimento di Fisica, University of Udine, Udine, Italy

Eilish McLoughlin, School of Physical Sciences, Dublin City University, Dublin, Ireland

This book series covers various aspects of physics teaching and learning across all educational levels and learning environments. The individual volumes cover a wide array of topics, including innovative approaches and pedagogical strategies for physics education, the development of effective methods to integrate multimedia into physics teaching and learning, pioneering lab experiments, and the utilization of web-based interactive activities. Selected research-based contributions will be prominently featured throughout.

The GIREP, the International Research Group on Physics Teaching, collaborates on this publication series, incorporating selected papers from globally recognized experts and monographs. Book proposals from other sources are also encouraged.

"Challenges in Physics Education" is targeted towards professionals, teachers, researchers, instructors, and curriculum developers, aiming to enhance physics teaching and learning and, consequently, the overall perception of physics within society.

Proposals for publication should be submitted to Marina Forlizzi (marina.forlizzi@springer.com).

Dagmara Sokołowska · Eilish McLoughlin
Editors

Embracing Changes Together in Physics Education

Springer

Editors
Dagmara Sokołowska
Marian Smoluchowski Institute of Physics
Jagiellonian University
Kraków, Poland

Eilish McLoughlin
School of Physical Sciences
Centre for Advancement of STEM
Teaching and Learning
Dublin City University
Dublin, Ireland

ISSN 2662-8422 ISSN 2662-8430 (electronic)
Challenges in Physics Education
ISBN 978-3-032-27384-0 ISBN 978-3-032-27385-7 (eBook)
https://doi.org/10.1007/978-3-032-27385-7

This Springer imprint is published by the registered company Springer Nature Switzerland AG
The registered company address is: Gewerbestrasse 11, 6330 Cham, Switzerland

Preface

The rapidly evolving landscape of the twenty-first century—shaped by economic transformation, technological development, and profound societal change—calls for new perspectives on physics education and physics teacher preparation. Educational systems that once advanced through gradual changes are now challenged to move more boldly, rethinking curricula, pedagogies in both classroom and online environments, and approaches to teacher professional learning. Global developments—from climate change to migration, pandemics, rapid digitalization, and the growing influence of artificial intelligence—continue to influence and shape the context in which education policies and systems are implemented.

In these rapidly changing times, physics educators at all levels of education, researchers in physics education, and the policy makers share a common responsibility, namely, to reveal the beauty and relevance of physics, to nurture the children's natural curiosity during their educational journeys, and to prepare young people for futures that are still unfolding. It is critical that connections among the diverse communities involved in physics education are strengthened to facilitate meaningful dialogue on the future of physics teaching and learning in schools and universities. *Embracing Changes Together in Physics Education* requires sustained collaboration, reflection, and innovation in our field.

The 4th World Conference in Physics Education (WCPE) was held in the historic city of Kraów, Poland, from 26 to 30 August 2024. The conference theme *Embracing Changes Together* was explored in a blended format, enabling participation both on-site and online. The aim of the conference was to bring together participants involved in physics education from across the globe and from different perspectives and levels. The objective was to engage delegates in meaningful discussions on various aspects of education in a world of rapid changes and new challenges, to increase mutual understanding and deepen learning and engagement of young people and society in physics.

This prestigious international conference was organized by the Faculty of Physics, Astronomy and Applied Computer Science, Jagiellonian University, Kraków, Poland, in cooperation with the International Research Group on Physics Teaching (GIREP vzw), IUPAP International Commission in Physics Education (C14), and the Faculty of Physics and Astronomy, University of Wrocław, Poland. The conference attracted over 800 contributors from 46 countries worldwide, among which 389 individuals participated in live (344) or hybrid sessions (45). Most delegates were from higher education (239); however, the conference also engaged a significant number of teachers (54), and undergraduate and postgraduate students (89). The scientific programme offered a diverse range of 288 contributions reflecting the rich and multifaceted nature of physics education research and practice. Of these, 36 contributions were presented in hybrid format and 252 were presented in person in Kraków.

The volume is organized in five parts, each discussing key aspects of physics education:

1. *Enhancing Physics Education for Young Learners* discusses innovative pedagogical approaches focused on developing our youngest learners' interest and engagement in physics education. The first chapter highlights the importance of cultivating a "feeling for physics" and embracing epistemic empathy and responsive reaching approaches. The second chapter promotes the role of dynamic modelling activities to empower student learning in physics education. The last three chapters provide examples of pedagogical approaches adopted in physics classrooms, i.e., blended learning, flipped classroom, and use of Arduino and Scratch programming, and discuss the influence on the development of student understanding of physics concepts, i.e., thermoelectric phenomena, optics, and electric circuits.
2. *Building Physics Competences at the University Level* examines pedagogical approaches and considerations for teaching and learning physics at the undergraduate level. The first chapter highlights that by addressing cognitive, affective, and metacognitive as dimensions of learning, physics educators can create enriching experiences that empower students to become lifelong learners and proficient problem solvers in physics. The second chapter investigates the prior knowledge of incoming first-semester physics students and compares differences between incoming students in 2013 and 2023. The third chapter reports on the design and influence of a student-centered collaborative learning approach in a first-year undergraduate quantum physics class. The fourth chapter examines students' difficulties with using graphical and algebraic representations in the context of non-constant acceleration kinematics and indicates the need to develop teaching materials that promote deeper understanding of kinematic concepts.
3. *Advancing Physics Education for a Modern Society* considers opportunities and challenges facing physics teachers and learners in today's world. The first chapter presents the use of thematic analysis to inform the design and evaluation of teaching and learning sequences. The second chapter reflects on the current state of research activities and teaching experiences on open questions of learning goals, assessment, and the interplay between the two in physics lab courses at all levels of education. The third chapter promotes a historical and pedagogical reconstruction of old quantum theory and early quantum mechanics approach to enhancing a deeper understanding of quantum physics. Over the past few decades, there has been a growing interest in understanding the evolution of research and

practices in physics education. In chapter four, the authors use natural language processing to conduct a review of 55 years of literature on physics education.
4. *Addressing Climate Change Through Physics Education* considers the urgent need to provide students with relevant learning experiences that foster scientific understanding and informed action towards addressing global grand challenges. The first chapter examines how physics-based climate change education can address such challenges by grounding instruction in core physical principles. The second chapter presents four examples focused on addressing dimensions of climate literacy in physics education. The third chapter promotes the development of student agency through empowering students with the knowledge, ethical grounding, and practical skills necessary to act as agents of change. The fourth chapter explores the current practices employed by STEM researchers when communicating climate change to the general public, as well as how these practices are formulated after a professional development program focused on strategies for enhancing public engagement with climate action.
5. *Supporting Physics Teacher Professional Learning* reports on innovative approaches to transformative teacher professional learning. The first chapter presents three collaborative visions for physics education, embracing domains from cognitive science to artificial intelligence. The second chapter offers insights into the role and development of teachers' pedagogical content knowledge for teaching physics. The third chapter promotes the importance of interdisciplinary collaboration between physics educators across levels and regions. The fourth chapter presents the development and use of digital teaching scenarios on advanced STEM topics in teacher education. The final chapter examines teachers' views of the role of educational simulations in physics education.

This volume highlights the diversity of physics education research and practice presented at the 4th WCPE conference in 2024. We greatly appreciate the high-quality contributions of researchers, teachers, students, and other contributors to this volume. These works not only reflect the dedication and expertise of the global physics education community but also offer valuable insights and practical tools that can inspire further advancements in this field.

Kraków, Poland — Dagmara Sokołowska
Dublin, Ireland — Eilish McLoughlin

Contents

Supporting Physics Teacher Professional Learning

Enhancing Physics Education for Young Learners

Cultivating a Feeling for Physics: The Importance of Epistemic Empathy and Responsive Teaching

Lama Ziad Jaber

Abstract Inspired by Evelyn Fox Keller's 1984 book "*A Feeling for the Organism: The Life and Work of Barbara McClintock,*" this chapter argues for the importance of cultivating a *feeling* for physics—an embodied, affective, and epistemic connection to the discipline. Drawing on narratives from learners and scientists, I illustrate how emotions are deeply entangled with the epistemic work of physics. *Feeling* an idea, for instance, is often central to constructing it. Scientific sensemaking, then, is not purely cognitive but inherently emotional, infused with curiosity, confusion, joy, and wonder. Building on these insights, I argue that responsive teaching—teaching that attends to and elevates learners' conceptual, emotional, and embodied experiences—is essential for cultivating a feeling for physics. Central to such teaching, I suggest, is teachers' *epistemic empathy* or their capacity to tune into and appreciate the emotional and intellectual experiences of learners within epistemic work. I explore how teacher education can support the development of epistemic empathy, while also acknowledging the tensions and complexities it entails. Ultimately, this chapter calls for a vision of physics education that is not only intellectually rigorous, but also emotionally invigorating, transformative, and deeply humanizing.

Keywords Affect · Emotion · Feeling · Epistemic empathy · Responsive teaching · Scientific sensemaking · Student engagement · Teacher education · Physics education · Transformative and humanizing science learning

1 Introduction

The first part of my chapter's title, *Cultivating a Feeling for Physics,* is inspired by Evelyn Fox Keller's seminal work *A Feeling for the Organism*, in which Keller (1984) chronicles the life and scientific contributions of cytogeneticist Barbara McClintock. In this beautifully written account, Keller traces McClintock's intellectual journey

L. Z. Jaber (✉)
School of Teacher Education, Florida State University, Tallahassee, FL, USA
e-mail: ljaber@fsu.edu

D. Sokołowska and E. McLoughlin (eds.), *Embracing Changes Together in Physics Education*, Challenges in Physics Education,
https://doi.org/10.1007/978-3-032-27385-7_1

while bringing to life the deeply personal and relational dimensions of her scientific inquiry. In doing so, Keller shows the ways in which scientists come to know through intimate, sustained engagement with their phenomena of interest, what she compellingly describes as "a feeling for the organism."

This notion of developing a "feeling" for the discipline—an embodied, affective, and epistemic connection to the work of science—has profoundly shaped my work as a researcher and educator. In this chapter, I discuss how my work takes up and extends Keller's provocation by asking: *What might it mean for learners, not just seasoned scientists, to develop a "feeling for physics"? And what does this ask of us as science educators and teacher educators?*

I begin by situating these questions within broader historical and contemporary narratives that have shaped the field of physics education, and education more broadly, highlighting how dominant paradigms have often sidelined or devalued feelings and emotions, treating them as secondary or even a hindrance to cognition. I then share the personal and scholarly commitments that anchor my work, including the experiences that have shaped my orientation toward science education as an affective, relational, and humanizing endeavor. From there, I explore what it might look like, both for scientists and for students, to cultivate a meaningful affective connection to physics. Drawing on examples from scientific practice and from science classrooms, I highlight how affect, including curiosity, uncertainty, wonder, and joy, is central to scientific sensemaking, what my colleagues and I have referred to as *epistemic* affect. Such affect reflects the emotional dynamics that emerge in the pursuit of scientific explorations (Davidson et al., 2020; Jaber & Hammer, 2016a), urging us to design learning environments that recognize, value, and nurture learners' emotional experiences within epistemic activity.

I then argue that designing such learning environments requires a shift in pedagogical practice and, as such, a reimagining of teacher learning. I introduce the notion of *epistemic empathy*—teachers' capacity to tune into and genuinely appreciate students' emotional and intellectual experiences as they engage in sensemaking about scientific phenomena and the world around them (Jaber, 2021). Epistemic empathy, I suggest, is central to teachers' learning to enact responsive science teaching, and it must therefore be intentionally cultivated within teacher education. In the final sections of the chapter, I discuss both the transformative possibilities and the complex tensions involved in this work. Ultimately, my hope is that this chapter invites readers to imagine and work toward a vision of physics education that is not only intellectually rigorous, but also emotionally invigorating, transformative, and deeply humanizing.

2 Situating the Discussion: Historical and Present Considerations

Historically, physics education has been shaped by epistemological understandings and norms that privilege objectivity, abstraction, and a detachment from lived experience (Barad, 2007; Haraway, 1988; Harding 1991; Prescod-Weinstein, 2020). This dominant orientation to physics—and more broadly science—education has contributed to the reproduction and continued marginalization of certain ways of knowing, being, and engaging the world that do not align with such norms (Archer et al., 2012; Gonsalves, 2018; Kayumova & Dou, 2022; Medin & Bang, 2014; Warren et al., 2020). Research has documented that when learners, particularly those from historically marginalized communities, do not see their perspectives, experiences, and epistemologies reflected and valued in the science classroom, they may come to view science as alienating, inaccessible, and even worse, dehumanizing (Hazari & Rodriguez, 2024; Lee et al., 2020; McGee, 2020; Ong et al., 2011; Rosa & Mensah, 2016; Wilson et al., 2015).

At the same time, research continues to document that learning environments that make space for and honor students' lived experiences and their emotional, familial, linguistic, and other backgrounds can offer a powerful counter-narrative, one that affirms students' identities and expands their participation in meaningful and transformative ways (Bang et al. 2013; Calabrese Barton & Tan, 2010; Chappell & Varelas, 2020; Kang, 2022; Luehmann et al., 2024; Morales-Doyle 2017; National Academies of Sciences, Engineering, and Medicine, 2020). Designing such inclusive and transformative learning spaces has never been more urgent than in present times, where collective efforts and diverse perspectives are becoming critical for our ability to tackle the various crises threatening our existence. We are indeed living through overlapping global crises: climate injustice, threats to public health, misinformation and distrust in science, and rising authoritarianism and ideological polarization, all of which threaten scientific inquiry, and more broadly, our ways of being with one another and with the world.

These realities demand that we reimagine how we approach instruction, to move beyond narrow, instrumental framings toward more expansive and imaginative views of science and science learning as a form of *world crafting*. As science educators and researchers, we must ask: What kind of world do we hope to shape through science? What forms of knowing and relating do we wish to cultivate and to what ends? How might science education become a means for healing, justice, and collective thriving? These questions invite us to reconceptualize science/physics teaching and learning as fundamentally relational and transformative endeavors grounded in deep care for one another and for the planet. They urge us to reimagine science classrooms not just as sites for learners to understand the world *as it is* but rather to envision *what it could become*. It is this vision that motivates my work as I ask: *How can science classrooms be transformative and empowering spaces where all students can see themselves as agentive and capable contributors to the work of science, and ultimately, to more hopeful, joyful, and just futures?*

2.1 Words of Gratitude and a Tribute

Before turning to the ideas I share in this chapter, I want to begin with some words of gratitude. The work presented here reflects a collective labor of love and joy that stems from many relationships, friendships, conversations, and communities that have shaped my intellectual journey and who I am as a scholar. I am deeply thankful for the mentors, teachers, students, colleagues, elders, family members, communities, and friends, who have walked alongside me on this journey, each offering their wisdom, generosity, and care. This chapter, and the broader arc of my work, is a reflection of the support and love of this village. While it is impossible to name all the people who have shaped my journey, throughout the chapter, I mention some names that were particularly influential for shaping the ideas that I discuss here.

But first, I would like to pay tribute to a remarkable human being and scholar whose legacy has left an indelible mark on the field of physics education: Dr. Edward F. (Joe) Redish, who passed away on Saturday, August 24, 2024, just days before I presented an earlier version of this paper at the fourth World Conference on Physics Education. Although our paths crossed only briefly during the early years of my doctoral journey at the University of Maryland, I consider myself deeply fortunate to have known him.

Joe's presence was magnetic. His intellectual curiosity, his boundless energy, and his sheer joy for ideas were infectious. But it was his warmth, humility, and unwavering respect for learners that left the deepest impression on me. Joe modeled what it means to be both a rigorous thinker and a deeply humane educator. He taught us that to teach physics well, we must center our attention on the learners themselves:

"If we are to make serious progress in reaching a larger fraction of our students, we will have to shift our emphasis from the physics content we enjoy and love so well to the students themselves and their learning. We must ask not only what do we want them to learn, but what do they know when they come in and how do they interact with and respond to the learning environment and content we provide" (Redish 1994, p. 803).

Joe persistently reminded us that learning begins not with explanation, but with attention, *with a willingness to see and hear* students as sensemakers in their own right: "Listen to your students. Understand how they are interpreting and understanding (or misunderstanding) what you are teaching. Respect their views and what they bring to class and respond by adjusting your instruction to match" (Redish, 2015). That idea of listening deeply and respecting learners' intellectual and emotional experiences has been, and continues to be, foundational to my work as a researcher and educator. At its core, my vision for teaching is about *listening*. Listening with care, with humility, and with respect, something that was further honed through my own learning journey and my research work with many wonderful colleagues and mentors, and especially Dr. David Hammer, as I discuss next.

2.2 Grounding my Commitments

As I reflect on my journey in science education, I realize that my orientations and commitments are deeply rooted in my experiences in the classroom. My entry into educational research did not begin from an abstract or theoretical place, but from within the walls of my own classroom, where I taught science in a range of contexts in Lebanon, including with children and families historically underserved by the school system.

As a middle school science teacher, I had the opportunity to collaborate with Dr. Hagop Yacoubian, a researcher studying student sensemaking about the nature of science and scientific inquiry. Through this collaboration, I was invited to engage in practices that would ultimately transform my understanding of teaching. I had the opportunity to think more deeply about my students' thinking, gaining insight into their rich intellectual and emotional lives that they brought to the learning environment. This experience opened a window into the world of educational research and ignited a curiosity that continues to guide my work.

These early encounters with research prompted a shift in my professional path. I moved to the USA to pursue my doctoral studies in science education, where I was fortunate to study with my mentor and advisor, Dr. David Hammer. David's generous support and thoughtful guidance have been foundational to my development as both a scholar and an educator. Since those formative years, I have been drawn to the textures of classroom life, particularly the generative moments when learners engage with one another and with ideas, when they grapple with uncertainty, ask questions, express wonder, and negotiate meaning. I became fascinated by how emotions in moments of sensemaking, such as curiosity, frustration, confusion, and joy, shape the nature of learners' participation and understanding of the natural world and of themselves. These questions have led me to explore how teachers can design responsive, empathetic, and humanizing learning environments that recognize and nurture learners' affective and intellectual lives, within and beyond the classroom.

I do this work not only as a researcher and educator, but also as a mother of a young child who recently entered the public school system in the USA. This personal perspective brings urgency and clarity to my commitments as I experience the stakes of education both through research and through the everyday realities of parenting, schooling, and society. These layered roles—teacher, researcher, teacher educator, and parent—inform my dedication to cultivating educational spaces where learners' ideas, emotions, and identities are recognized and valued, and where their humanity and dignity are affirmed.

3 A Feeling for Science: Examples from Sandra and Marya

A central thread in my scholarship has been the study and enactment of responsive teaching, an instructional approach that invites educators to attend closely and with care to the substance of students' ideas and to build instruction from students' thinking, feelings, and sensemaking (Hammer et al., 2012; Robertson et al., 2016; Watkins et al., 2020). Thanks to David's support, as a doctoral student, I had the opportunity to join the *Responsive Teaching* project, a multi-year research project supporting elementary and middle school teachers in implementing responsive teaching practices. As a professional learning community, we worked together as researchers and teachers to center students' thinking in ways that make instructional decisions grounded in what students were saying and doing in the moment, with an eye toward broader curricular and disciplinary objectives.

In these classrooms, teaching was not an "anything goes" approach but involved deliberate attention and thoughtful improvisation within a broader pedagogical and disciplinary vision. While the teachers remained attentive to canonical and disciplinary goals, those goals were pursued by intentionally responding to the emergent ideas and curiosities students brought into the classroom.

What I saw in those classrooms was fascinating! These learning environments became spaces where students raised their own questions, listened to one another's ideas with curiosity, actively constructed and debated knowledge claims, and, more importantly, did so by bringing their full selves, including their feelings and emotions, into the inquiry. As I discuss below, this project laid the foundation for my research on epistemic affect and teacher empathy.

3.1 "Science is the Other Person that I'm Hooked Into!"—Illustrations from Sandra

To illustrate what students' experiences looked and felt like in practice in the *Responsive Teaching* project, I introduce you to Sandra, a Hispanic student who participated in the project from Grade 4 through Grade 7. While I had the privilege of speaking with Sandra about her experiences at different times throughout the project, here I share only a glimpse of her reflections from an interview when she was in sixth grade (for more, see Jaber & Hammer 2016a).

As we sat down to talk, knowing our conversation would center on science, Sandra offered an unprompted and striking description of her relationship with science: "Me–Science [gesturing the connection], like when you like somebody or something, it's like you and somebody else? I'm that somebody, and science is the other person that I'm hooked into!" Sandra spoke of science as a relationship, one marked by connection, affinity, and joy. When I asked her what made science feel so special to her, she explained: "It's just like a way that science is to me. See, when I have science, it really, it's like something that makes me happy when we do science. It's something that makes me really happy."

During our interview, Sandra and I watched video clips from her science classrooms, recorded as part of the *Responsive Teaching* project. One clip captured a lively discussion in which students tried to make sense of what happened to a puddle of water that had evaporated. As Sandra reflected on that moment, she described how experiences in her science classes over the years had fostered a deep sense of joy and connection to the work of science:

"WOW! … There were so many different responses flying back and forth, back and forth [...] And just by that one question, all those opinions came flying out. That's why I like science, because one question can bring up EVERYTHING that you think about. That's what I think, it can bring up EVERYTHING […] When we talked about the puddle, most kids, they start with the same thing: 'evaporation, evaporation.' But then everybody's mind exPANded into different things of what happened to the water. […] When you have a group discussion, it's what you think. And when somebody says something, you can say: 'Oh yeah, I wanna add on that.' That's what I like about science. In science, you can do that!"

In these powerful reflections, we hear Sandra delighting in the collective process of inquiry, the exchange of ideas, and the generativity of questions in science. She experiences herself as someone who has ideas in science, someone whose thinking matters. Her reflections illustrate what is possible when students are invited to be active sensemakers and when their emotional and epistemic experiences are taken seriously in the science class.

Sandra was 11 years old at the time of this interview, an age when many children, especially those from communities historically excluded from science, begin to feel disenchanted with the discipline. Research shows that students like Sandra—girls, students of color, multilingual learners—often come to see science as distant, intimidating, or irrelevant to their lives. And yet, here was Sandra, joyfully expressing a deep connection to science and a sense of belonging within it (Archer et al., 2012; Brickhouse et al., 2000; National Academies of Sciences, Engineering, and Medicine, 2020). Her story is both an affirmation and a call: *Wouldn't it be powerful if all students were given opportunities to experience science in this way?* Next, I share the story of another student, Marya, whose experience with responsive teaching occurred at the undergraduate level.

3.2 *"I've Always Been Intimidated by Physics, But in this Class, I Was More Fascinated than Intimidated"—Illustrations from Marya*

I met Marya when she was taking a physics course designed by Dr. David Hammer, a course that aimed to help students experience physics not as a collection of formulas to memorize, but as a way of making sense of the world. I met Marya thanks to

Jen Radoff, who at the time was a teaching assistant in the course. Jen noticed a significant shift in Marya's engagement and ways of participating throughout the course. At the start of the course, Marya expressed anxiety and intimidation about physics. She stood out to her instructors as someone who felt deeply uncertain about her ability to succeed. But over the course of the semester, there was an evident shift in Marya's engagement and her feelings about physics and about herself as a physics learner. Through a learning environment that welcomed her questions, supported her sensemaking, and honored her experiences, Marya came to engage in physics with increasing confidence and joy. By the end of the term, she had even decided to minor in physics.

Jen invited me to meet with Marya and reflect on her experiences in physics at the end of the course. In our reflective interview, Marya recounted her journey in the course, noting the significant shift she has experienced, as illustrated in this composite excerpt (for more, see Radoff et al., 2019):

"I've always been intimidated by physics… Physics this semester was a big trigger for me with anxiety. As an outsider, it just looks really complex… Do I really have the brains for that? I'm not sure. […].

But I reached a point in this class where I was more *fascinated* than intimidated […] It started being like: 'If I don't know the answer, then oh goody! We have another problem to solve! […] Ooh, we have this big problem, you know, and it's like so complex. And it's scary but it's also exciting because, "Let's see if we could figure this out," you know? And when you do, it's so rewarding in the end because like it's just, I don't know, it's such a high when you figure something out, you're just so excited and just like, I dunno- you see the smile on my face […].

You're so excited, it's like a kid walked into a candy store, and you're like, "You know what? Who cares? The anxiety can just like take a back seat because physics is just too awesome to pass up."

Marya's story is compelling not only for the transformation it reveals, but for what made that transformation possible. In this course, Marya experienced a learning environment that allowed her to bring her full self into the learning process. She was encouraged to express vulnerability, confusion, excitement, and uncertainty. Her instructors, David and Jen, responded with deep care, empathy, and attentiveness to her ideas and emotions. In turn, Marya was able to redefine her relationship with physics, from viewing herself as an "outsider" to the discipline to deciding to pursue a minor in physics.

4 What Can the Field Learn from Sandra and Marya?

As I reflect on Sandra and Marya's stories with science, I consider several insights from their experience to inform the field of science/physics education. These include: (1) the central role of affect in scientific sensemaking; (2) the importance of responsive teaching in cultivating learners' affect and their "feeling" for the discipline; and

(3) the need to design teacher education that fosters epistemic empathy as a core dimension of responsive teaching and teacher learning.

4.1 Epistemic Affect as Central to Scientific Sensemaking

Sandra and Marya's stories vividly illustrate that science is not a purely cognitive endeavor but one deeply infused with feelings that both arise from and shape inquiry. Their experiences offer compelling evidence that emotions are not add-ons or distractions to learning; they are rather inherent to doing science, woven into learners' efforts and expectations to make sense of phenomena. From the joy of formulating ideas to the unease of encountering contradictions, from anticipating the pleasure of new understanding to learning to persist through confusion and uncertainty, these affective dynamics are not peripheral to science; they are *within* it. They are therefore *epistemic* in that they emerge within and help drive intellectual work and are deeply entangled with the conceptual and epistemological dynamics of inquiry. As such, they are a core part of doing and learning science.

Understanding epistemic affect as central to science stands in stark contrast to prevailing stereotypes regarding scientific fields, especially physics, as emotionally detached and disconnected from culture and identity. Yet the experiences of learners like Sandra and Marya, alongside accounts and reflections from scientists, show us that science learning is deeply personal, relational, and affective (see for example El Halwany, 2021; El Halwany & Adams 2025; Jaber et al. 2023; Krishnamoorthy, et al., 2025; Lanouette, 2022, 2024; Pierson, et al., 2023; Pierson et al., 2024; Zembylas, 2004; Zhang & Kuo, 2025). Ethnographies and autobiographical accounts of scientists make clear that science is indeed infused with feelings and emotions that shape and propel the intellectual lives of scientists (e.g., Girod, 2007; Keller, 1984; Wolpert & Richards, 1997). For example, Einstein recalled being overwhelmed with "nervous conflicts" during the development of relativity: "When the Special Theory of Relativity began to germinate in me, I was visited by all sorts of nervous conflicts... I used to go away for weeks in a state of confusion" (as cited in American Institute of Physics, n.d.). Barbara McClintock described her corn plants as "friends" and took great pleasure in knowing them intimately: "I know every plant in the field. I know them intimately…The more I worked with [chromosomes], the bigger and bigger they got … I was part of the system… these were my friends" (Keller, 1984). Robin Wall Kimmerer reminds us that "doing science with awe and humility is a powerful act of reciprocity with the more-than-human world" (Kimmerer, 2013, p. 252). These emotions are not tangential or incidental to scientific inquiry; they are inseparable from the intellectual work of doing science.

Echoing this argument, Megan Bang and colleagues (2017) note:

"The work that scientists do reflects who they are, what they care about, and how they experience the world. It should be no different for students. It is in recognizing who our students are—the ideas, perspectives, and sense-making resources they bring, the issues they care about—that we can expand our efforts to engage them deeply." (p. 55).

If we take seriously that affect is part and parcel of scientific inquiry, we must also take seriously and honor the emotional lives of our students as they learn science to support them to experience science as a joyful, meaningful, and emotionally invigorating process. Doing so requires that we push back against prescriptive teaching that reinscribes narrow, exclusionary definitions of success and strips away pleasure and relevance from learning, as I discuss next.

4.2 Cultivating a Feeling for Science through Responsive Teaching

If learning involves developing a deep and meaningful connection with the discipline, then recognizing and nurturing the beginnings of science in our students' contributions must involve tuning into and making room for their emotions as they grapple with ideas, wrestle with uncertainty, and make meaning of the world (Jaber & Hammer, 2016b). In Sandra and Marya's classrooms, their science teachers invited students to share not only their ideas and knowledge, but also their uncertainties and fascinations, their questions and vulnerabilities, creating spaces where curiosity, confusion, and wonder were valued as central to science learning. Their teachers also worked intentionally to attend and respond to learners' varied conceptual, experiential, cultural, emotional, linguistic, and other funds of knowledge as resources for teaching and learning. Such responsive teaching promoted meaningful and empowering learning opportunities for Sandra, Marya, and their peers, supporting them to recognize that their ways of thinking and feeling matter—a recognition that is particularly consequential for learners whose cultures, communities, and lived experiences have been and continue to be marginalized within educational systems.

Sandra and Marya's stories indeed highlight the power of responsive science classrooms that nurture students' disciplinary drives and creativity, and that embrace rather than suppress their multiple ways of knowing, being, and *feeling* in science. Their stories remind us that students thrive when they are listened to, when their ideas are taken seriously, and when their emotions are not treated as distractions but rather as central to their learning. Importantly, they remind us that science teaching must support learners in cultivating personal and expansive relationships with the discipline. As my colleagues and I have argued:

"[A] physics class has the potential to be transformative, not just by developing powerful understandings of content, but by shaping students' affectively laden experiences of the world in idiosyncratic, personal, and emotionally saturated ways. Setting those as a goal of a course invites us to reimagine the discipline" (Jaber, Atkins, Elby, & Suárez, 2023, p. 14–10).

By intensively listening and responding to learners' ideas and emotions, a physics class can reshape students' identities and relationships with the discipline and more broadly with the world—not just through their developing rich content knowledge but importantly through fostering affectively rich experiences that connect to their lives and sense of self. This is especially important today in light of the ongoing and well-documented challenges facing the field of science, and especially physics education, including student disengagement, persistent underrepresentation, gender disparities, teacher shortages, and enduring cultural narratives that render physics distant and exclusive (Archer et al., 2012; Avraamidou, 2020; Gonsalves, 2018; Ong, 2023).

This vision of responsive teaching cannot be achieved through isolated practice of teaching components, scripted routines, or "best practices." Instead, it requires a reimagining of science teaching as embedded within "discretionary spaces" (Ball, 2018), where teachers constantly make decisions based on their "pedagogical judgment" (Horn, 2019) in highly contextualized settings and improvisational interactional moments. It calls for educators to position students as agentive and brilliant sensemakers, to listen with genuine interest and curiosity to their reasoning and emotions, and to respond in ways that honor and affirm their full humanity (Krist, 2024; Luehmann et al., 2024; Patterson Williams & Sheth, 2025; Robertson & Atkins Elliot 2017; Vea & Jaber, 2025). It also necessitates an understanding of classrooms as embedded within larger systems of power whereby societal narratives and ideologies inevitably infiltrate classroom interactions (Calabrese Barton & Tan, 2020; Madkins & McKinney de Royston, 2019; Morales-Doyle, 2019). These insights have important implications for theorizing and designing for teacher learning, as I discuss in the next section.

4.3 The Role of Epistemic Empathy in Responsive Teaching

As I consider what these insights mean for teacher education, I ask myself: How can teachers be supported to enact responsive teaching practices that honor and elevate learners' varied ways of knowing, being, and feeling, and that cultivate intellectual joy, wellness, and a sense of belonging? How can teacher education programs help physics (and more broadly science) teachers reimagine the discipline to foster personally meaningful and affectively rich experiences for learners in the classroom?

These are the kinds of questions that became central to my recent research on teacher learning as I work to support teacher candidates and novice teachers to enact responsive teaching practices in their science classrooms. In my role as a teacher educator, I consider how teacher education can be designed to foster teacher professional judgment and political clarity in order to support responsive teaching practices that are equitable and humanizing. I argue that a key aspect of this work pertains to cultivating teachers' epistemic empathy—i.e., their ability to tune into and value learners' intellectual and emotional experiences within epistemic activities—particularly in moments when students' and teachers' frames of reference are incongruent (Jaber 2021; Jaber et al., 2018; Jaber, Davidson, & Metcalf, 2024; Jaber, Dini, & Hammer, 2022).

Such empathy allows teachers to decenter their own perspectives to see how, *to the learners*, their ideas, feelings, and meaning-making repertoires make sense. It also highlights the need to move beyond a focus on conceptual ideas to attend to learners as whole beings with emotions, hopes, and dreams. In this way, learners' affect becomes an instructional target that teachers notice, cultivate, and respond to, not as separate from, but as deeply entangled with, cognition (Davidson et al., 2022; Herrick et al., 2025; Jaber & Hammer, 2016a; Jaber et al., 2021; Pierson et al., 2023; Vea, 2020).

As a goal for teacher learning, epistemic empathy expands conversations about teacher attention and teacher noticing to include teachers' affective attunements—their care for and curiosity about students' ways of being and feeling. This perspective on empathy redefines notions of teacher care and relationality beyond merely building relationships and community, though important, toward deeply caring about and connecting with the conceptual and emotional complexity of students' thinking and learning experiences. Unlike generalized notions of empathy, which are often framed outside the realm of cognition or disciplinary knowledge, epistemic empathy is inseparable from learners' sensemaking. It asks teachers to move beyond their familiar ways of thinking and to learn to see how students' ideas and emotions make sense to them. It challenges technocratic approaches that reduce teaching to a set of prescribed practices and instead centers teaching as a complex intellectual, relational, and emotional endeavor.

Epistemic empathy manifests in the ways teachers notice and interpret students' reasoning and affect, how they make sense of rather than judge their contributions, and how they respond with curiosity and care, especially in moments when students' sensemaking diverges from the expected norms. Epistemic empathy allows teachers to see such divergent moments as productive, broadening their conceptions of "what counts" as intellectually generative for sensemaking and "who counts" as capable of meaningful participation in the discipline. It also asks teachers to attend to what it feels like to struggle with a concept, to take intellectual risks publicly, to voice unconventional ideas, or to connect disciplinary concepts to one's lived experience. At the heart of this work is a deep commitment to approaching learners as human beings whose intellectual and emotional lives are inextricably intertwined; a commitment to hone one's capacity to recognize and value the complex ways students come to know, to feel, and to make sense of the world. In this sense, epistemic empathy is not only a way to reimagine physics (and more broadly science) teaching, but also a call to humanize it.

5 Implications for Teacher Education

5.1 *Fostering Epistemic Empathy as an Aspect of Teacher Learning in Teacher Education*

In my current work, along with a team of co-researchers, teacher educators, and co-designers, we are investigating through a design-based research project how epistemic empathy can be cultivated within teacher education, and how it can support more responsive, inclusive, and expansive forms of classroom practice. Together, we work to cultivate nonjudgmental and humanizing spaces of learning where we engage in deep interrogation of taken-for-granted assumptions about teaching and learning. In doing so, we center three interrelated epistemological and axiological commitments that we see as essential for cultivating epistemic empathy as a target of teacher learning (Metcalf & Jaber, in preparation). These commitments include:

- Fostering asset-based orientations that position all learners as brilliant, capable sensemakers and transformative change agents in the classroom and beyond (Mejia et al., 2018).
- Interrogating settled expectations of "success," "intelligence," and "objectivity" in science that continue to marginalize learners who do not align with prevalent normative framings (Bang et al., 2013; Mignolo, 2009) toward valuing disciplinary heterogeneity or different ways of knowing, being, feeling, and participating as fundamental for scientific sensemaking (Kayumova & Dou, 2022).
- Promoting political clarity that centers critical understandings of systemic inequity in order to support teachers in recognizing how dominant conceptions of science are historically rooted in powered structures (Madkins & McKinney de Royston, 2019; Warren et al., 2020; Ladson-Billings, 2007).

These commitments guide teachers to understand the deeper epistemic and affective work students are engaged in, and to recognize how their actions as educators, including what they attend to, how they respond, and what they value, can either reproduce or interrupt patterns of exclusion and harm.

These commitments drive our design of educative experiences to cultivate teachers' capacity to see differently, to listen differently, and in turn, to teach differently. As such, we have worked to iteratively curate and (re) design a series of learning experiences within our teacher education program for STEM pre-service teachers (PSTs) that include:

1. Readings and discussions aimed at troubling and broadening assumptions of what counts as science and who is capable of doing science.
2. Science explorations where PSTs engage as learners in sensemaking about science phenomena to surface, examine, and reflect on their own emotions, expectations, and past/current experiences as learners within the discipline.

3. Video-based analysis of student thinking, centered on rich, unconventional examples of sensemaking that can expand teachers' ideas of what it may look like to think, act, feel, and be in science.
4. A field placement where PSTs observe weekly and ultimately facilitate a *Learning-to-Listen* capstone event: A recorded whole-class discussion around a generative science question designed to elicit and respond to student thinking. PSTs then use the video recording and transcripts to reflect on the intellectual and emotional aspects of students' experiences as well as their own teaching, including the tensions, joys, and dilemmas that come with this work.

Through ongoing reflections, class discussions, and "coffee chats" (or informal conversations with PSTs throughout the course), PSTs come to see and articulate powerful shifts in how they understand their roles as teachers, their students' experiences, and the work of teaching and learning science more broadly. They come to name and interrogate the patterns that limit who and what we value in science education, and they come to understand science teaching as way more than teaching content and practices, but rather as about world-crafting, to create classrooms and futures where all learners can feel rightfully present, experience joy, and thrive.

Across a range of enactments and reflections in this design-based project, we found that these educative experiences hold promise in supporting science teachers to develop more empathic, asset-based orientations to learners (e.g., Jaber, 2021; Jaber et al., 2022; Jaber et al., 2024; Davidson et al., 2024). Importantly, this empathy often translated into classroom practice in teachers' enactments of more responsive pedagogies, where they engaged with students' thinking, emotions, and lived experiences in intellectually meaningful and humanizing ways.

While epistemic empathy can support deeper, more meaningful, and more equitable engagement in science, this work is not without its challenges. The teachers in our project made meaningful shifts in their orientations and practices, but they also grappled with difficult, layered tensions in their efforts to empathize with learners. As I discuss next, these tensions complicate how we understand and approach empathy in teacher education, surfacing the intellectual, emotional, and sociopolitical work it entails.

5.2 *Grappling with Tensions Embedded in the Work of Empathy*

While epistemic empathy can serve as a powerful pedagogical resource, the work of empathy is laden with tension. To illustrate the kinds of tensions teachers face, I present the following challenges and dilemmas that surfaced across various enactments in the project, which raise deeper questions about teacher education and empathy (see Jaber et al., 2024 and Metcalf, 2024 for more details):

1. *Interpreting Diverse Ways of Knowing:* Teachers spoke about the intellectual and emotional labor of making sense of multiple lines of student reasoning, especially

when those ideas were in tension with their own frames or were presented and communicated in unfamiliar or unconventional ways.
2. *Thinking on One's Feet in Moments of Uncertainty:* Empathizing with students in real time challenged teachers to respond flexibly and vulnerably, often without knowing exactly where the lesson or discussion might lead.
3. *Honoring Students' Ideas While Advancing Disciplinary Goals:* Teachers grappled with how to respond to students' ideas and emotions in ways that were respectful and supportive but also rigorous and aligned with their disciplinary goals.
4. *Grappling with and Navigating Various Commitments:* Teachers navigated various commitments that influenced their practice, including epistemological, ethical, and political commitments. These factors shaped how they framed the purpose of teaching, the role of empathy, and what constituted meaningful learning.

One powerful reflection that illustrates some of these tensions came from Leslie, a White biology preservice teacher, who was grappling with her own sense of agency, as both a classroom teacher and a participant in larger systems of schooling. Leslie noted:

"When we have issues with test-score-driven curriculum affecting how students learn—especially students from marginalized communities—how do we fix these issues when we are only a speck in the system? This may be a deficit perspective, but the state of our country today makes this concern ever more apparent. How do we find ways to safely work against the system? Or will it ever be possible to change and improve it altogether?"

Leslie's reflection captures the kind of critical consciousness and emotional wrestling that lies at the heart of this work. She was grappling with her own sense of agency, specifically her political agency, as she questioned oppressive systems and structures that she saw as both restricting her actions and as deeply harmful for students and their communities, especially those from marginalized communities. Yet, she approached this tension committed to dreaming of and asking *what could be possible*, claiming her own role and agency to either work within the system or to push back toward systemic change.

Another powerful quote that reflects different dynamics around the challenge of empathy comes from Carmen, a Black preservice teacher who was grappling with the tension between her commitment to connect with and care for students and the risk of harm embedded in the work of empathy (Metcalf, 2024; Jaber & Metcalf, 2025). In her reflections, Carmen shared:

"I've had to stay away from students I don't identify with, or students that I feel...triggered by based on past experiences with people who look like them [...] I think back to how I was treated by my White male peers while growing up, and the fact that I do not know what their parents say about Black people at home. There are many closeted racists [...] When I was in middle school or elementary school, I was treated a certain way because I'm Black, and I just carry that with me. How do I get past my own experiences? For the sake of teaching this child, who I may not feel comfortable around? Like, how do you navigate that?"

Carmen's quote highlights the disproportionate weight that teachers of color and those from historically marginalized communities often carry as they navigate relationships with students across powered lines of political, racialized, and gendered difference. Her reflection helps us see that empathizing across such powered lines of difference involves significant emotional labor and vulnerability, especially for teachers with marginalized racial, ethnic, and gender identities.

More broadly, these reflections trouble idealized educational discourse that has depoliticized the work of empathy, framing it as unproblematic and universally good. They instead compel us to understand the work of empathy as always situated and configured in powered social, cultural, and institutional spaces that are fraught with tensions. As Mirra (2018) argued,

"The very ideas of empathy and perspective-taking are politicized and inequitable… Those with more power often fail to consider their perspectives as just that—perspectives—because they are so accustomed to seeing them represented as the norm. By contrast, those with less power are constantly forced to navigate others' views to survive, while their own are minimized, othered, or ignored" (Mirra, 2018, p. 6).

It is therefore important for us as educators (and more broadly as humans) to confront and disrupt our assumptions about the neutrality of empathy, especially in socio-politically violent and polarized contexts. Relatedly, as teacher educators, we must wrestle with how to support teachers, especially those with marginalized identities, as political allies and advocates (Jaber & Metcalf, 2025). We must also grapple with how to listen and respond to teachers who may be inadvertently perpetuating deficit narratives of students and communities, by both making space for teacher sensemaking while also pushing back against these harmful and dehumanizing narratives.

These insights raise critical and urgent questions for us as teachers, teacher educators, and researchers:

- How do we cultivate empathy across lines of epistemological and ideological difference?
- How do we grapple with the gendered, racialized, and sociopolitical dimensions of empathy?
- How do we navigate the emotional, ethical, and intellectual demands of this work, especially when epistemological, axiological, and ethical, tensions arise?
- And how do we do all of this in ways that recognize and affirm learners' and teachers' humanity, while supporting rigorous and meaningful disciplinary engagement in science?

These questions matter way beyond our classroom learning spaces. They are deeply consequential for how we live and learn together as human beings, especially in a world increasingly shaped by polarization, dehumanization, and a refusal to listen across lines of difference. In such a context, empathy becomes not just a teaching tool, but an ethical and moral commitment, a commitment to listening with humility, honoring multiple perspectives, and withholding judgment to see the merits in one another across lines of difference.

6 Conclusion: Reimagining Physics Education through Empathy and Affect

This chapter has explored how cultivating a "feeling for physics" invites us to reimagine what it means to learn, teach, and *feel* with science. Drawing inspiration from both scientists and learners, I have argued that scientific inquiry is not solely a rational or cognitive endeavor but one that is deeply emotional. Feelings of curiosity, wonder, joy, frustration, and confusion are not peripheral to the learning process; they are central to how meaning is made, and how knowledge is constructed.

By highlighting the role of epistemic empathy in science teaching, I have made the case that responsive teaching requires more than simply noticing students' ideas; it demands that we care about how students are thinking *and* feeling, and that we create space for their full humanity in the classroom. This means honoring the diverse experiences, identities, and emotions that students bring into physics classrooms, and designing environments where learners' varied affective, relational, and experiential backgrounds and contributions are seen as assets, not as distractions from doing and learning science.

Ultimately, a physics education grounded in empathy and responsiveness is not only intellectually rigorous but also emotionally invigorating and socially transformative. It supports learners in developing deep and personal connections to the discipline, promoting their sense of belonging and agency. This vision is rooted in care, justice, and collective responsibility, reminding us that learning is not an isolated act but one deeply intertwined with people, communities, and histories. By emphasizing the relational and affective dimensions of science and science learning, this vision compels us to rethink science (and physics) education toward crafting more sustainable, hopeful, and joyful ways of being with one another and with the world.

References

American Institute of Physics. (n.d.). The great works I Einstein

Archer L, DeWitt J, Osborne J, Dillon J, Willis B, Wong B (2012) Science aspirations, careers and gender: the problems of fixing the leaky pipeline. Palgrave Macmillan

Avraamidou L (2020) "I am a young immigrant woman doing physics and on top of that I am Muslim": identities, intersections, and negotiations. J Res Sci Teach 57(3):311–341

Ball DL (2018, April) Just dreams and imperatives: the power of teaching in the struggle for public education. In: Presidential address at the 2018 American Educational Research Association annual meeting, New York NY

Bang M, Warren B, Rosebery A S, & Medin D (2013). Desettling expectations in science education. Hum Dev, 55 (5:6), 302–318

Barad K (2007) Meeting the universe halfway: quantum physics and the entanglement of matter and meaning. Duke University Press
Brickhouse NW, Lowery P, Schultz K (2000) What kind of a girl does science? The construction of school science identities. J Res Sci Teach 37(5):441–458
Calabrese Barton A, Tan E (2010) We be burnin'! Agency, identity, and science learning. J Learn Sci 19(2):187–229
Calabrese Barton A, Tan E (2020) Beyond equity as inclusion: a framework of "rightful presence" for guiding justice-oriented studies in teaching and learning. Educ Res 49(6):433–440
Chappell MJ, Varelas M (2020) Ethnodance and identity: black students representing science identities in the making. Sci Educ 104(2):193–221
Davidson S, Jaber LZ, Metcalf A (2024) Learning to listen: cultivating pre-service teachers' attunement to student thinking. J Sci Teach Educ 35(5):504–523. https://doi.org/10.1007/s11191:021-00276:1
Davidson S, Jaber LZ, Southerland S (2020) Emotions in the doing of science: exploring epistemic affect in elementary teachers' science research experiences. Sci Educ 104(6):1008–1040
Davidson S, Jaber LZ, Southerland S (2022) Cultivating science teachers' understandings of science as a discipline. Sci Educ 31:657–683. https://doi.org/10.1007/s11191:021-00276:1
El Halwany S, Adams JD (2025) Affective politics of belonging to stem: some conceptual and methodological considerations. Science Education
El Halwany S, Bencze L, Hassan N, Schaffer K, Milanovic M, Zouda M (2021) Becoming alive within science education (research): thinking with life history (ies), bodies and stickiness. Cult Stud Sci Educ 16:85–107
Girod M (2007) A conceptual overview of the role of beauty and aesthetics in science and science education. Stud Sci Educ 43(1):38–61. https://doi.org/10.1080/03057260708560226
Gonsalves AJ (2018) Exploring how gender figures the identity trajectories of two doctoral students in observational astrophysics. Phys Rev Phys Educ Res 14(1):010146
Haraway D (1988) Situated knowledges: the science question in feminism and the privilege of partial perspective. Fem Stud 14(3):575–599. https://doi.org/10.2307/3178066
Harding S (1991) Whose science? Whose knowledge? Thinking from Women's lives. Cornell University Press, Ithaca NY
Hazari Z, Rodriguez I (2024) Examining the relationship between disciplinary counterspaces for undergraduate women and sense of belonging: a physics exemplar. J Res Sci Teach 61(10):2336–2361
Herrick IR, Lawson M, Matewos A (2025) "How do these data make you feel?" the emergence of emotional pathways in community science data talks about climate justice issues. Science Education
Horn I S (2019). Supporting the development of pedagogical judgment: connecting instruction to contexts through classroom video with experienced mathematics teachers. In International handbook of mathematics teacher education: Vol 3 pp 321–342. Brill
Jaber LZ (2021) "he got a glimpse of the joys of understanding"—the role of epistemic empathy in teacher learning. J Learn Sci 30(3):433–465
Jaber LZ, Hammer D (2016a) Engaging in science: a feeling for the discipline. J Learn Sci 25(2):156–202
Jaber LZ, Hammer D (2016b) Learning to feel like a scientist. Sci Educ 100(2):189–220
Jaber LZ, Atkins LE, Elby A, Suárez E (2023) Affect in physics learning: entanglement with cognition and learning goals. In: Taşar MF, Heron P (eds) Handbook of research on physics education volume 1: learning physics (pp. 14:1–14:26). AIP Publishing Melville New York
Jaber LZ, Davidson S, Metcalf A (2024) "I loved seeing how their brains worked!"—examining the role of epistemic empathy in responsive teaching. J Teach Educ 75(2):141–154. https://doi.org/10.1007/s11191:021-00276:1
Jaber LZ, Dini V, Hammer D (2022) "well that's how the kids feel!" epistemic empathy as a driver of responsive teaching. J Res Sci Teach 59(2):223–251

Jaber LZ, Hufnagel E, Radoff J (2021) "this is really frying my brain!": how affect supports inquiry in an online learning environment. Res Sci Educ 51:1223–1246
Kang H (2022) Teacher responsiveness that promotes equity in secondary science classrooms. Cogn Instr 40(2):206–232
Kayumova S, Dou R (2022) Equity and justice in science education: toward a pluriverse of multiple identities and onto-epistemologies. Sci Educ 106(5):1097–1117
Keller EF (1984) A feeling for the organism: the life and work of Barbara McClintock. W H Freeman, San Francisco
Kimmerer R W (2013). Braiding sweetgrass: indigenous wisdom, scientific knowledge and the teachings of plants. Milkweed editions
Krishnamoorthy R, Duncan RG, Tan E (2025) "That's just gonna make them upset": youth authoring emerging epistemic ideals through rightful presence. Science Education
Krist C (2024) Striving for relationality: teacher responsiveness to relational cues when eliciting students' science ideas. Cogn Instr 1:36
Ladson-Billings G (2007) I used to love science… and then I went to school: the challenge of school science in urban schools. Using culture as a starting point, xiii-xix, Teaching science to every child
Lanouette K (2022). Emotion, place, and practice: exploring the interplay in children's engagement in ecologists' sampling practices. Sci Educ, 106(3):610–644
Lanouette K (2024) Elevating configurations of data and emotion: dynamics of coproduction, collaboration and competition. Sci Educ
Lee MJ, Collins JD, Harwood SA, Mendenhall R, Huntt MB (2020) "If you aren't white Asian or Indian, you aren't an engineer": racial microaggressions in STEM education. Int J STEM Educ 7(1):48
Luehmann A, Zhang Y, Boyle H, Tulbert E, Merliss G, Sullivan K (2024) Toward a justice-centered ambitious teaching framework: shaping ambitious science teaching to be culturally sustaining and productive in a rural context. J Res Sci Teach 61(2):319–357
Madkins TC, McKinney de Royston M (2019) Illuminating political clarity in culturally relevant science instruction. Sci Educ 103(6):1319–1346
McGee EO (2020) Black, brown, bruised: how racialized STEM education stifles innovation. Harvard Education Press
Medin DL, Bang M (2014) Who's asking? native science, western science, and science education. MIT Press
Mejia JA, Revelo RA, Villanueva I, Mejia J (2018) Critical theoretical frameworks in engineering education: an anti-deficit and liberative approach. Educ Sci 8(4):158
Mignolo W D (2009). Epistemic disobedience, independent thought and decolonial freedom. Theory Cult Soc, 26 (7:8), 159–181
Mirra N (2018) Educating for empathy: literacy learning and civic engagement. Teachers College Press
Morales-Doyle D (2017) Justice-centered science pedagogy: a catalyst for academic achievement and social transformation. Sci Educ 101(6):1034–1060
Morales-Doyle D (2019) There is no equity in a vacuum: on the importance of historical, political, and moral considerations in science education. Cult Stud Sci Educ 14:485–491
National Academies of Sciences Engineering, and Medicine (2020) Promising practices for addressing the underrepresentation of women in science, engineering, and medicine: opening doors. The National Academies Press
Ong M, Wright C, Espinosa L, Orfield G (2011) Inside the double bind: a synthesis of empirical research on undergraduate and graduate women of color in science, technology, engineering, and mathematics. Harv Educ Rev 81(2):172–209
Patterson Williams A, Sheth MJ (2025) Cultivating inner witnesses attuned to the emotional experiences of students of color in science. Sci Educ
Pierson AE, Brady CE, Lee SJ (2023) Emotional configurations in STEM classrooms: braiding feelings, sensemaking, and practices in extended investigations. Sci Educ 107(5):1126–1162

Pierson AE, Brady CE, Lee SJ, Shuler D, Sengupta P, Clark DB (2024) Intrinsic and instrumental care in pen pal letters: recognizing care in STEM classrooms. Sci Educ 108(6):1608–1636
Prescod-Weinstein C (2020) Making black women scientists under white empiricism: the racialization of epistemology in physics. Signs J Women Cult Soc 45(2):421–447
Radoff J, Jaber LZ, Hammer D (2019) "It's scary but it's also exciting": evidence of meta-affective learning in science. Cogn Instr 37(1):73–92
Redish EF (2015). My teaching philosophy
Redish EF (1994) Implications of cognitive studies for teaching physics. Am J Phys 62:796–803
Robertson AD, Elliott LJA (2017) "All students are brilliant": a confession of injustice and a call to action. Phys Teach 55(9):519–523
Robertson AD, Scherr RE, Hammer D (2016) Responsive teaching in science and mathematics. Routledge
Rosa K, Mensah FM (2016) Educational pathways of black women physicists: stories of experiencing and overcoming obstacles in life. Phys Rev Phys Educ Res 12(2):020113
Vea T (2020) The learning of emotion in/as sociocultural practice: the case of animal rights activism. J Learn Sci 29(3):311–346
Vea T, Jaber LZ (2025) Centering affect and emotion toward dignity and justice in science education. Science Education, Online First
Warren B, Vossoughi S, Rosebery A S, Bang M, Taylor E V (2020). Multiple ways of knowing*: re-imagining disciplinary learning. In Handbook of the cultural foundations of learning pp 277–294. Routledge
Watkins J, Jaber LZ, Dini V (2020) Facilitating scientific engagement online: responsive teaching in a science professional development program. J Sci Teach Educ 31(5):515–536
Wilson D, Jones D, Bocell F, Crawford J, Kim MJ, Veilleux N, Floyd-Smith T, Bates R, Plett M (2015) Belonging and academic engagement among undergraduate STEM students: a multi-institutional study. Res High Educ 56(7):750–776
Wolpert L, Richards A (1997) Passionate minds: the inner world of scientists. Oxford University Press
Zembylas M (2004) Young children's emotional practices while engaged in long-term science investigation. J Res Sci Teach 41(7):693–719
Zhang M, Kuo E (2025) "Everyone's struggling:" coping with institutionalized hierarchies of competence through emotional resonance. Science Education

Empowering Learning Through Dynamic Modelling Activities to Enhance Physics Education

Zuzana Ješková and Katarína Kozelková

Abstract Computer modelling is a powerful tool in physics and science education, enabling students to explore complex phenomena, develop inquiry skills, and enhance conceptual understanding. This paper discusses the role of modelling in science education, emphasizing its integration into the modelling cycle and the impact of digital tools in facilitating this process. The use of graphical modelling software allows students to build and manipulate dynamic models intuitively, bridging the gap between abstract mathematical concepts and real-world applications. The paper reviews international efforts to incorporate computer modelling into curricula and highlights the challenges of implementation, particularly in teacher training and digital competency development. Practical modelling activities are outlined, demonstrating how students can progress from simple to complex models across various physics topics. The findings underscore the importance of expanding modelling-based instruction to foster problem-solving, computational thinking, and a deeper understanding of scientific principles in secondary education.

Keywords Physics education · Computer modelling

1 Introduction

Inquiry-based science education emphasizes active learning, where students engage in the process of scientific inquiry to explore concepts and phenomena. Through inquiry, students answer a research question by applying a scientific method based on experimentation or a modelling approach. Experimentation is a primary means of inquiry, when students pose questions, design, and conduct investigations through experiments, and draw conclusions based on evidence. Modelling, as another essential aspect of the inquiry approach, allows students to represent and manipulate abstract concepts, explore relationships between variables, and make predictions

Z. Ješková (✉) · K. Kozelková
Faculty of Science, Pavol Jozef Šafárik University in Košice, Košice, Slovakia
e-mail: zuzana.jeskova@upjs.sk

D. Sokołowska and E. McLoughlin (eds.), *Embracing Changes Together in Physics Education*, Challenges in Physics Education,
https://doi.org/10.1007/978-3-032-27385-7_2

about the system's behaviour. By integrating experimentation and modelling into learning, students are provided with authentic opportunities to explore the world around us.

2 Modelling in Physics and Science Education

As defined by Hestenes (1987) 'a model is a surrogate object, a conceptual representation of a real thing. The models in physics are usually mathematical models, which is to say that physical properties are represented by quantitative variables in the models'. Blum and Leiss (2005) further expanded on the concept of modelling by describing it as a cyclical process known as the modelling cycle Fig. 1a. Greefrath (2011) later modernized the original seven-step model (1. Understanding, 2. Simplifying, 3. Mathematising, 4. Working mathematically, 5. Interpreting, 6. Validating, 7. Exposing) by integrating the digital tools Fig. 1b. As described by the authors, the modelling process begins with understanding and analyzing a real-world problem or question. In this step, the complexity of the situation is often reduced by focusing on its most critical aspects and translating these into a mathematical form. This simplification enables the construction of a mathematical model that represents the problem in terms of variables and equations. Once the model is formulated, mathematical computations are performed. This phase can be greatly enhanced by using computational tools and software, when mathematical terms are translated into the language of the computer, using a coding or programming environment. After obtaining results, the numerical or graphical outputs must be translated back into mathematical language and then connected to the original situation to interpret these results in the context of the real-world problem. This involves evaluating whether the results make sense and how they correspond to observable phenomena. The model's predictions are then tested against empirical data, which may lead to further refinement of the model or adjustments to its assumptions.

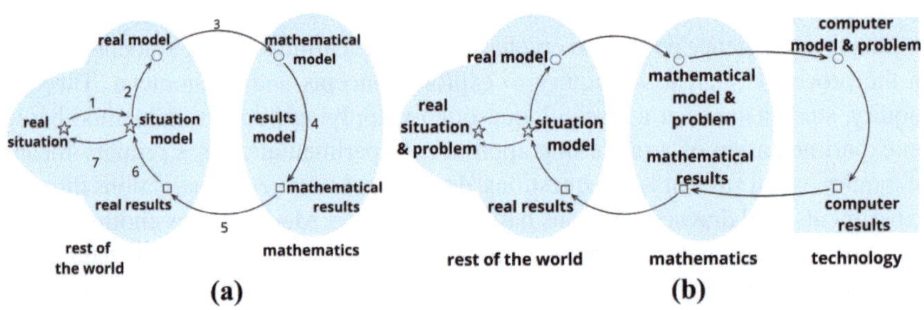

Fig. 1 **a** Modelling cycle by Blum and Leiss (2005), **b** Modelling cycle extended by digital tools by Greefrath (2011)

Table 1 Categories of icons in COACH learning environment

Type	Characteristics and examples	Icon
State variable	Quantities that change in time, e.g. water in a bathtub, distance, speed, heat, number of trees in a forest, a country population, etc.	
Flow variable	Quantities that represent rate of change of state variable, e.g. volume of water flowing to the bathtub in a minute, speed as a rate of change of distance, acceleration as a rate of change of speed, etc.	
Auxiliary variable	Quantity that is calculated based on another quantity using a formula, e.g. acceleration is calculated from force and mass of an object	
Constant	Variables that are defined by constant values, e.g. mass of an object, its crossectional area, air density, etc.	
Connector	The variable at the end of a connector, to which a connector's arrow is attached, depends on the variable at the start of the connector	

3 Integration of Digital Tools into the Modelling Cycle

One of the most significant contributions of digital tools is their ability to handle the computational phase of the modelling cycle. They not only streamline calculations but also allow for handling complex systems that would otherwise be impractical to solve manually. In this phase, mathematical terms are translated into the language of the computer, using coding or programming environments. Computer models or computer modelling (often used interchangeably with computational models and modelling), instead of solving equations analytically, usually use numerical methods to approximate solutions and simulate the evolution of a system over time. This involves discretizing the mathematical equations into a set of computational steps that can be solved iteratively using algorithms. Additionally, the results of modelling—often referred to as dynamic models since they evolve over time—can be further enhanced through visual representations.

Several software tools are available for students, including MATLAB, Python, or GeoGebra. Moreover, computer software, such as COACH7 (Learning by doing the CMA way n.d.) or STELLA (Stella Online n.d.), enables users to develop dynamic models of phenomena not only in a text mode using mathematical equations but also in an iconographical mode when the variables and the relationships between them are represented by graphic symbols (icons) (Table 1).

4 Implementation of Computer Modelling in Education

Computer modelling can be implemented in physics education either through the use of ready-made simulations or by engaging students in developing their own models. While ready-made simulations, such as PhET interactive Simulations (Phet n.d.), are widely used due to their accessibility and ease of integration into the classroom, the practice of having students develop their own models remains less common. Nevertheless, encouraging students to develop their own models can foster a deeper understanding of underlying physical principles and computational thinking.

There have been a number of educational initiatives that highlight the importance of integrating computer modelling into high school science curricula. In the United States, the National Science Foundation (NSF) developed the Modelling Instruction Curriculum, which emphasizes 'developing and using models' as a core practice in science education. This curriculum also identifies 'computational thinking' as one of the fundamental skills to be incorporated into K−12 science instruction (Cavallero et al. 2014). Hurt (2023) developed a framework for computational thinking for science including modelling. Vieyra et al. (2024) created a framework for the integration of computational modelling into physics education at high school level based on combining instructional modelling practices, physical lab materials, and computer programming activities. The authors claim that the successful implementation requires that teachers expand the vision of their discipline beyond the traditional view of science presented in most curricula.

In Europe, the integration of computer modelling into science education varies significantly between countries. In majority of countries, computer modelling is not a mandatory part of a science curriculum, even though they incorporate software tools like Python, Stella, or COACH to model physical phenomena using numerical methods. On the other hand, in the Netherlands, computer modelling is an integral component of science-oriented programs in secondary education. Research indicates that incorporating modelling activities into the curriculum enhances students' conceptual understanding of scientific principles and their ability to engage in scientific inquiry (van Buuren and Heck 2019; Heck and Ellermeijer 2014). In Germany, the integration of computer modelling into science education is not uniformly mandated; however, there have been many different attempts to implement computational modelling and there are notable instances where such approaches have been adopted to enhance the learning experience in physics (Weber and Wilhelm 2020). Currently, there is an ongoing initiative, The Computational and Mathematical Modelling Program (CAMMP), which aims to raise public awareness of the societal importance of mathematics and simulation sciences. CAMMP engages students and teachers in problem-solving activities that involve real-world challenges drawn from everyday life, industry, and research. Participants use mathematical modelling and computational tools to analyze and address these problems (Welcome at CAMMP n.d.).

In Slovakia, similarly to other countries, even though it is not explicitly included in the curriculum, there are attempts to implement computer modelling in regular physics classes. The results of research on the consistent implementation show that students are capable of following the designed learning scenario and can solve designed problems with the help of computer modelling tools (Timková and Ješková 2018). Based on these positive results, a teaching and learning unit on modelling and a corresponding learning scenario was designed and tested at school (Ješková and Kireš 2019). Large national projects with a strong focus on computer science, science, and mathematics education in Slovakia such as IT Academy (IT Akadémia n.d.) running in 2016–2022, followed by the national project DiTEdu (Digital transformation of Education and School) 2023–2028, provide strong support and encompass these efforts (DiTEdu—Digitálna transformácia vzdelávania a školy n.d.).

One of the most critical factors influencing the successful implementation of computer modelling in education is the teacher's own modelling competencies and confidence in this field (Vieyra et al. 2024). Teachers' digital competencies play a crucial role in effectively integrating digital tools into learning environments. The DigCompEdu framework (Digital Competence Framework for Educators), developed by the European Commission's Joint Research Centre (Redecker 2017), provides a detailed description of digital competencies required for educators. It establishes pedagogical guidelines for using ICT in education, including an area specifically focused on developing students' digital skills in alignment with the DigComp 2.2 framework (Digital Competence Framework for Citizens) (Vuorikari et al. 2022). However, while DigCompEdu provides a broad and general foundation, existing competency frameworks often lack subject-specific guidance. To address this gap, the DiKoLAN framework (Digital Competencies for Teaching in Science Education) was developed specifically for pre-service science teachers. It outlines seven key areas of digital competencies relevant to science education. One of these areas is Simulation and Modelling, which defines the competencies and skills necessary for performing computer-aided modelling and utilizing digital simulations effectively, applying simulations in a targeted and audience-specific manner to support knowledge acquisition and communication and understanding the limitations and potential of models and simulations in the learning process (Kotzebue et al. 2021). By providing a subject-specific approach to digital competencies, DiKoLAN helps future science educators develop the necessary skills to integrate modelling and simulations into their teaching effectively.

5 Modelling Activities

Computer modelling can be integrated into existing physics and science curricula at various levels to explore a wide range of topics and phenomena. In physics, a common starting point is modelling simple motion, such as uniform or uniformly accelerated motion. However, computer modelling becomes particularly powerful when using numerical methods to investigate more complex and realistic scenarios, such as those involving non-constant forces acting on a moving object. Beyond motion, computer modelling can be applied to simulate a variety of other phenomena, including oscillations, thermodynamics, electric circuits, radioactive decay and many other phenomena. Furthermore, computer modelling extends well beyond physics, offering rich possibilities in STEM-related subjects, e.g. simulating real-world engineering problems, biological processes such as population dynamics, spread of diseases, climate change, and many others.

We have been developing and integrating computer modelling activities into physics education at the upper secondary school level, in collaboration with physics teachers who test these activities in their classrooms after participating in professional development courses in this field. These activities have been developed through a design-based research approach, where iterative feedback from classroom implementation played a key role in refining both the content and structure of the modelling tasks. Feedback from implementation revealed several important insights. On the positive side, teachers reported that students were highly engaged when working on real-world problems, especially when they could compare their models with experimental or video measurement data. The interactive and visual nature of modelling where variables are represented by icons helped students better understand complex phenomena and fostered meaningful classroom discussions. However, the implementation also revealed challenges. Some students initially found it difficult to design the model structure corresponding to the more complex physical situation, in particular. They also struggled with interpreting models or understanding variable relationships in more abstract contexts, such as heat transfer. These issues led us to revise the activities by clarifying instructions, introducing intermediate steps and additional scaffolding, and refining the use of analogies.

Based on the feedback gathered during the design and classroom implementation of modelling activities, we identified the following key principles that emerged as essential for effective modelling activity design:

- Problem-based Approach: Each modelling activity begins with a problem that students are tasked with solving using a modelling approach. This process is closely integrated with experimentation, and, where possible, the model is compared to experimental data, which may be collected through video measurements or real-time data collection.
- Progression from Simple to Complex: Activities are structured to move from simpler to more complex problems. Students begin with a basic model and progressively elaborate and extend it by introducing new variables and relationships. This gradual development allows students to create increasingly sophisticated models that more closely mirror real-life situations.
- Use of Analogies: Many phenomena, even across different fields of physics, follow similar formal relationships between variables. Therefore, analogies can be effectively employed in the design of models across various scientific disciplines. This approach helps students recognize the connections between different types of problems and systems, fostering a deeper understanding of underlying principles.

In the following sections we present examples of modelling activities and modelling paths designed in iconographical mode within the COACH learning environment (Learning by doing the CMA way n.d.). Moreover, COACH combines software and hardware tools to create a versatile STEM platform that not only enables computer modelling but also supports a wide range of scientific activities, including data logging from real experiments, video analysis, animations, and data processing. This integration allows for the combination of experimental and modelling approaches within a single environment, effectively reinforcing the scientific method in problem-solving. Another advantage of this platform is also its availability for use in many schools, making it accessible for implementation in physics and STEM education.

5.1 *From Bathtub Through Linear Motion to Heat Exchange*

In Table 2 a modelling unit with five subsequent modelling subunits can be seen. The presented learning scenario represents a structured modelling path that includes a sequence of modelling activities designed to align with the three principles outlined above. Each of the five subunits is specified in terms of individual activities, along with examples of corresponding graphical models and experiments or video measurements (Ješková and Kireš 2019). The full set of resources is accessible at https://drive.google.com/drive/folders/155_4gF8QJbTxcrqXHZKkX lAA3qDx8hpC?usp=drive_link.

The modelling path begins with the Introduction to Modelling subunit, where a simple model of filling a bathtub is used to develop an analogous model for a cyclist's motion. This activity models kinematic variables such as distance, speed, and acceleration.

In the second part, titled How Force Affects Motion, the force variable is incorporated into a model of a falling ball. Initially, a constant gravitational force is applied (model of a heavy ball), followed by the addition of a speed-dependent drag force (model of a light ball). This model is then extended to simulate a skydiver's fall.

The next part introduces a drag force with changing air density to model the fall of Felix Baumgartner, who broke the speed of sound during his descent from a height of 39 km. In the subsequent part, a decreasing mass variable is introduced to model rocket motion.

Finally, the unit concludes with an activity on Heat Transfer, which broadens the scope of modelling by applying the principles to contexts beyond motion, such as the heat transfer in a house.

5.2 *How to Keep a House Warm? Heat Exchange Modelling Activity*

While motion modelling is one of the most common contexts for introducing modelling concepts—often due to its straightforward nature and direct visual representation—in Table 2 we have focused on presenting only the final exemplary graphical models for these activities. In contrast, the modelling activity 'How to Keep a House Warm' challenges students to apply their knowledge and skills in a more complex and abstract domain: heat transfer. Modelling heat transfer is more

Table 2 Modelling unit and its subunits with corresponding modelling activities and experiments

Activity	Model I	Model II	Experiment
Intro to modelling – How is the water tank filled? – How does a cyclist move (uniform and uniformly accelerated motion)	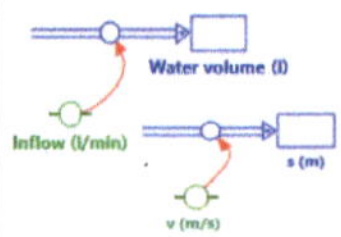	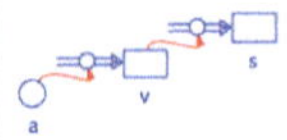	Video measurement of cyclist – Moving uniformly – Speeding up – Slowing down
How force affects motion – Fall of a heavy ball – Fall of a light ball – Fall of a sky diver	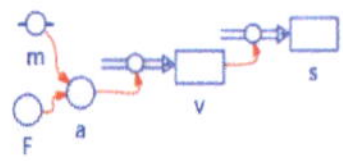	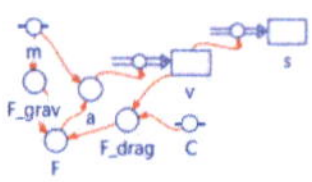	Video measurement of fall of – A heavy ball – A light ball Video recording: – Sky diver with data
How Felix achieved the speed of sound – Free fall – Fall with drag force – Fall with drag force and changing air density	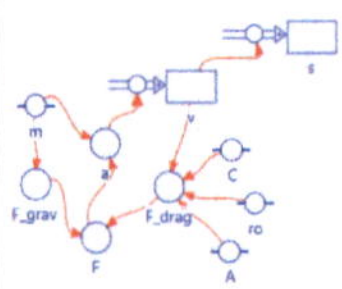	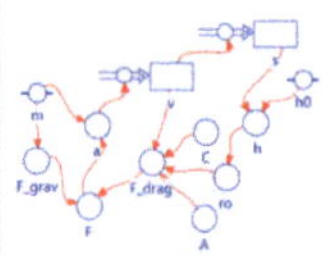	Video recording: – Felix's fall – Data from the event
How rocket starts – Rocket in free space – Rocket in constant gravitational field – Rocket with drag force – Rocket with changing air density	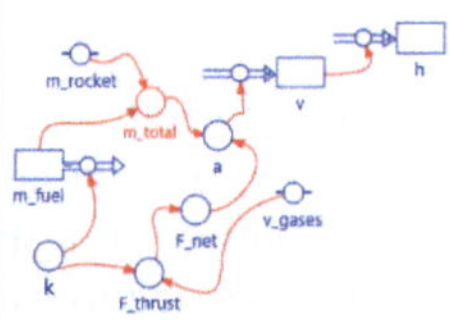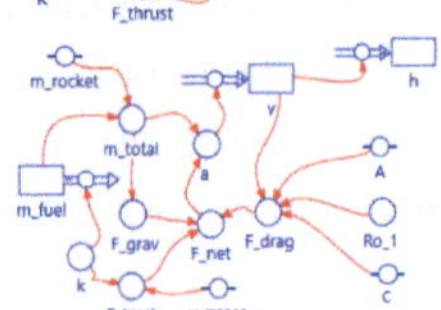		
How to keep a house warm			

demanding than modelling motion because it requires an understanding of multiple concepts of heat and temperature, the principles of heat transfer, energy inputs or outputs, temperature gradient, or insulation properties. This makes it an excellent opportunity for students to deepen their understanding of modelling principles while engaging with a real-world problem.

The activity addressing the main research problem, 'How to Keep a House Warm', begins by exploring the question of energy consumption in households. Students are introduced to the fact that the largest share of household energy consumption is typically attributed to heating. This observation serves as a starting point for exploring the factors influencing energy efficiency and heat retention in homes.

The activities gradually progress through different stages with increasing complexity, introducing additional variables at each step and building on the previous model. They involve the following parts:

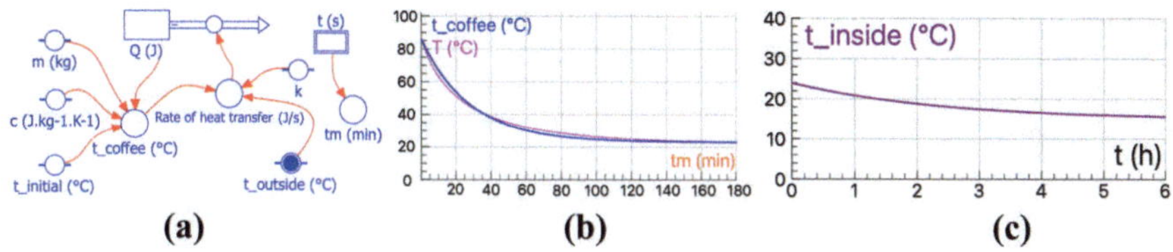

Fig. 2 **a** Model of coffee cooling, **b** Comparison of modelling (blue) and experimental (pink) results, **c** Results of modelling temperature in a house

5.2.1 Modelling of Cooling a Cup of Coffee in Analogy with a House Without Heating

The model of cooling is built on the fundamental principle of heat transfer: heat flows from a warmer object (coffee) to a cooler environment (surroundings) at a rate proportional to the temperature difference between the two. Students first simulate the cooling process of a cup of coffee. The model predicts the rate of heat loss based on the temperature difference, as shown in Fig. 2a. The model is then validated using real-time measurements of the coffee's cooling curve, shown in Fig. 2b. This comparison helps students understand how closely the model approximates real-world phenomena.

After constructing the coffee-cooling model, students extend their understanding by modelling the temperature drop in a house on a cold day after the heating system is turned off. In this analogy, the coffee represents the house, and the surrounding environment mirrors the cold outdoor temperature. The results are presented in Fig. 2c, which illustrates the initial indoor temperature, the outside temperature and the time required for the house to cool down to match the outside temperature. Students then modify the model by adjusting the constant in the rate of heat transfer variable and identify its origin. They recognize the role that insulation plays in slowing heat loss and maintaining a comfortable indoor environment.

5.2.2 Modelling a House Temperature with Added Heating or Cooling System

This activity builds on the principles of heat transfer and introduces students to the dynamics of maintaining a comfortable indoor temperature using heating or cooling systems. Students simulate how the temperature of a house changes when external heating or cooling systems are applied, highlighting the role of thermostats in regulating the temperature. During cold weather, a heating system is used to keep the house at a comfortable temperature. The rate of heating (energy supplied by unit time) determines how fast the heat is added. At the same time, there is a heat loss due to lack of insulation. Then the thermostat is introduced into the model to switch on or off the heating depending on whether the indoor temperature falls below or exceeds a predefined threshold. Similarly, on summer days, air conditioning is used

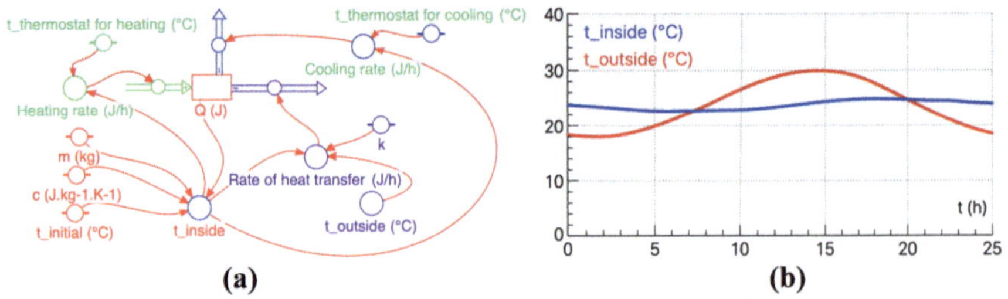

Fig. 3 **a** Model of temperature in a house equipped with heating and cooling system. The outside temperature is set to a constant value, **b** Results of modelling of outside (red line) and inside temperature (blue line) during a summer day

to cool the house. The cooling system removes heat from the house at a specific rate of cooling and the thermostat turns the air conditioning on or off. In Fig. 3a there is a model with both heating and cooling systems and thermostats for heating and cooling with threshold temperatures added.

5.2.3 Simulation of Temperature Fluctuations During a Day

The previous model Fig. 3a can be elaborated with the changing outside temperature simulating summer day time and night time variations from 18 to 30 °C. The heating system (threshold temperature 20 °C) as well as the cooling system (threshold temperature 35 °C) is switched off, and the inside temperature is oscillating a little due to good house insulation Fig. 3b.

5.3 From Mechanical Oscillations to Electric Circuits

5.3.1 Spring-Mass System Oscillations

The motion models described in Sect. 5.1 can be extended to other mechanical systems. A model of oscillatory motion can be constructed in a similar way by applying Newton's second law considering all the forces acting on the system, including the restoring force exerted by the spring, the damping force opposing motion, and the external driving force. Using this approach, models of undamped, damped, or driven oscillations can be designed Fig. 4a. These models can be tested for various parameters and compared with the real-time experimental results obtained from the real-time measurement of a mass-spring oscillator. In our experimental setup, a magnet attached to the CD (to increase damping) is mounted to the spring Fig. 4b. A surrounding coil carrying an alternating current generates a magnetic field that serves as the driving force influencing oscillations. The displacement of the oscillator is recorded using an ultrasonic motion detector, while the voltage of the alternating current generator is measured using a voltage sensor within the COACH system. Initially, the mass-spring oscillator vibrates at its natural frequency in the absence of an external driving force. At a specific moment, indicated in the experiment, the alternating current generator is switched on at a defined frequency. The experimental results for different driving frequencies can be thoroughly analyzed, allowing for the investigation of how the system responds to various external forces. In particular, when the driving frequency approaches the natural frequency of the oscillator, resonance occurs, leading to a significant increase in the amplitude of oscillations.

Figure 4c displays a displacement-time diagram for an oscillator with a natural frequency of 997 mHz. At the moment marked by the green arrow, the alternating current generator was activated at a frequency of 1031 mHz. The experimental results

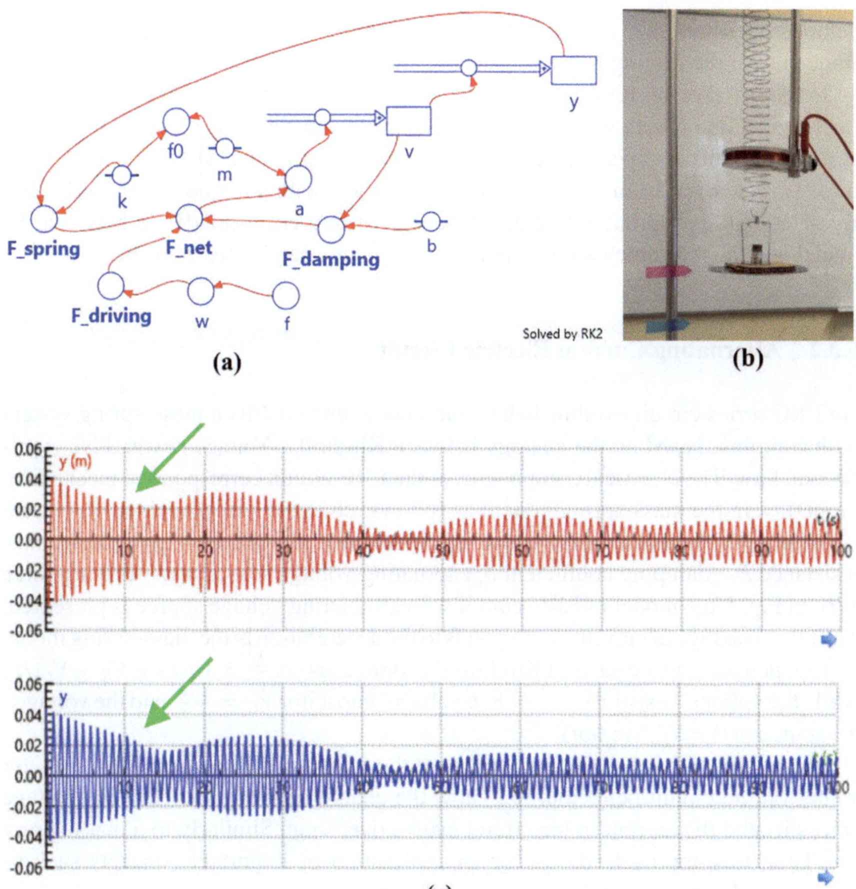

Fig. 4 **a** Model of driven oscillations, **b** Experimental setup, **c** Comparison of model results (blue) with real-time measurement results (red)

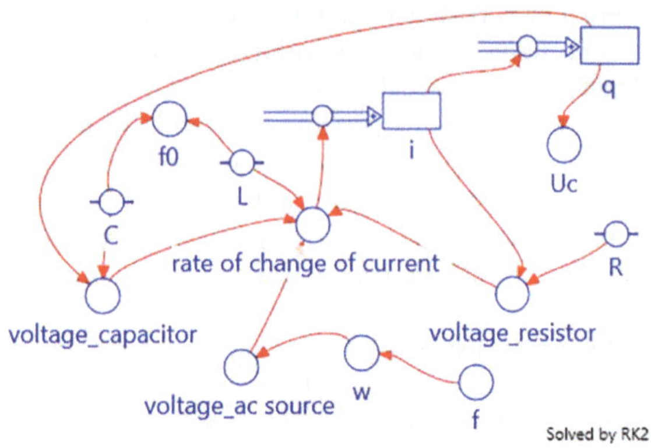

Fig. 5 Model of LRC circuit behaviour

demonstrate that, after a transient period, the mass-spring system oscillates at the frequency of the driving force.

In the corresponding computer model, the mass and spring constant of the experimental oscillator were entered, and the damping constant was adjusted to achieve agreement with the experimental data for damped oscillations. After 10 s, the driving force was changed from zero to an alternating force with a frequency of 1031 mHz. By selecting appropriate parameter values, a strong correspondence between the model and the real-time experimental measurements can be observed Fig. 5c.

5.3.2 Alternating Current Electric Circuits

An LRC series circuit exhibits behaviour analogous to a driven mass-spring system with damping, based on the analogy between Kirchoff's Voltage law and Newton's Second Law. The electrical components of the LRC circuit correspond to mechanical properties in the mass-spring system as follows: charge *q*—displacement *y*, current *I*—velocity *v*, inductance *L*—mass *m*, capacitance *C*—*1/k* (*k*—spring constant), resistance *R*—damping coefficient *b*, alternating voltage source *u(t)*—driving force *F(t)*. In Fig. 5 the model of LRC circuit with alternating voltage source is presented. The rate of change of current corresponds to the acceleration in the mass-spring model and is calculated on the basis of Kirchhoff's voltage law as $\frac{\Delta i}{\Delta t} = (u(t) + V_R + V_c)/L$, while the voltage_resistor $V_R = -R.\, i$, voltage_capacitor $V_C = -\frac{q}{C}$ and the voltage_ ac_source $u(t) = U_0 \sin(\omega t)$.

The model enables students to simulate the effect of different circuit parameter values on the circuit behaviour, e.g. how the charge on a capacitor (or the voltage across it) and the current in the circuit evolve over time. Similarly to a mass-spring oscillator, this approach allows for an exploration of important concepts such as resonance, damping, and also the phase relationships.

Figure 6a shows the voltage across the capacitor and the current in the LRC circuit without an applied external voltage. In Fig. 6b the external voltage was added to the

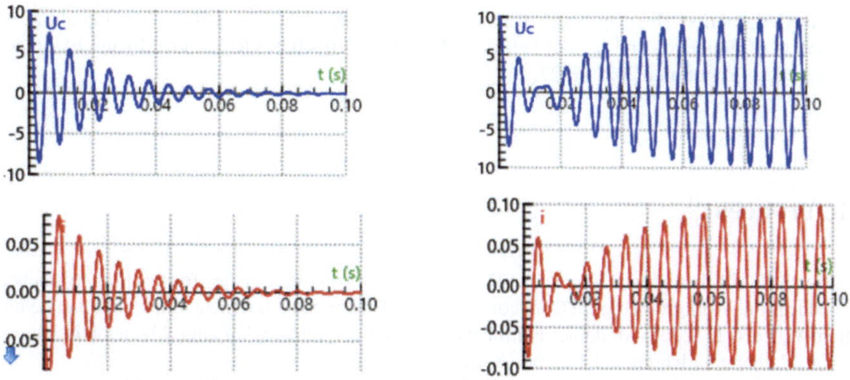

(a) (b)

Fig. 6 **a** Damped oscillations, **b** Resonance in LRC circuit

circuit and set to the natural frequency of the circuit leading to the resonance when the current oscillates with the highest amplitude.

6 Conclusion

The integration of computer modelling into physics and science education enriches students' learning experiences by combining theoretical knowledge with practical applications. While challenges remain in fully embedding these methodologies into curricula worldwide, the positive impact of modelling activities on students' conceptual understanding, inquiry skills, and computational thinking underscores the importance of continuing to develop and promote their use in science education. Future efforts should focus on expanding teacher training, developing comprehensive modelling activities with high-quality teaching materials, and designing curricula that integrates modelling as a component of science instruction. This is particularly important for modern STEM education, where students must be equipped not only with subject-specific knowledge but also with the skills to explore and understand complex, real-world problems—such as climate change, pandemics, energy systems, or sustainable development. Fostering interdisciplinary connections—linking physics, biology, environmental science, chemistry and engineering, and modelling helps students see beyond the boundaries of individual subjects and appreciate the broader context in which scientific and technological challenges exist. In this way, modelling becomes a common language in STEM education, promoting transferable skills such as problem-solving, critical thinking, and the ability to analyze interconnected systems.

Acknowledgments This paper was made in the framework of the National Project Digital Transformation of Education and School DitEdu ITMS2014+: 401402DVR6, funded by the EU under the Slovakia Programme 2021–2027.

References

Becker S. B.-J. (dátum neznámy) (n.d.) DiKoLAN—digital competencies for teaching in science education. Workgroup digital Core competencies. Dostupné na Internete. https://dikolan.de/en/

Blum W, Leiss D (2005) Modellieren im Unterricht mit der Tanken-Aufgabe. Mathematik lehren. In: Mathematik lehren, vol 128, pp 18–21

Cavallero M, Burk J, Aiken J, Thoms B, Douglas S, Scanlon E, Schatz M (2014) Integrating numerical computation into. Phys Teach 52:38–42. https://doi.org/10.1119/1.4849153

Christine R (2017) European framework for the digital competence of educators: DigCompEdu. In: Yves P (ed) Dostupné na Internete. https://publications.jrc.ec.europa.eu/repository/handle/JRC107466

Computational And Mathematical Modeling Program: Welcome at CAMMP. (dátum neznámy) (n.d.) Dostupné na Internete. https://www.cammp.online/english/

DiTEdu—Digitálna transformácia vzdelávania a školy. (dátum neznámy) (n.d.) Dostupné na Internete. www.ncdtv.sk

Greefrath G (2011) Using technologies: new possibilities of teaching and learning modelling—overview. In: Kaiser G, Blum W, Ferri R, Stillman G (eds) Trends in teaching and learning of mathematical modelling, ICTMA 14. Springer, Dordrecht

Heck A, Ellermeijer A (2014) Realizing authentic inquiry activities with ICT. In: Tasar M (ed) Proceedings of the world conference on physics education. Pegem Akameni, Istanbul, pp 775–786

Hestenes D (1987) Toward a modeling theory of physics instruction. Am J Phys 55(5):440–454. https://doi.org/10.1119/1.17449

Hurt TG (2023) The computational thinking for science (CT-S) framework: operationalizing CT-S for K–12 science education researchers and educators. Int J STEM Educ 10(1). https://doi.org/10.1186/s40594-022-00391-7

IT Akadémia vzdelávanie pre 21.storočie. (dátum neznámy) (n.d.) Dostupné na Internete. http://itakademia.sk/

Ješková Z, Kireš M (2019) Modeling activities in the framework of an interdisciplinary subject informatics in science and mathematics. AIP Conf Proc 2152:030011. https://doi.org/10.1063/1.5124755

Kotzebue L, Meier M, Finger A, Kremser E, Huwer J, Thoms L-J, Thyssen C (2021) The framework DiKoLAN (digital competencies for teaching in science education) as basis for the self-assessment tool DiKoLAN-grid. Educ Sci 11:775. https://doi.org/10.3390/educsci11120775

Learning by doing the CMA way: A unique approach to STEM Education, to enhance students' ability to reason and be independent learners. (dátum neznámy) (n.d.) Dostupné na Internete. http://cma-science.nl

PhET: Interactive Simulations for Science and Math. (dátum neznámy) (n.d.) Dostupné na Internete. https://phet.colorado.edu/

Redecker C (2017) European framework for the digital competence of educators: DigCompEdu. Luxembourg: Publications Office of the European Union. https://publications.jrc.ec.europa.eu/repository/handle/JRC107466

Stella Online Powerful modeling and diagramming capabilities on any web browser. (dátum neznámy) (n.d.) Dostupné na Internete. https://www.iseesystems.com/store/products/stella-online.aspx

Timková V, Ješková Z (2018) Computer modeling in physics teaching. In: Sokolowska D, Michelini M (eds) The role of laboratory work in improving physics teaching and learning. Springer, Krakow, pp 121–130

van Buuren O, Heck A (2019) Computer modelling in physics. In: McLoughlin E, van Kampen P (eds) Concepts, strategies and models to enhance physics teaching and learning. Springer, pp 87–100. https://doi.org/10.1007/978-3-030-18137-6

Vieyra R, Megowan-Romanowicz C, Fisler K, Lerner B, Politz J, Krishnamurthi SR (2024) Expanding models for physics teaching: a framework for the integration of computational modeling. Edus Sci 14(8):861. https://doi.org/10.3390/educsci14080861

Vuorikari RK (2022) DigComp 2.2: the digital competence framework for citizens—with new examples of knowledge, skills and attitudes. Publications Office of the European Union, Luxembourg

Weber J, Wilhelm T (2020) The benefit of computational modelling in physics teaching: a historical overview. Eur J Phys 41:043003

Exploring Thermoelectric Phenomena by the Method of Blended Learning

Marián Kireš and Antónia Juhásová

Abstract Blended learning combines online learning in guided home self-study with active student inquiry in the school's learning lab. During the workshop, the teacher will alternately be in the position of a student trying to understand the physical nature of thermoelectric phenomena but also in the position of a teacher discovering the supporting elements of flipped learning and habitat rotation in blended learning. Peltier cell stations will be available to participants through different activities. At the end of the workshop, participants will use personal experience to evaluate the suitability of blended learning for active student physics learning.

Keywords Blended learning · Physics education · Thermoelectric phenomena · Flipped learning · Station rotation · Classroom environment · Learning zones

1 Introduction

Thanks to the Internet, the media, social networks as well as book and magazine sources, the student has enormous access to the latest knowledge. The student creates his own set of concepts with varying depth and correctness of understanding. In information overload, we perceive two essential cognitive problems: superficial processing of information mainly at the level of memorization and insufficient development of inquiry skills.

We understand the essence of education as giving meaning to a child's new experiences, in cooperation with others (Harlen 2015). The sustainability of the quality of education depends on conceptual decisions, the deployment of adequate learning methods, preparedness and professional support of teachers (Baeta and Pedro 2017). Education is a difficult but strategic investment for the future. Already today in school, we need to prepare students for jobs and technologies that do not yet exist (OECD

M. Kireš (✉) · A. Juhásová
Faculty of Science, Pavol Jozef Šafárik University in Košice, Košice, Slovakia
e-mail: marian.kires@upjs.sk

D. Sokołowska and E. McLoughlin (eds.), *Embracing Changes Together in Physics Education*, Challenges in Physics Education,
https://doi.org/10.1007/978-3-032-27385-7_3

2018). From the perspective of current challenges in education, we see the implementation of blended learning and the creation of an engaging learning environment for students as one of the suitable solutions (Kireš et al. 2024).

Among various blended learning models, we have selected flipped classroom (Ramirez and Buitrago 2022) and station rotation (Blended Learning n.d.) with the integration of learning zones (Colton et al. 2020) for the pilot phase of teacher innovation in education. Our goal is to gain initial positive experiences from teachers with an innovative approach to physics education. We place emphasis on the teacher's personal conviction about the benefits of the teaching method and their enthusiasm for innovation. We specifically focus on the development of selected student skills, individualized approaches, and the opportunity to choose from multiple learning paths. We understand learning zones as part of a flexible space in which the student finds conditions for personal self-development. From the teacher's perspective, our goal is to understand the expected change in their approach to supporting and guiding (mentoring) active student learning.

From the perspective of the importance of significance, our starting point is the educational goals (especially the targeted development of selected skills), followed by the application of the teaching method and the adaptation of a flexible learning environment.

2 Theoretical Background

In line with OECD recommendations, building an engaging learning environment is essential for achieving the individual success of each student. Schools that emphasize high expectations, support, and active student involvement contribute to the development of well-rounded citizens with a broad range of skills necessary for life in modern society (OECD 2020). In a successful, student-centered learning environment, students have greater control over their learning, allowing them to set their own educational goals, consider strategies, and make decisions to achieve these goals. This engaging process makes learning more meaningful and fosters higher-order thinking skills (Govindaraj and Silverajah 2017). To ensure that teachers can engage every student in a student-centered learning process, Dzakiria et al. (2013) suggest implementing blended learning, as it increases the interactivity of learning and reduces feelings of isolation among students.

It is necessary to reflect on the prevailing reality of school practice, in which the dominant arrangement of classroom space is determined by the requirement that every pupil in the classroom can see and hear the teacher and has a desk available. Teachers encountered this type of learning environment long ago when they were students. Such a learning space is predisposed to the transmission of teaching methods, the dominant role of the teacher, and a unified approach to the attainment of knowledge. The development of students' skills becomes a secondary outcome, the achievement of which is not specifically monitored.

2.1 Blended Learning Method

The blended learning method, which combines face-to-face instruction with online learning activities, is a promising way to develop students' scientific literacy and skills needed for the twenty-first century. It helps to create a dynamic and stimulating learning environment (Tucker 2022). The Clayton Christensen Research Institute (Christensen Institute n.d.) from the education sector offers a generally accepted definition of blended learning in the context of education. According to this definition, it is a formal educational program that combines online learning with some degree of control over the time, place, way, and pace of learning with face-to-face teacher-supervised learning. The learning methods used in a course or subject are linked together to create a "coherent learning experience" (Staker and Horn 2012). However, to achieve these benefits, it is essential to design high-quality learning activities within this method that incorporate the seven essential elements: Teacher, Pace, Preparation, Active Participation, Personal Interaction, Personalization, and Environment. By integrating these elements, blended learning can help students develop the knowledge and skills they need for future success, including communication, collaboration, critical thinking, and creativity. Therefore, blended learning can be a key factor in achieving this goal (Graham 2019).

There are a variety of blended learning models that differ in their physical and pedagogical focus, so it is important to specify both elements when conducting research. However, the learning method is generally considered more important than the environment in terms of student learning (Graham and Halverson 2023). However, it is important to recognize that each BL model has its own advantages and disadvantages. With careful planning and implementation, BL can be a successful addition to any curriculum. BL models have gained popularity due to their flexibility and ability to address the individual learning needs of learners. However, the success of these models depends on the availability and accessibility of the technology, which can be challenging for teachers and learners.

In general, we define the following basic models of blended learning:

- *Station rotation*: students rotate between stations with different sub-activities according to a fixed schedule set by the teacher. Some of these stations may also include online activities. The students do not have to complete all of the prepared stations.
- *Lab rotation*: Students rotate between stations dealing with complex problem solving according to a set schedule. Online learning takes place in a computer lab of the students' choice.
- *Individual rotation*: Students rotate between personalized activities based on a schedule set by the teacher or adapted by the software. They do not have to go through all the stations, but only those that are relevant to their individual needs.
- *Flipped classroom*: Learning takes place in the home environment, at the time and pace the student prefers, and with the help of online materials. Classroom time is then dedicated to clarification of more complex topics, practical exercises and project work, with the teacher acting as a facilitator.

- *Flex model*: Students work independently on curriculum-based activities, with the opportunity to customize the pace of learning. Teachers provide support as needed, giving students more control over their own learning.
- *Model À la carte*: Students take an online course outside their normal timetable that supplements and deepens their knowledge in a specific field. The choice of courses is often recommended by the teacher based on the learner's needs and interests. Students choose from an educational menu on offer, much like in a restaurant.
- *Enriched virtual*: The majority of the course takes place online, but students have the opportunity to meet regularly with the teacher in face-to-face meetings, consultations, and participation in group activities, which enhances their learning experience (Blended Learning n.d.).

According to the new blended learning classification, the Station Rotation, Lab Rotation and Flipped Classroom approaches are classified as hybrid innovations that strike a balance between traditional teaching methods and online learning. On the other hand, the Flex, À La Carte, Enriched Virtual and Individual Rotation models are considered more disruptive compared to the traditional system, as they deviate from traditional teaching methods (Aksak et al. 2023). Govindaraj and Silverajah (2017) emphasize the extensive use of the rotation model as a successful learning approach. The popularity of the rotational model in blended learning is due to its ability to support classroom learning in a convenient, direct, adaptable and self-paced manner (Divayana 2019). In this model, students follow a predetermined schedule that alternates between online and face-to-face classes.

To gain initial experience, we opted for the flipped classroom method of self-study at home and station rotation in the classroom. The idea of the flipped classroom is to motivate and introduce students to the concept to be taught in class in order to arouse their curiosity and interest in the classroom activity. In self-study at home, students are introduced to a new topic before the actual lesson with the help of learning materials prepared by the teacher, such as videos, articles, presentations, and simulations. This process involves guided self-study where students use a variety of learning resources and work independently to master the easier parts of a problem. In addition, they have the opportunity to work together in small groups of peers online, which encourages peer learning, collaboration and discussion. Learning takes place in a home environment, which allows for time flexibility and individual adjustment of the learning pace. The aim of these activities is to motivate learners and help them to recognize ambiguities or misunderstandings about the topic. Formative assessment (Ganajová and Kireš 2021) using a selected tool at the end of self-study at home gives feedback to both the learner and the teacher. It is an integral part of the learning process and supports personalized learning. The teacher then selects appropriate learning activities for the station rotation model in the school environment to promote further development and deeper understanding of the topic. During the lesson, the teacher focuses on guiding and encouraging students to actively participate in activities that require deeper engagement with key concepts, critical thinking, and peer discussion. Classroom activities also encourage continuous feedback between teacher and students, which improves and deepens understanding.

The station rotation model provides the opportunity to customize the content of the curriculum, the way it is taught, and the forms in which students perform classroom activities. Students are divided into groups and rotate gradually between stations, with each station focusing on a different type of activity. It is not necessary for all students to complete all activities, and the teacher can give them a choice. However, in the longer term, the teacher monitors student engagement in the learning activities and guides the completion of the student's skills portfolio.

However, the use of blended learning as an innovative learning method naturally requires a suitable learning environment that can be adapted to the current needs of the students. One option is to divide the learning space into learning zones according to the Future Classroom Lab concept (European Schoolnet n.d.).

2.2 *Future Classroom Lab*

The Future Classroom Lab (FCL) is an innovative flexible learning environment divided into learning zones (Fig. 1) focused on active student learning, student skill development, and the promotion of different pedagogical approaches and forms, and the meaningful use of technology. The standard model FCL, European Schoolnet, includes 6 zones: Interact, Investigate, Develop, Exchange, Create, Present. These zones are not strictly partitioned and activities within them can overlap (European Schoolnet n.d.). The number of zones depends on the needs of the school, the teacher or the curriculum of the country.

The station rotation model in the Future Classroom Lab environment offers several advantages. Each learning zone enables the development of specific skills tailored to the individual needs of the learner. The description of the zones is already based on our experience, which we have developed based on the Future Classroom Lab concept (European Schoolnet 2013).

Interact is a place where a lively and interactive discussion takes place between the teacher and the students. The aim is to formulate questions that help students to better understand the topic while summarizing the background knowledge they have acquired during their self-study at home. This zone provides a bridge between the self-study at home and the face-to-face activities, allowing students to transition smoothly from theoretical knowledge to practical activities. The zone also includes the use of digital formative assessment tools that allow the teacher to monitor students' progress and adapt lessons to their needs.

Investigate is a zone aimed at promoting active discovery and problem solving where students plan and carry out activities using a research method. This process encourages their critical thinking and ability to approach different problems analytically. Within this zone, interdisciplinary themes can be explored, linking different fields of knowledge and giving students a broader perspective. Interaction with peers

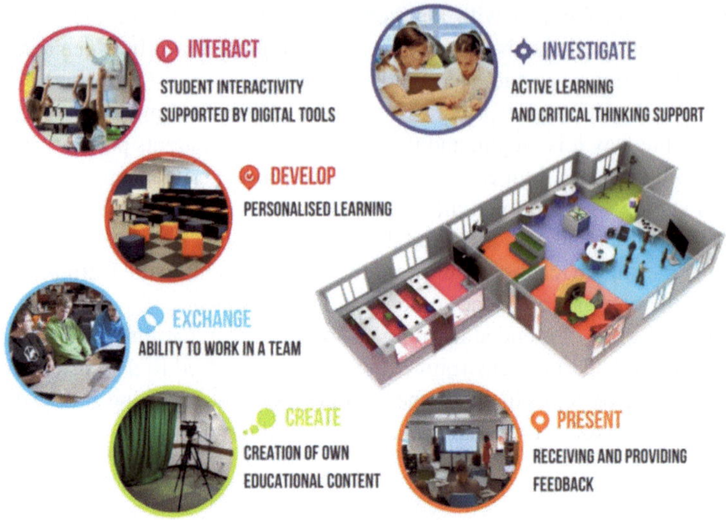

Fig. 1 European Schoolnet's standard FCL model in Brussels (European Schoolnet n.d.)

plays a key role as students work together to solve problems, which encourages teamwork and the sharing of ideas. The zone also uses digital technologies to create and process the collected data so that students can better analyze the results and use modern technologies in research.

Exchange focuses on developing teamwork, where students collaborate and take responsibility for tasks within the group. Brainstorming is a key element that encourages finding innovative solutions to various problems. In this zone, forms of face-to-face and synchronous collaboration alternate with online and asynchronous activities so that students learn to communicate flexibly and collaborate remotely. The zone also encourages project planning, where students propose new solutions and create concrete plans together, which they then divide up and implement. In this way, they develop their organizational and communication skills, which are essential for successful teamwork.

Develop focuses on promoting students' personal development through informal communication that creates a space for free self-expression. Self and peer assessment allow students to reflect on their progress and provide feedback to their peers. The informal design of the zone helps students to find a balance between learning and relaxation. Individual self-study encourages consolidation of knowledge and independence. In addition, students have the opportunity to participate in counseling sessions with a teacher to help them address specific problems and provide guidance according to their individual needs.

Create provides space for students to create their own content and products, which encourages the development of their skills and creativity. It strengthens students' autonomy, independence, and responsibility through their own project design and implementation. The Create Zone focuses on designing for real life and allows students to experience the practical application of their skills in everyday life. Technology and digital tools are also actively used in this zone, giving students a wide range of opportunities to create and work on different products.

Present focuses on presenting their own results, presenting the students' work, and arguing their findings. This zone also includes scientific discussion that encourages critical thinking and the ability to defend one's own views. Students have the opportunity to share and communicate their findings with others, learning to respect different perspectives and opinions. By interacting with a wider audience, they can gain new perspectives and improve their communication skills. The focus is on constructive feedback to help students improve their results and performance. A summary of the results at the end comprehensively concludes the whole process and provides a clear overview of what has been achieved.

The advantage of using multiple zones within the classroom is also one of the possible solutions to the relatively common problem of lack of aids in schools, as it is not necessary to have multiple aids available for all students (groups) at the same time.

2.3 Educational Content

When designing blended learning activities in a stimulating learning environment, we emphasize several key aspects. We strive to make activities engaging and stimulating while encouraging active learner participation. Ensuring that activities are appropriate for the pupils' age allows them to develop both personally and intellectually. We focus on the practical applicability of acquired knowledge and skills so that pupils remain motivated and see value in what they are learning. The accessibility of professional information is also crucial, which is why we select teaching methods that facilitate comprehension and long-term retention. Additionally, we ensure that the learning space is appropriately adapted and actively use feedback to better understand individual learners' needs. Based on this feedback, we adjust teaching activities to best align with students' abilities and requirements.

The targeted development of students' skills requires the use of active learning methods. Implementing these methods places new demands on the learning environment, including its layout and flexibility. A motivated, enthusiastic, and well-prepared teacher who understands the benefits of innovative educational approaches plays a key role in this process.

3 Methodology

For grammar school students aged 15–18, we have chosen the rarely presented topic of thermoelectric phenomena. Specifically, we focus on the Seebeck and Peltier effects, as they are both intriguing and highly relevant due to their widespread applications in modern engineering. Our primary goal was to monitor students' engagement in blended learning and their work in learning zones. We provide pupils with personalized learning opportunities, but with their active involvement.

3.1 Study Methodology

The activity combines two blended learning models: Flipped Classroom (guided self-study at home) and Station Rotation (eight stations within a six-zone FCL environment). At home, students engage in self-paced study using a teacher-prepared learning resource. The e-learning material (https://lnk.sk/csl0) includes a structured explanation of phenomena, practical applications, a video (Fyzikálne vzdelávanie ODF PF UPJŠ 2024) demonstrating a student project, and a feedback form. This form helps the teacher assess students' conceptual understanding and identify areas where additional support may be needed.

In the school laboratory, students come with their own research questions, which they validate through small-group work within the station rotation model, structured across different learning zones. They follow guided exploration using prepared worksheets, with the primary goal of discovering new knowledge through hands-on experience. The station rotation model promotes cooperative learning, where each group contributes to the overall outcome. Depending on time availability, pupils complete all or part of the assigned tasks in each learning zone. At the beginning of the activity, in the Interact zone (Fig. 3, middle), all students participate in a brief review of their self-study. This includes revisiting key concepts, as well as addressing any uncertainties. Students then can experiment, either in pairs or individually, while receiving a frontal explanation of its working principles. Finally, they receive organizational instructions for further activities in each zone (Table 1).

The teacher guides the pupils to work independently in groups and at the end, together with the students, formulates conclusions from the findings.

3.2 Curricular Methodology

Through home online self-study, students learned about thermoelectric phenomena and became familiar with basic practical applications. They obtained basic information about the Seebeck and Peltier effects. As part of the feedback on self-study,

Table 1 Instructions for working in different learning zones

Learning zone	Interact	Investigate	Exchange	Develop	Create	Present
Participants	All	3 + 3 + 3	3	3	3	In groups
Activity	Frontal repetition, instructions	3 stations	1 station	1 station	1 station	Presentation of group results
Time for activity	15 min	15–20 min	15–20 min	15–20 min	15–20 min	20 min(3 min for each group)

students formulated their own research questions and chose stations on which they will work at school. In the *Investigate* zone there are 3 stations for students:

How much electrical voltage can we get from the temperature difference?

At this station, students investigate how the electric voltage on a Peltier cell depends on the temperature difference between its two sides. Their task is to set up the apparatus and verify whether the generated voltage is sufficient to light up an LED or power a small electric motor, helping them better understand the applications of thermoelectric phenomena.

And yet it moves (thanks to Peltier).

The station (Fig. 2, right) is designed for students who are more interested in the qualitative explanation of thermoelectric phenomena rather than measurements. Students observe why a stove fan starts rotating when placed on a hot surface or stove. They investigate how increasing the temperature of the surface affects the fan's rotation speed and how the temperature difference influences airflow velocity. Their task is also to explain the working principle of the stove fan.

What is the power of the Peltier cell?

At the station (Fig. 2, left), students investigate the power of a Peltier cell. They place ice in one container of the holder and pour hot water into the other, creating a temperature difference across the Peltier cell. Voltage and current sensors are connected to the cell's terminals and linked to the LabQuest measurement panel to display these values. After integrating a rheostat into the circuit, students start

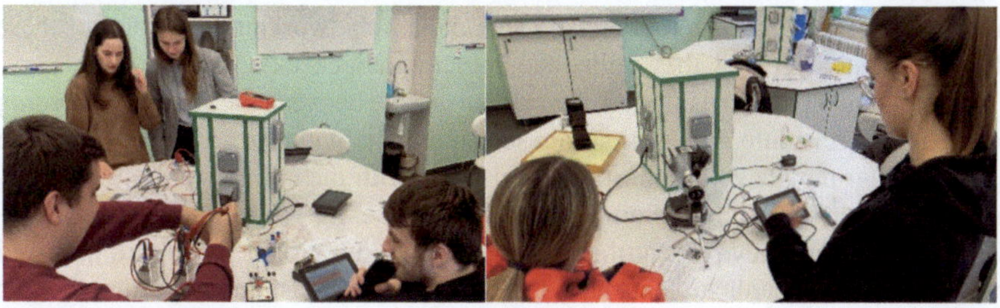

Fig. 2 Investigate zone: measuring the electrical voltage on the Peltier cell (left) and observing the speed of the stove fan (right)

Fig. 3 Create zone (left), Interact zone (in the middle), Develop zone (right)

the measurement and gradually increase the resistance. Throughout the experiment, they observe how the relationship between voltage and current changes under load. Once the measurement is complete, they determine the power of the Peltier cell based on the recorded values. Finally, they formulate a response to the research question regarding the power of the Peltier cell.

The *Exchange* zone offers students the opportunity to work on a project that connects thermoelectric phenomena with various applications in technical fields, such as space technology, survival in nature, smart homes, a healthy lifestyle, and industrial use. Their task is to choose one of these topics, explore the use of thermocouples in the selected area, engage in discussions, and collaboratively design new, practical applications. This station helps students develop analytical thinking, critical reasoning, and the ability to propose innovative solutions based on their acquired theoretical knowledge.

The *Create* zone (Fig. 3, left) is ideal for students who enjoy experimenting, building, and applying their knowledge to create functional devices. Their task is to connect 10 Peltier cells in series, cooling one side with ice and heating the other with tea candles, to determine whether the generated voltage is sufficient to charge a mobile phone. As part of the task, they must connect the Peltier cells to a charger and monitor its functionality. Finally, they create an educational video documenting the entire process and demonstrating how the device works.

The *Develop* zone (Fig. 3, right) is designed for students who want to go beyond a basic understanding of thermoelectric phenomena and deepen their knowledge. It is also suitable for those who are not yet fully familiar with the topic and want to strengthen their understanding of the concept. At this station, students have access to real-world devices that utilize Peltier cells, such as a dehumidifier, a car cooler, and a stove fan. Their task is to understand and explain the working principles of these devices and analyze how they utilize the Peltier and Seebeck effects.

Depending on the time allotted, the students rotate between the stations according to the agreed schedule. The stations are selected by individual registrations from the students after they have familiarized themselves with the online learning materials as part of their self-study at home. In the final phase, all groups move to the "Present" zone, where they present their results to the other groups in a short, prepared presentation. As not all students have to visit all the stations, it is important that the results presented by the other groups make sense to them.

The last part of the activity is dedicated to the evaluation by the students using an online form. We use formative assessment tools that we combine with the extension (Extended) in terms of the 5E model IBSE (ESTABLISH Consortium n.d.).

Our target groups were two groups of physics teacher students (12 students in total) and one group of physics teachers (44 teachers in total) with several years of teaching experience. The participants of the training were introduced to the blended learning method and the station rotation with the use of learning zones during the interactive lecture. In the practical part of the workshop, participants took on the role of students exploring thermoelectric phenomena in a Future Classroom Lab environment using the blended learning method. An online survey was conducted immediately after the learning activity. From an action research perspective, we seek answers to the following research questions:

Q1: Based on their first personal experiences with blended learning and the use of learning zones, to what extent would teachers be willing to implement them in their own teaching practice?

Q2: What kind of support do teachers expect when implementing blended learning?

Q3: Which learning zones would trained teachers incorporate into the arrangement of a school physics lab and what potential obstacles do they see to using these zones?

4 Results

In the sample of teachers surveyed, student teachers (0 years of experience) account for 21% (Table 2). 11% of teachers with less than 10 years of experience are considered novice teachers and 45% of teachers with more than 21 but not more than 36 years of experience are considered experienced teachers.

Immediately after completing a training course on the use of blended learning (BL), 51% of physics teachers are determined (definitely yes, probably yes) to try blended learning in their lessons (Table 3). Of the teachers with a positive attitude, 50% belong to the group of pre-service teachers and 48% to the group of experienced teachers. Teachers with less teaching experience tend to be reluctant or to repeat the teaching methods they were taught as students. Teachers with more than 36 years of experience have established teaching methods that they are reluctant to change toward the end of their careers. In terms of the extent of blended learning use, selected educational topics or thematic units dominate.

Table 2 Length of respondents' teaching experience

Teaching experience (years)	0	1–5	6–10	11–15	16–20	21–25	26–30	31–35	36 and over
Frequency	12	3	3	3	4	11	2	12	6

Table 3 Determination to use blended learning in physics lessons

Absolutely yes	Rather yes	I will consider	Rather not	Absolutely not
21%	**30%**	32%	16%	0%
As the prevailing method	In selected thematic units	On selected topics	In non-formal education	Not yet at all
2%	**39%**	**38%**	9%	5%

In order to support the implementation of blended learning in physics education, teachers expect the provision of methodological materials, worksheets and, to a considerable extent, the provision of teaching aids for the active participation of students (Fig. 4). It is encouraging to see that there is an interest in including the learning zones of inquiry and collaboration in the design of the school learning environment (Fig. 5, left). The main barriers to the use of these learning zones, according to teachers, are the lack of teaching aids and flexibility in the design of physics classroom layout (Fig. 5, right). In most schools, the laboratory furniture is usually permanently installed.

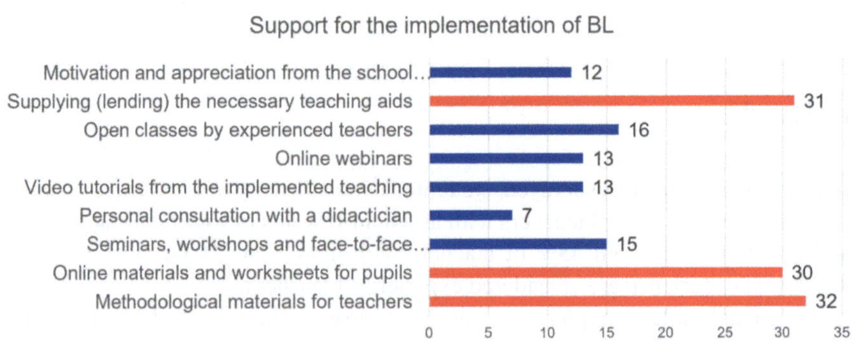

Fig. 4 Teachers' expected support for the implementation of blended learning

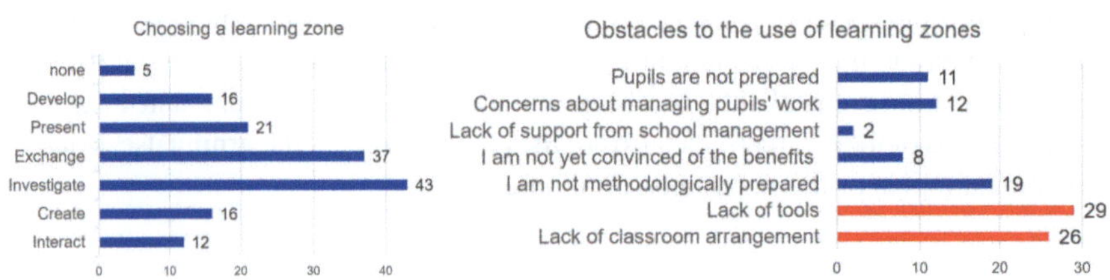

Fig. 5 The interest in and obstacles to the use of learning zones

5 Conclusion

Through interactions with physics teachers of different age groups, we are increasingly convinced of the need for training and practical support for innovation in teaching towards blended learning and Future Classroom Lab. Teachers recognize the need for changes in education due to the demands placed on the graduate profile. A strong motivation is the need to increase students' interest in physics as a science. It is evident to teachers that the development of research skills and scientific literacy in students can only be achieved through innovative teaching methods and active student engagement in a stimulating learning environment. Teachers may become convinced of the relevance of blended learning in Future Classroom Lab environments based on improved student outcomes in skills, conceptual understanding as well as engagement with the learning content. We will continue to support this focus with the data gained from action research and offer specialist guidance and help.

Acknowledgments The work was created and supported as part of the grants: National Project Digital Transformation of Education and School DitEdu ITMS2014+: 401402DVR6, funded by the EU under the Slovakia Programme and the Slovak Research and APVV-22-0515 Blended learning of Physics in the Future Classroom Lab.

References

Aksak K, İstek Kılınç H, Okur MR (2023) The rotation model in blended learning. Asian J Distance Educ 18(2):63–74. https://www.asianjde.com/ojs/index.php/AsianJDE/article/view/704
Baeta P, Pedro N (2017) Future classroom: analysis of educational activities developed by teachers and students. In: ICERI2017 proceedings, pp 130–139. https://doi.org/10.21125/iceri.2017.0069
Blended Learning (n.d.-a). Blended learning. https://www.blendedlearning.org/
Blended Learning (n.d.-b). Blended learning models. https://www.blendedlearning.org/models/
Christensen Institute (n.d.) Education research. Clayton Christensen Institute. https://www.christenseninstitute.org/research/education/#topics
Colton S, Smith C, Sourdot L (2020) Designing a future classroom laboratory for exploring the science of teaching and learning. Int J Des Learn 11:36–46. https://doi.org/10.14434/ijdl.v11i3.25860
Divayana D (2019) The implementation of blended learning with Kelase platform in the learning of assessment and evaluation course. Int J Emerg Technol Learn (iJET) 14(17):114–132. https://doi.org/10.3991/ijet.v14i17.8308
Dzakiria H, Don MS, Abdul Rahman HD (2013) Action research on blended learning transformative potential in higher education: learners' perspectives. Bus Manag Res 1(2):125. https://doi.org/10.5430/bmr.v1n2p125
ESTABLISH Consortium (n.d.) ESTABLISH project. http://www.establish-fp7.eu/project.html
European Schoolnet (2013) Future classroom lab learning zones. https://edufor.pt/icl/documentos/FCL_LearningZones-Description_ING.pdf
European Schoolnet (n.d.) Future classroom lab. https://fcl.eun.org/
Fyzikálne vzdelávanie ODF PF UPJŠ (2024). Netradičná nabíjačka mobilu [Video]. YouTube. https://www.youtube.com/watch?v=lcQFJZgZufQ
Ganajová M, Kireš M (2021) Štyri rozmery vzdelávania ako východisko k tvorbe kurikula. Didaktika 6

Govindaraj A, Silverajah VSG (2017) Blending flipped classroom and station rotation models in enhancing students' learning of physics. In: ACM international conference proceeding series, pp 73–78. https://doi.org/10.1145/3175536.3175543
Graham CR (2019) Current research in blended learning. In: Moore MG, Diehl WC (eds) The handbook of distance education, 4th edn. Routledge, pp 173–188. https://doi.org/10.4324/9781315296135-15
Graham CR, Halverson LR (2023) Blended learning research and practice. In: Zawacki-Richter O, Jung I (eds) Handbook of open, distance and digital education. Springer, pp 1–17. https://doi.org/10.1007/978-981-19-2080-6_68
Harlen W (ed) (2015) Working with big ideas of science education. Science Education Programme of IAP. https://www.ase.org.uk/bigideas
Kireš M, Juhásová A, Karolčík Š, Hubeňáková V (2024) Learning environment in the digital age for personalised STEM education. In: International conference on emerging elearning technologies and applications (ICETA), pp 332–337. https://doi.org/10.1109/ICETA63795.2024.10850782
OECD (2018) The future of education and skills: education 2030. https://www.oecd.org/content/dam/oecd/en/publications/reports/2018/06/the-future-of-education-and-skills_5424dd26/54ac7020-en.pdf
OECD (2020) Education at a glance 2020: OECD indicators. OECD Publishing. https://doi.org/10.1787/69096873-en
Ramirez M, Buitrago CR (2022) In-class flip: a student-centered approach to differentiated learning. ISTE, ISBN 9781564849588
Staker H, Horn MB (2012) Classifying K–12 blended learning. Innosight Institute. http://www.christenseninstitute.org/wp-content/uploads/2013/04/Classifying-K-12-blended-learning.pdf
Tucker CR (2022) The complete guide to blended learning: activating agency, differentiation, community, and inquiry for students. Solution Tree Press, ISBN 978–1–954631-33-5

Flipped Classroom: Effects on the Conceptual Understanding in Electric Circuit Teaching

Wolfgang Lutz and Thomas Trefzger

Abstract On the basis of a design-based research approach, a teaching concept suitable for the application of flipped classroom was developed for teaching simple electric circuits, the Würzburg Teaching Concept (WTC). The WTC combines digital learning materials with in-class hands-on activities and was adapted from the electric pressure curriculum. In a large-scale field study involving over 2,000 students, the effects on conceptual understanding were assessed. On the one hand, the study examined which effect the teaching concept had when applied within the frame of traditional teaching. On the other hand, it compared the effects the concept had applied to different methods, i.e., traditional teaching and flipped classroom. Results show that the WTC improves conceptual understanding in the non-science-oriented branch whereas combining the WTC and the flipped classroom method improves learning in all school branches.

Keywords Electric pressure curriculum · Digital learning · Flipped classroom

1 Background and Motivation

The teaching subject Physics is considered particularly difficult and challenging by many pupils (Merzyn 2010). Difficulties in understanding occur in particular regarding topics, in which physical concepts and correlations are barely illustratable and do not match pupils' experiences. One example for this is electric circuit teaching as most experimental effects can only be proven indirectly by use of measuring devices. To illustrate their explanations, teachers often refer to models which do not match learners' experiences and thus seem abstract and complicated (Frank and Trefzger 2024). Moreover, numerous misconceptions derived from daily life impede

W. Lutz (✉) · T. Trefzger
Physik und ihre Didaktik, Julius-Maximilians-Universität Würzburg, Würzburg, Germany
e-mail: wolfgang.lutz@uni-wuerzburg.de

T. Trefzger
e-mail: thomas.trefzger@uni-wuerzburg.de

D. Sokołowska and E. McLoughlin (eds.), *Embracing Changes Together in Physics Education*, Challenges in Physics Education,
https://doi.org/10.1007/978-3-032-27385-7_4

pupils' development of conceptual understanding of the correlations between voltage, resistance, and current (Wilhelm and Hopf 2018). One possible explanation for this is the constructivist learning theory according to which new information is always linked to existing cognitive structures (De Corte 2000). Misconceptions can thus impede the reception, processing, and interpretation of new information (Schecker and Duit 2018). Hence, the research interest of this article is to develop methods to facilitate pupils' conceptual understanding of electric circuit teaching. This was achieved through the following three aspects.

1.1 Promotion of Conceptual Understanding Using the Electric Pressure Curriculum

According to Minstrell (1982), taking into account the everyday experiences pupils bring to class is an important starting point for constructing correct conceptual understanding. Within diSessa (1993), it is estimated that each concept is based on a variety of independent building blocks of the cognition structure that are valid in themselves. These are called *"phenomenological primitives"* (*p-prims*). For developing conceptual understanding, it is useful to activate favourable p-prims and to avoid unfavourable p-prims (Weinert 2000) and to use a suitable didactical concept to form a coherent mental structure from the previously loosely connected and unstructured p-prims. Such an approach has proven particularly valuable for teaching complex contents like the concept of voltage.

Thus, due to the consideration of misconceptions, electric circuit teaching uses analogy-based teaching concepts applying models based on the learners' experiences, like the Electric Pressure Curriculum (Burde 2018b; Burde and Wilhelm 2021). The idea behind the Electric Pressure Curriculum is to compare the conduction electrons in a metal with free moving gas particles. In analogy to gas pressure, "electrical pressure" dependent on the electron density is introduced in the conductor, which illustrates the electrical potential. Electrical voltage is interpreted as electric pressure difference. This approach is used in the Electric Pressure Curriculum (Burde 2018a) to explain air flow on the basis of air pressure differences. This principle experienced as intuitive by learners is then applied to teaching electric circuits so that voltage is understood as cause for electron flow. The objective is to establish a qualitative relationship between voltage, current, and resistance before quantitatively specifying this connection afterwards. Pupils taught according to the Electric Pressure Curriculum, showed significantly better conceptual understanding than the comparison group (Burde 2018b).

1.2 Promotion of Conceptual Understanding Through Activation

Many pupils' conceptual understanding remains impeded after physics classes, even if they were based on the Electric Pressure Curriculum (Burde 2018b; Rhöneck 1986). This is due to the fact that pre-concepts have proven to be highly resistant to change as they continue to seem plausible to pupils despite teaching efforts (Driver et al. 1985; Duit and Treagust 2003). This resistance is rooted in beliefs developed from pre-teaching theories and seemingly logical connections of concepts and rules (Vosniadou and Skopeliti 2014). Didactical research has shown that complete reconstruction of these conceptions is neither possible (Duit 1998), nor desirable (Duit and Treagust 2003). So, the objective should rather be to promote the conscious coexistence of every-day and physically correct termini and experiences so that learners can switch between these different perspectives in a reflected way.

It is important to provide learners with a space in which they can apply their knowledge in as many and as diverse contexts as possible. However, there are frequent reports of teaching contexts characterised by lack of cognitive activation of pupils, constructivist-oriented teaching offers, and a context-based everyday-based orientation of teaching (Fischer et al. 2003; Duit 2006). In this context, the term activation does not refer to unconscious and unreflected actions but to targeted stimulation of mental processes that promote thought processes and understanding (Hasselhorn and Gold 2017). Learning environments can support learners' cognitive activation, e.g. by providing problem-oriented and challenging but still solvable tasks, or the opportunity for inquiry-based learning, or by supporting a discussion-oriented communication culture in class (Lipowsky 2020).

1.3 Promotion of Conceptual Understanding Through the Flipped Classroom Method

The effectiveness of a teaching offer depends on the teaching materials as well as their didactical application in the learning process and the way they are used by learners (Bergmann and Sams 2012; Perkhofer-Czapek and Potzmann 2016). In this context, the teacher is a significant actor who shapes the didactical framework for integrating the materials in class. In order for pupils to develop deeper understanding, the teacher can use targeted impulses and efficient class management to support cognitive processes and to maximise time-on-task. At the same time, the teacher can create a motivating and supportive teaching environment in which pupils experience competence, autonomy, and social integration (Deci and Ryan 1993). In physics classes, however, a teacher-based approach prevails in which teachers spend a lot of time for explanations while pupils have too little time to practice and experiment on their own which also negatively impacts individual support (Duit and Wodzinski 2010; Prenzel et al. 2002). With the USA-originated flipped classroom teaching method, knowledge transfer, which is otherwise firmly anchored in the classroom, is shifted to a home preparation phase. The pupils concern themselves with the subject content of the next lesson at home and develop the theoretical foundations at their own pace (Bergmann and Sams 2012). The time in the classroom can then be focussed on cooperative forms of learning and on knowledge application. The change in classroom management opens up new opportunities to diagnose individual learning difficulties in pupils and to constructively support the learning process with appropriate suggestions and assistance (Nicolaisen 2016; Perkhofer-Czapek and Potzmann 2016). Learners are supported in working on cognitively complex tasks, while they acquire the theoretical foundations in a self-organised manner and at their own learning pace (Gilboy et al. 2015). The role of the teacher is changing from knowledge transfer to individualised learning support (Roehl et al. 2013; Yildirim and Kiray 2016), which enables the teacher to provide more targeted support for the pupils' learning process. The central potential of the flipped classroom method therefore lies in the different use of attendance time for intensification, cooperation, and individual support, which, according to the theory of good teaching (Duit and Wodzinski 2010), should have a positive effect on working behaviour and conceptual understanding.

2 The Würzburg Teaching Concept

Against the theoretical background described above, a design-based research project was initiated in Würzburg in cooperation with Goethe University, the University of Tübingen, and Freie Universität Berlin. The aim of the project was to further develop the existing Electric Pressure Curriculum (Burde 2018a) to increase the comprehensibility of the content, to promote active work phases more strongly, and to enable the application of the flipped classroom method. As part of this development, animations specifically adapted to the Electric Pressure Curriculum were created (Lutz et al. 2020), in order to better visualise the temporal changes of physical variables in dynamic processes, such as the movement of electrons, within the framework of model concepts. This is supposed to reduce the learners' cognitive load so that they have more capacity for information intake and processing (Sweller et al. 2011).

To promote active learning, the teaching materials from the Electric Pressure Curriculum have been expanded to include numerous other interactive tasks and supplemented by a number of student experiments, some of which are also digital. Digitalisation in particular enables learners to apply and network their acquired knowledge in a variety of ways. The SAMR model (Puentedura 2006), the Cognitive Load Theory (Sweller et al. 2011), the cognitive theory of multimedia learning (Mayer 2005), and the integrative model of text and image comprehension (Schnotz and Bannert 2003) were considered for the development of the digital teaching materials.

A central element of the materials for the self-learning phase of the flipped classroom method are the explanatory videos, which were designed on the basis of Kulgemeyer's quality criteria (Kulgemeyer, 2020). The explanatory videos are characterised by the integration of self-developed animations to illustrate the models, an appropriate video length (5–8 minutes), and good comprehensibility of the content (Lutz et al. 2020, 2022, 2021). The entire Würzburg Teaching Concept (WTC) and its teaching materials are available for free use on the website www.phytet.de.

3 Study Description

3.1 Research Questions and Hypotheses

Based on the theory presented, the WTC addresses central elements for promoting conceptual understanding in the field of electricity. In addition to the teaching materials, differences in the development of conceptual understanding depending on the teaching method can also be assumed for the flipped classroom method due to the more active involvement of the learners in the lesson and the more individualised support from the teacher. Hence, the following research questions and hypotheses were derived for the survey:

Research Question 1: Does the WTC in traditional lessons lead to an improvement in learners' conceptual understanding of electricity compared to a comparison group without specific teaching materials?

The WTC constitutes a further development of the Electric Pressure Curriculum. Burde (2018b) found, that traditionally-taught learners showed a significantly higher conceptual understanding than learners in the comparison group. Compared to the original concept, the WTC offers the possibility of better visualising physical processes and reducing the cognitive load of the learners through the use of animations. At the same time, an expanded repertoire of tasks, digital offerings, and numerous opportunities for independent experimentation are aimed at activating learners to a greater extent. The aim is to increase learners' active engagement with the content and thus support the development of a profound conceptual understanding. Hypothesis 1 is based on this extension:

H1: Through the WTC, traditionally-taught learners achieve a higher level of conceptual understanding in electric circuit teaching than those of the comparison group.

Research Question 2: Does the use of the flipped classroom method change the effect of the WTC in terms of learners' conceptual understanding?

Through the flipped classroom method, lessons can be increasingly learner-centred (Bishop and Verleger 2013) and focus more on the promotion of cooperative learning, which leads to increased active involvement of learners in class. By shifting the transfer of knowledge to a self-learning phase, there is more time for discussion, independent experiments, and the co-operative completion of challenging learning tasks in the classroom, which can support deeper learning processes (Bloom et al. 1956). By working independently in the self-learning phase and applying what has been learnt in the classroom, the method can support the learners' experience of competence (Deci and Ryan 1993). In addition, the teacher's role as a learning companion (Roehl et al. 2013) enables more individualised support tailored to learning needs. The explanatory videos developed in the WTC provide additional media support for the learning process (Mayer 2005). Regarding the flipped classroom method, there is empirical evidence of a positive effect of the method on the development of declarative knowledge of electrodynamics in German-taught physics lessons (Finkenberg 2018). Hypothesis 2 is derived from this:

H2: In the WTC, the flipped classroom method increases learners' conceptual understanding compared to traditional teaching.

3.2 Method and Sample Group

This article is based on data from a field study conducted in the 2021/22 school year at a total of 60 Bavarian *Gymnasien* (German schools of secondary education) with 122 classes and 1581 female and 1318 male pupils in Year 8. The classes taking part in the survey were divided into two intervention groups and one control group.

In the first intervention group, electricity was taught entirely according to the WTC and the flipped classroom method (G_{flip}). In the second intervention group, the WTC was also applied, but the teachers used the prescribed materials to teach in a traditional way (G_{trad}). The control group (G_{cont}) was not given any specific guidelines regarding the teaching materials or methods and could be individually organised by the teachers. The decision as to whether a class was assigned to one of the two intervention groups or the control group was made by the respective teacher.

In the Bavarian school system, a special influencing factor must be taken into account. At the end of Year 7, pupils choose different branches that will shape their future school career. One of these options is the science and technology branch (NTG), which is characterised by a stronger focus on physics, among other things. Compared to the general physics lessons in all other branches (nNTG), the NTG branch has one additional profile lesson per week in which the students deepen their knowledge, work on more advanced topics, and carry out more independent experiments. Classes are often halved for these lessons, allowing for more individualised support from the teacher. In addition to the extra teaching time, pupils who consciously choose the NTG branch are on average more interested in physics and have a more pronounced physics-related self-concept (Elsholz et al. 2024; Lutz et al. 2023). Due to the different external framework conditions, separate analyses are calculated for the branches.

Incomplete information provided by the participants in the questionnaires and absences at one of the two test times led to a reduction in the data, which is evenly

Table 1 Pupils' gender distribution within the groups

	NTG			nNTG		
Pupils	G_{flip}	G_{trad}	G_{cont}	G_{flip}	G_{trad}	G_{cont}
Female	247	138	78	284	268	154
Male	372	154	112	152	77	43

distributed across all participating classes and can be attributed solely to statistical reasons. The sample size remaining for the analysis comprises 1169 female pupils and 910 male pupils, distributed across the groups according to Table 1. On the one hand, the uneven group sizes result from the individual group-related decisions of the teachers, and on the other hand from the fact that the NTG branch is chosen more often by male pupils than by female pupils.

3.3 Measuring Instrument

A valid and reliable multiple-choice test by Urban-Woldron and Hopf (2012) was used to assess conceptual understanding. The test, which was developed in Vienna, consists of 22 items relating to the conceptual understanding of current, resistance, and series and parallel circuits. Nine of the items have a two-stage structure in which the students must not only answer a question (first stage) but also have to provide an explanation (second stage). An item is only scored as correct if both the answer and the explanation are correct. Figure 1 shows an example item of the test instrument, with the correct answers highlighted accordingly.

Item 3
The resistor R_1 has a low resistance and is replaced by R_2, which has a higher resistance.

a) How does this affect the electric current in the circuit?

- ☐ **It increases.**
- ☒ **It gets smaller, but will not be zero.**
- ☐ **It stays the same.**
- ☐ **No more current flows.**

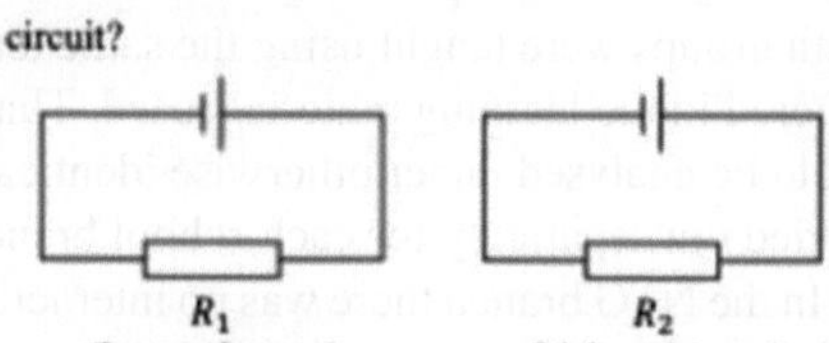

b) How do you explain your answer?

- ☐ **The battery is strong enough to push any electric current through the new resistor.**
- ☒ **The battery can only push a smaller current through the new resistor.**
- ☐ **R_2 with a high resistance consumes more current than R_1 with low resistance.**
- ☐ **It is the same battery, hence the current does not change.**

Fig. 1. A two-stage example item from the test instrument. The correct answers are marked. (Translated in accordance with Burde and Wilhelm 2020)

3.4 Modelling

Conceptual understanding was estimated in the pre- and post-test as a latent variable using item response theory (IRT) (e.g. Neumann 2014), R software (R Core Team 2022), and the test analysis modules (TAM) package (Robitzsch et al. 2020). Only data sets in which at least 10 of the 22 items were completed in both the pre-test and post-test were taken into account. This restriction aimed to ensure that the estimates of personal abilities were based on a sufficient number of responses and that the estimation errors were minimised. The one-parameter logistic model (1PL), the two-parameter logistic model (2PL) model, and the three-parameter logistic model (3PL) model were compared with each other in order to find the most suitable model for the data set. The 1PL model only considers item difficulty and assumes the same discrimination ability for all items. In the extended 2PL model, different discrimination abilities of the items are permitted, resulting in increased degrees of freedom. The 3PL model also takes into account guess probabilities for the items. Following the likelihood-ratio test ($\chi^2 = 1382$, $\Delta df = 22$, $p < 0.001$), the 2PL model (AIC: 144788; BIC: 145076) is significantly more suitable for the data than the 1PL model (AIC: 146126; BIC: 146282). Since many items in the test used have two stages and are only rated as correct if they are answered correctly at both stages, the guess probability for a randomly correct answer to an item is negligible. For these reasons, a 2PL model was used to analyse the data in this study. The expected a posteriori (EAP) reliability estimation have moderate estimation accuracies in both the pre-test 0.62 and the post-test 0.77 and are sufficient for group comparisons.

4 Results

The effect of the WTC was first analysed in traditional lessons. For this, groups G_{trad} and G_{cont} were compared against each other separated by the different branches. Both groups were taught using the same teaching method (traditional teaching), but differed in the learning materials used. That way, the specific influence of the WTC could be analysed under otherwise identical conditions. The statistical analysis was carried out separately for each school branch using a mixed ANOVA.

In the NTG branch there was no interaction between groups and time ($F(1, 480) = 0.150$, $p = 0.698$, $\eta_p^2 < 0.001$). This indicates that there is no significant difference in learning gains between the two groups. However, there was a significant main effect for the groups with a small effect size ($F(1, 480) = 3.932$, $p = 0.048$, $\eta_p^2 = 0.008$), which indicates a difference in the learning prerequisites in the groups. Regardless of group affiliation, learners showed a significant improvement in their conceptual understanding over time with a strong effect size ($F(1, 480) = 1124$, $p < 0.001$, $\eta_p^2 = 0.7$).

In the nNTG branch there was a significant interaction with a small effect size between groups and time ($F(1, 540) = 6.821$, $p = 0.009$, $\eta_p^2 = 0.012$). The learning gain in the treatment group with the WTC was higher over time than in the control group. To further analyse the main effects, the significance level was set at 0.025 using a Bonferroni correction both for the group and for time. It was also applied to all further observations. There were significant differences between the groups with small effect sizes in both the pre-test ($F(1, 540) = 5.84$, $p = 0.016$, $\eta_p^2 = 0.011$) and the post-test ($F(1, 540) = 12.8$, $p < 0.001$, $\eta_p^2 = 0.023$). Over time, both groups showed significant learning gains with high effect sizes (G_{trad}: $F(1, 344) = 543$, $p < 0.001$, $\eta_p^2 = 0.612$ | G_{cont}: $F(1, 196) = 258$, $p < 0.001$, $\eta_p^2 = 0.568$) (Fig. 2 and Table 2).

In a further step, the effect of the teaching method in the WTC was analysed by comparing the G_{trad} and G_{flip} groups. Both groups were taught using the same materials but different teaching methods: While group G_{trad} was taught traditionally, group G_{flip} was taught using the flipped classroom approach. The use of the same teaching materials made it possible to specifically analyse the influence of the teaching method on conceptual understanding. The statistical analysis was also carried out separately for each school branch using a mixed ANOVA.

In the NTG branch there was a significant interaction with a small effect size between groups and time ($F(1, 909) = 6.396$, $p = 0.012$, $\eta_p^2 = 0.007$). Learning gains in the flipped classroom model were higher over time than in traditional lessons. In the subsequent analysis of the main effects, no significant differences were found

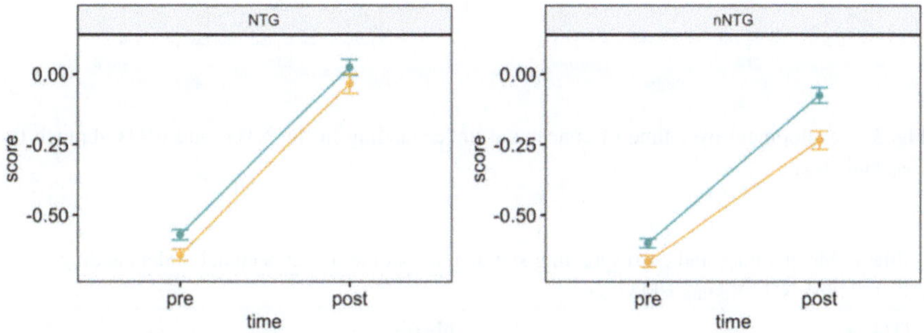

Fig. 2 Development over time of conceptual understanding in the NTG- and nNTG-branch for G_{trad} and G_{cont}

Table 2 Mean values and corresponding standard errors (se) for conceptual understanding in the pre- and post-test for G_{trad} and G_{cont}

NTG				nNTG			
Time	Group	Mean	Se	Time	Group	Mean	Se
Pre	G_{trad}	–0.570	0.024	Pre	G_{trad}	–0.600	0.0222
	G_{cont}	–0.642	0.029		G_{cont}	–0.665	0.0294
Post	G_{trad}	0.025	0.024	Post	G_{trad}	–0.0750	0.0222
	G_{cont}	–0.033	0.029		G_{cont}	–0.234	0.0294

between the groups, neither in the pre-test ($F(1, 909) = 0.312$, $p = 0.577$, $\eta_p^2 < 0.001$) nor in the post-test ($F(1, 909) = 2.96$, $p = 0.086$, $\eta_p^2 = 0.003$). Both groups showed significant learning gains with strong effect sizes (G_{flip}: $F(1, 618) = 1392$, $p < 0.001$, $\eta_p^2 = 0.692$ | G_{trad}: $F(1, 291) = 685$, $p < 0.001$, $\eta_p^2 = 0.702$).

In the nNTG branch there was a significant interaction with a small effect size between groups and time ($F(1, 779) = 3.951$, $p = 0.047$, $\eta_p^2 = 0.005$). The learners in the flipped classroom group also achieved higher learning gains than those in traditional lessons. In the further analysis of the main effects, no significant differences were found between the test times (pre: $F(1, 779) = 0.177$, $p = 0.674$, $\eta_p^2 < 0.001$|post: $F(1, 779) = 3.43$, $p = 0.064$, $\eta_p^2 = 0.004$). The learning gains of both groups over time were significant with strong effect sizes (G_{flip}: $F(1, 435) = 843$, $p < 0.001$, $\eta_p^2 = 0.660$ | G_{trad}: $F(1, 344) = 543$, $p < 0.001$, $\eta_p^2 = 0.612$) (Fig. 3 and Table 3).

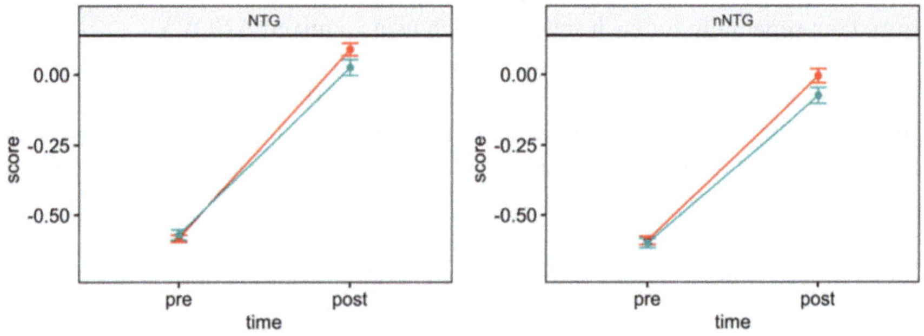

Fig. 3 Development over time of conceptual understanding in the NTG- and nNTG-branch for G_{flip} and G_{trad}

Table 3 Mean values and corresponding standard errors (se) for conceptual understanding in the pre- and post-test for G_{flip} and G_{trad}

NTG				nNTG			
Time	Group	Mean	Se	Time	Group	Mean	Se
Pre	G_{flip}	–0.583	0.018	Pre	G_{flip}	–0.590	0.021
	G_{trad}	–0.570	0.026		G_{trad}	–0.600	0.023
Post	G_{flip}	0.089	0.018	Post	G_{flip}	–0.006	0.021
	G_{trad}	0.025	0.026		G_{trad}	–0.075	0.023

5 Discussion

The analysis of the effect of the WTC on the conceptual understanding of electricity in traditional lessons shows a differentiated picture in the various school branches. While greater learning gains were observed in the nNTG branch among pupils who were taught using the WTC, there was no significant difference between the NTG branch and the control group. Hypothesis H1 could therefore be confirmed in the nNTG branch and must be rejected in the NTG branch.

One possible explanation for this result could lie in the different learning requirements of the pupils. Due to the deliberate choice of a focus on natural sciences, it can be assumed that the learners in this branch have higher mathematical and physical competences as well as a pronounced capacity for abstraction. This is reflected, among other things, in a higher self-concept of the learners (Elsholz et al. 2024). In addition, pupils in the NTG branch show a greater interest in physics than their fellow pupils in the nNTG branch (Lutz et al. 2023). Based on these prerequisites, it can be assumed that learners in the NTG branch can familiarise themselves more easily with complex topics due to a greater willingness to learn and better skills. It therefore seems less important for learners in the NTG branch how the learning content is didactically reconstructed. Pupils in the nNTG branch, on the other hand, benefit from the didactic approach of the WTC.

The other results show that the flipped classroom method led to a significant increase in learning gains in both school branches with a similar effect size, so that hypothesis H2 can be confirmed. These results are consistent with an earlier study on the use of the flipped classroom method in eleventh-grade electricity lessons at Bavarian *Gymnasiums* (higher secondary school) (Finkenberg 2018). It is interesting to note that the positive effects were evident in both the NTG- and nNTG branches, despite the different learning prerequisites. This suggests that not only high-performing or particularly interested students, but also lower-performing and less motivated learners benefit from the flipped classroom method. At the same time, however, it should not be forgotten that the flipped classroom method—like any other teaching method—does not appeal to all students and does not automatically lead to improvements in learning. The factors influencing learning success in the flipped classroom remain unclear in this study.

6 Conclusion and Outlook

The results of this study show that the WTC in traditional lessons only leads to higher learning gains in the nNTG branch. In the NTG branch, however, there was no significant difference to the control group. This indicates that the teaching concept does at least not represent a disadvantage for the learners. The teaching concept can unfold its full potential in combination with the flipped classroom method. The results show that learners of both school branches were able to increase their performance through the flipped classroom method. The change in the use of attendance time for co-operative learning phases and in-depth applications as well as the more intensive support from the teacher therefore contributed to the promotion of conceptual understanding. However, it should be noted that the effect sizes in this study are small. Learning is a complex process of which the teaching materials and the teaching method are only two small levers. Decisive for learning success is not only the learning offer provided, but also how this offer is utilised by the learners in the end. Factors such as interest, self-concept, cognitive abilities, and the discipline to actively engage with the learning content play a central role and determine the extent to which pupils benefit from a particular teaching concept. A further analysis will therefore investigate the extent to which learning success in the flipped classroom depends on these variables.

References

Bergmann J, Sams A (2012) Flip your classroom: Reach every student in every class every day, 1st ed. International Society for Technology in Education

Bishop J, Verleger MA (2013) The flipped classroom: A survey of the research. ASEE Annual Conference & Exposition 2013:23–1200

Bloom BS, Engelhart MD, Furst ED, Hill WH, Krathwohl DR (1956) Taxonomy of educational. Objectives: Handbook I-Cognitive Domain/BS Bloom, MD Englehart, EJ Furst, WH Hill, DR Krathwohl. David McKay Co Inc., New York

Burde J-P (2018b) Konzeption und Evaluation eines Unterrichtskonzepts zu einfachen Stromkreisen auf Basis des Elektronengasmodells. Logos Verlag Berlin. https://doi.org/10.30819/4726

Burde J-P, Wilhelm T (2020) Teaching electric circuits with a focus on potential differences. Phys Rev Phys Educ Res 16. https://doi.org/10.1103/PhysRevPhysEducRes.16.020153

Burde J-P, Wilhelm T (2021) Unterrichtskonzeptionen zu elektrischen Stromkreisen. In: Wilhelm T, Schecker H, Hopf M (eds) Unterrichtskonzeptionen für den Physikunterricht: Ein Lehrbuch für Studium, Referendariat und Unterrichtspraxis, pp 231–277. Springer. https://doi.org/10.1007/978-3-662-63053-2_8

Burde J-P (2018a) Eine Einführung in die Elektrizitätslehre mit Potenzial. Institut für Didaktik der Physik der Goethe-Universität, Frankfurt am Main

De Corte E (2000) Marrying theory building and the improvement of school practice: A permanent challenge for instructional psychology. Learn Instr 10(3):249–266

Deci EL, Ryan RM (1993) Die Selbstbestimmungstheorie der Motivation und ihre Bedeutung für die Pädagogik. Zeitschrift für Pädagogik 39(2):223–238

diSessa AA (1993) Toward an epistemology of physics. Cogn Instr 10(2–3):105–225. https://doi.org/10.1080/07370008.1985.9649008

Driver R, Guesne E, Tiberghien A (1985) Children's ideas in science, vol 105. Open University Press, Milton Keynes

Duit R (1998) Learning in science: From behaviourism towards social constructivism and beyond. Int Handb Sci Educ, 3–25

Duit R, Wodzinski CT (2010) Merkmale guten Physikunterrichts. Piko-Briefe. Der Fachdidaktische Forschungsstand Kurzgefasst. IPN Kiel

Duit R (2006) Initiativen zur Verbesserung des Physikunterrichts in Deutschland. PhyDid A-Physik Und Didaktik in Schule Und Hochschule 2(5):083–096

Duit R, Treagust DF (2003) Conceptual change: A powerful framework for improving science teaching and learning. Int J Sci Educ 25(6):671–688
Elsholz M, Lutz W, Trefzger T (2024) Das physikbezogene Selbstkonzept von Schüler:innen im Flipped Classroom
Finkenberg F (2018) Flipped classroom im Physikunterricht. Logos Verlag Berlin
Fischer HE, Klemm K, Leutner D, Sumfleth E, Tiemann R, Wirth J (2003) Naturwissenschaftsdidaktische Lehr-Lernforschung: Defizite und Desiderata. Z Für Didakt Naturwissenschaften 9:179–209
Frank F, Trefzger T (2024) Zum Einfluss der Nutzung von Modellanalogien auf das Konzeptwissen—Bericht zum Leistungsstand gymnasialer Mittelstufenschüler*innen in der Elektrizitätslehre
Gilboy MB, Heinerichs S, Pazzaglia G (2015) Enhancing student engagement using the flipped classroom. J Nutr Educ Behav 47(1):109–114
Hasselhorn M, Gold A (2017) Pädagogische Psychologie: Erfolgreiches Lernen und Lehren (4, aktualisierte. Kohlhammer, Stuttgart
Kulgemeyer C (2020) Erklären im Physikunterricht. In: Kircher E, Girwidz R, Fischer HE (eds) Physikdidaktik | Grundlagen. Springer, Berlin Heidelberg, pp 403–426
Lipowsky F (2020) Unterricht. In: Wild E, Möller (eds) Pädagogische Psychologie. Springer, pp 69–118. https://doi.org/10.1007/978-3-662-61403-7_4
Lutz W, Burde J-P, Wilhelm T, Trefzger T (2020) Digitale Unterrichtsmaterialien zum Elektronengasmodell. In: Nordmeier V, Grötzebauch H (eds) PhyDid B - Didaktik der Physik—Beiträge zur DPG-Frühjahrstagung Bonn 2020, pp 333–342. http://www.phydid.de/index.php/phydid-b/article/view/1098
Lutz W, Haase S, Burde J-P, Wilhelm T, Trefzger T (2021) Nutzung digitaler Materialien zum Elektronengasmodell im Homeschooling. Naturwissenschaftlicher Unterr Und LehrBildung Im Umbruch GDCP-Jahrestag., 137–140
Lutz W, Elsholz M, Haase S, Burde J-P, Wilhelm T, Trefzger T (2023) Flipped Classroom im Physikunterricht der Sekundarstufe I–Auswirkungen auf die Veränderung des individuellen Interesses im Bereich der E-Lehre, pp 35–49. https://doi.org/10.1007/978-3-662-66133-8_3
Lutz W, Haase S, Burde J-P, Wilhelm T, Trefzger T (2022) Erste empirische Ergebnisse zum Einsatz digitaler Materialien im Flipped Classroom zur E-Lehre und Optik. *Unsicherheit Als Element von Naturwissenschaftsbezogenen Bildungsprozessen.* GDCP-Jahrestagung 2021:260–263
Mayer RE (2005) Cognitive theory of multimedia learning. The Cambridge Handbook of Visuospatial Thinking/Cambridge University Press
Merzyn G (2010) Physik—Ein schwieriges Fach? In: Praxis der Naturwissenschaften—Physik in der Schule, vol 59, Issue 5, pp 9–12
Minstrell J (1982) Explaining the "at rest" condition of an object. The Physics Teacher 20(1):10–14
Neumann K (2014) Rasch-Analyse naturwissenschaftsbezogener Leistungstests. In: Krüger D, Parchmann I, Schecker H (eds) Methoden in der naturwissenschaftsdidaktischen Forschung. Springer, Berlin Heidelberg, pp 355–369
Nicolaisen T (2016) Unterstützung von Lernprozessen in Schule und Unternehmen. Coaching Als Individuelle Antwort Auf Gesellschaftliche Entwicklungen. Springer, pp 113–120
Perkhofer-Czapek M, Potzmann R (2016) Begleiten, Beraten und Coachen. Springer Fachmedien Wiesbaden. https://doi.org/10.1007/978-3-658-12828-9
Prenzel M, Seidel T, Lehrke M, Rimmele R, Duit R, Euler M, Geiser H, Hoffmann L, Müller C, Widodo A (2002) Lehr-Lernprozesse im Physikunterricht—Eine Videostudie. In: Bildungsqualität von Schule: Schulische und außerschulische Bedingungen mathematischer, naturwissenschaftlicher und überfachlicher Kompetenzen: Vol. Zeitschrift für Pädagogik, Beiheft, Issue 45, pp 139–156. Weinheim, Beltz. https://doi.org/10.25656/01:3944
Puentedura R (2006) Transformation, technology, and education
R Core Team (2022) R: A language and environment for statistical computing [Computer software]
Rhöneck von C (1986) Vorstellungen vom elektrischen Stromkreis und zu den Begriffen Strom, Spannung und Widerstand. Naturwissenschaften Im Unterr-Phys 34 (13):10–14

Robitzsch A, Kiefer T, Wu M, Robitzsch MA, Adams W, Rcpp L, LSAmitR RE (2020) Package 'TAM'. Test Analysis Modules–Version, 3, 26

Roehl A, Reddy SL, Shannon GJ (2013) The flipped classroom: An opportunity to engage millennial students through active learning strategies. J Fam Consum Sci 105(2):44

Schecker H, Duit R (2018) Schülervorstellungen und Physiklernen. In: Schecker H, Wilhelm T, Hopf M, Duit R (eds) Schülervorstellungen und Physikunterricht: Ein Lehrbuch für Studium, Referendariat und Unterrichtspraxis. Springer, pp 1–21

Schnotz W, Bannert M (2003) Construction and interference in learning from multiple representation. Learning and Instruction 13(2):141–156

Sweller J, Ayres P, Kalyuga S (2011) Intrinsic and extraneous cognitive load. In: Sweller J, Ayres P, Kalyuga S (eds) Cognitive load theory. Springer New York, pp 57–69

Urban-Woldron H, Hopf M (2012) Entwicklung eines Testinstruments zum Verständnis in der Elektrizitätslehre. Z Für Didakt Naturwissenschaften: ZfDN 18:201–227

Vosniadou S, Skopeliti I (2014) Conceptual change from the framework theory side of the Fence. Sci & Educ 23(7):1427–1445. https://doi.org/10.1007/s11191-013-9640-3

Weinert FE (2000) Lehren und Lernen für die Zukunft-Ansprüche an das Lernen in der Schule. Nachr Ges Zur Förderung Pädagogischer Forsch 2:4–23

Wilhelm T, Hopf M (2018) Schülervorstellungen zum elektrischen Stromkreis. In: Schecker H, Wilhelm T, Hopf M, Duit R (eds) Schülervorstellungen und Physikunterricht: Ein Lehrbuch für Studium, Referendariat und Unterrichtspraxis, pp 115–138. Springer. https://doi.org/10.1007/978-3-662-57270-2_6

Yildirim FS, Kiray SA (2016) Flipped classroom model in education. Res Highlights Educ Sci 2(6):1–8

A Scheduled Teaching Intervention for Newton's Disc, Programmable with Scratch, for Teaching and Learning Optics

Aristotelis Gkiolmas, Aikaterini Benisi, Anthimos Chalkidis, Artemisia Stoumpa, Ilias Boikos, Gianna Katsiampoura, Zografia Papanagiotou, George Koutromanos, and Elina Karagiannidou

Abstract This paper refers to a scheduled teaching intervention, based on Arduino and the Scratch programming environment, for programming the rotation of small Newton discs. The aim is for students in Primary or Secondary school classes, as well as for University students, to explore concepts of optics and issues related to elements of physics. In this context, concepts such as "colours" and "persistence of vision" are explored. The project has been published on electronic platforms, in Greek. The application and the corresponding teaching intervention in school classrooms can contribute to the feedback from students and teachers regarding learning outcomes.

Keywords Physical computing · Newton's disc · Scratch programming · Optics education

1 Introduction

As regards the concept of physical computing, it is a concept becoming more and more widely used and disseminated at all levels of education. In particular, physical computing is becoming an increasingly powerful tool for physics and informatics' education (Przybylla and Romeike 2014). Arduino, in particular, is already a popular device for teaching physics concepts (Organtini 2018), which is easy to use and

A. Gkiolmas (✉) · E. Karagiannidou
School of Primary Education, Aristotle University of Thessaloniki, Tower of the Faculty of Pedagogy, University Campus, Thessaloniki, Greece
e-mail: agkiolm@eled.auth.gr

A. Benisi · A. Chalkidis · A. Stoumpa · I. Boikos · G. Katsiampoura · Z. Papanagiotou · G. Koutromanos
Department of Pedagogy and Primary Education, National and Kapodistrian University of Athens, Athens, Greece

D. Sokołowska and E. McLoughlin (eds.), *Embracing Changes Together in Physics Education*, Challenges in Physics Education,
https://doi.org/10.1007/978-3-032-27385-7_5

relatively low cost. In education, projects with Arduino micro-controllers have often been used, in order to teach the properties of light (Saadi et al. 2019; Carvalho and Hahn 2016; Petry et al. 2016). However, a lack of educational proposals for using Newton's disc, in combination with educational robotics and Arduino devices, is definitely apparent. An exception to this is some efforts made with rotating LEDs (Saadi et al. 2019).

What has been apparent in the pre-existing literature is that combining Arduino with light measurements is mainly related to LEDs (Nieh and Chen 2023). On the other hand, Arduino has been used, for example, in physics projects, with the aim of teaching properties of light (Oprea 2018), but never related to the rotation of Newton's disc.

As can be seen, rotating a Newton's disc with Scratch programming is a project existing even in the international Scratch platform (for example https://scratch.mit.edu/projects/991316531), but there are no mentions in the literature about attempts to control its speed and check when "persistence of vision", or "additive colour mixing" appears, or what is the difference between six and seven colours (adding indigo) on the disc.

2 Theoretical Framework

The aim of the scheduled research is the interaction of students with programming environments (Scratch) and with educational robotics' devices, based on physical computing (such as Arduino). A sub-objective is the attempt to solve issues and problematic situations concerning light (with the main focus on Newton's disc), and another one is to achieve basic programming skills—particularly in Scratch programming—with a focus on physical computing, which means programming in order to control the comportment of an arrangement. Furthermore, the theoretical framework of the designed intervention as a whole was Constructionism (Papert and Harel 1991).

The experimental exploration of Newton's disc serves as the foundation for this teaching intervention, aiming to enhance computational learning. Computational thinking is an essential skill for everyone, not just those in computer science. It involves problem-solving, system design, and interpreting human behaviour through key computing concepts (Wing 2006). Most studies on assessing computational thinking typically rely on either student-developed or pre-structured programming tasks to gauge understanding of abstraction, logic, algorithms, and related ideas. Furthermore, educational methods like deconstructing, reverse engineering, and debugging have consistently been recognized as valuable tools for evaluating children's computational skills (Grover and Pea 2013).

There are some well-known papers about Primary School students' understanding of colour and light that focus on the misconceptions about primary colours (Martini et al. 2019) and students' apprehension of light concepts (Sahin et al. 2008). But both these—and other well-known studies—do not offer specific uses of Newton's disc. These exist only in school activities described on the Internet, and this consists a novelty in the current research.

On the other hand, it is recorded (Berns 2019) that the Newton's disc needs to spin at speeds ranging from 3000 to 5000 revolutions per minute (RPM) in order to achieve additive colour mixing ("see" white light). But these values are highly dependent on the characteristics of the disc, such as its size or the materials that it is made of Williamson and Cummins (1983). Additionally, the techniques of measuring the accurate rotation speed of Newton's disc, such as the stroboscopic method or the optical tachometers, photodetector arrays or even video analysis using high-speed cameras, are all considered as difficult to apply—especially in school praxis—and have limits in the validity of their results (Berns 2019; Sharma and Bala 2017; Wolfe 2007; Wyszecki and Stiles 2000).

3 Research Questions

The research questions that this educational research seeks to answer are: (i) what is the difference between the visual impression of Newton's classical disc (with indigo as the seventh colour, as found in many textbooks) and the theoretical scheme of the six-colour disc; and (ii) what kind of visual effects happen when the two aforementioned disc types are rotated, at various rotation speeds. There is literature evidence (Houstoun 1917) that the colours were chosen by Newton to be seven (instead of six), somehow arbitrarily, because he wanted the spectrum to have a similarity to the musical notes (Newton 1952). But this is an issue still undergoing investigation.

4 Methodology—Evidence from the Proposed Didactic Intervention

In the current teaching intervention activities students should work collaboratively, plan, test, and implement their own ideas, with the teacher/researcher acting as a facilitator and assisting the groups, if they need help.

The teaching sequence is designed—and pilotically applied—to students of the third grade of the Greek Gymnasium (Junior High School), who are about 15 years old in age.

In the first stage of the teaching intervention, the students in the classroom are to be familiarised—in groups—with various forms of small-scale Newtonian discs, controlled and driven by a motor, attached to an Arduino Uno micro-controller. The main components used, except for the Arduino, the breadboard, the resistor and the jumper wires, are:

- An OEM DC Motor for Arduino.

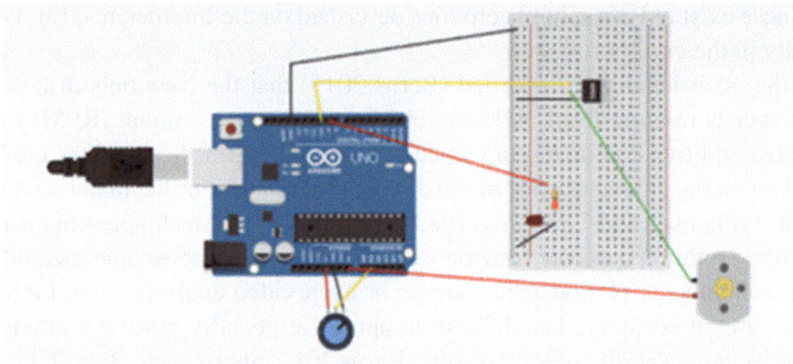

Fig. 1 The layout with the Arduino and the ONE motor, for turning one Newton disc, the classical one, with the seven colours

- An NPN Transistor (Model: TIP120).
- A Flyback Diode (Model: 1 N4007).
- An analogue rotation Potentiometer for Arduino (Model: DF Motor).
- The Digital Tachometer (Model: DT-2234C+, Manufacturer: OEM).

In Fig. 1, the first such device with ONE motor, designed to rotate ONE Newton disc, is depicted, as it will be used by the students in the teaching intervention.

Then, with the help of a digital revolution counter, students could measure the rotational speed of the disc, always watching out for possible technical difficulties and errors. Common mistakes to avoid, when measuring the rotational speed of a Newton's Disc with a digital tachometer, are:

- To place the reflective tape off-center or to use multiple reflective patches.
- To hold the tachometer at an angle or to not point it directly at the reflective surface.
- To let the disc wobble or not be securely mounted on its axis.
- To try to measure at very low speeds (Rounds per Minute, RPMs) (e.g., <100 RPM).
- To take measurements under strong ambient light (e.g., sunlight, reflective surfaces nearby).

The rotation speed can also be programmed from the Scratch code at one stage of the intervention. In particular, as shown in Fig. 2, a Scratch for Arduino code section is built, where the students may decrease or increase by 1 the rotational velocity ω of the motor, and, thus, of the Newton's disc. This way they can control it.

Also, with the aim of finding their correspondence, the experimentally measured values would be compared with the programmed ones. In Fig. 3, this device, a form of a digital tachometer, like the technique we described in the theory before, is shown, used by a student to measure the rotational speed of Newton's disc. In addition to this, the Arduino Arrangement is equipped with an LCD screen, where—by applying

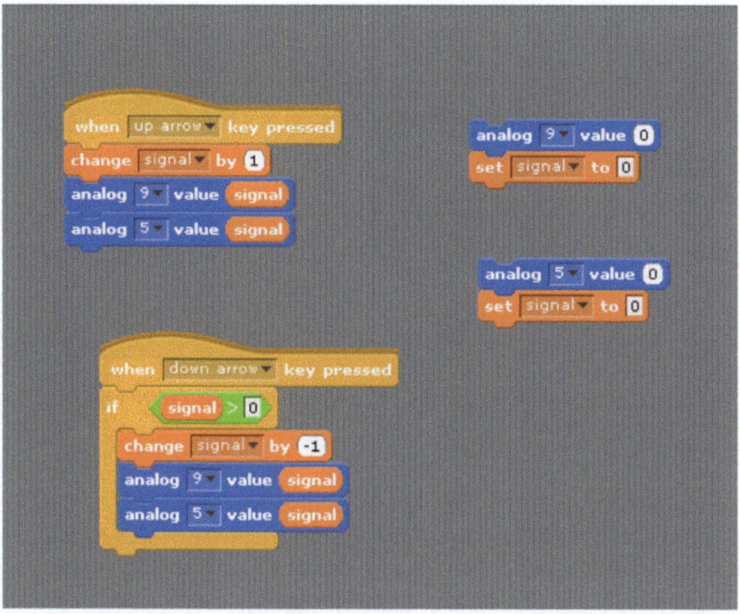

Fig. 2 A screenshot of the Scratch for Arduino code where the students may increase or decrease the rotational velocity of the disc by a "unit of 1"

a simple multiplication algorithm—the value of ω given in Scratch for Arduino is depicted as Rounds Per Minute (RPM), as the ones measured by the tachometer.

In the next stage of the intervention, the students, assisted by both the teacher and the researcher, should mount two Newton's discs on the Arduino, with the help of two motors. The one disc possesses the classic six (6) colours, while the other also possesses the seventh colour, indigo, as shown in Fig. 4. At the same time, the students are answering the Worksheets distributed to them. The images and

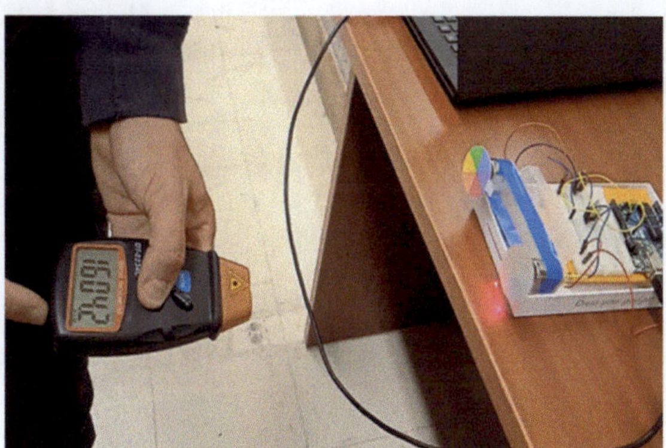

Fig. 3 The digital tachometer, used to measure the rotational speed of the Newton's disc

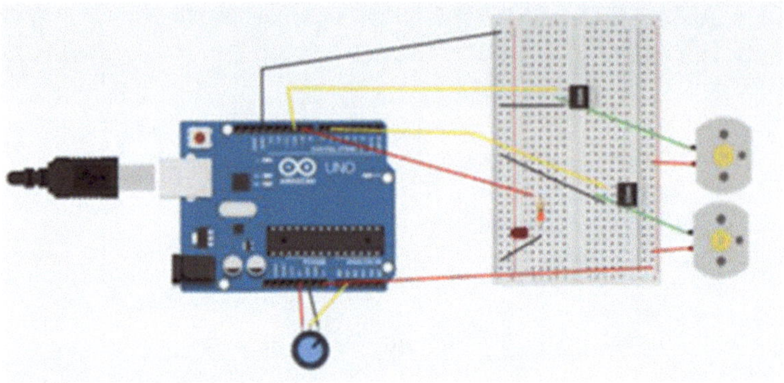

Fig. 4 The layout with the Arduino and the two motors turning the two Newton disks, one with six and one with seven colours

the descriptions/videos of the activities have also been uploaded to the "Github" web platform (Juviler 2021) (https://github.com/artemisia69/-.-O-Arduino.git), all written in the Greek language, with the aim of creating a community of users/teachers who will use it in their own classrooms, thus providing feedback and improvements.

In Fig. 5, one can see the actual arrangement with the two Newton's discs, mounted on the Arduino micro-controller and programmed through Scratch (as regards their speed of rotation). It was created by one of the groups of students, in a pilot implementation of the teaching sequence, that was tested in a class of students.

In the further, and last stages of the proposed teaching sequence, a set of teaching and learning activities takes place in the classroom, as depicted in Fig. 6.

Fig. 5 The arrangement with the two Newton's discs, one with the six colours, and one with the seven colours, mounted at the Arduino breadboard, and rotated trough two motors, created by a group of students

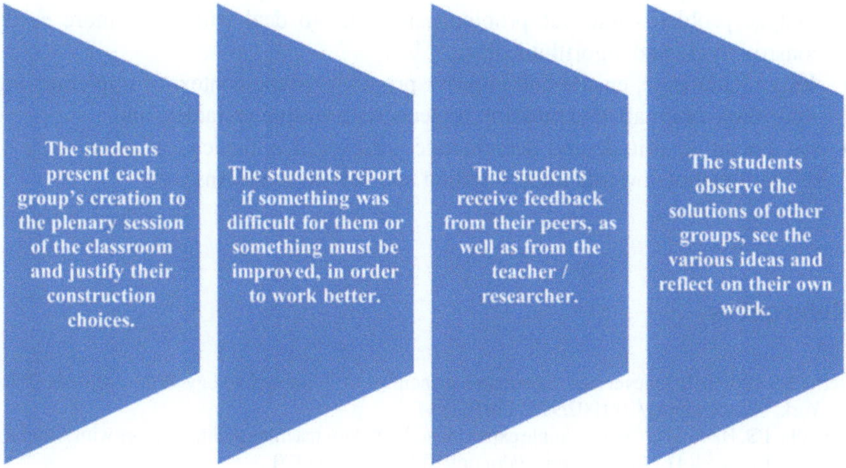

Fig. 6 The series of the educational activities within the classroom, which complete the teaching sequence

5 Expected Results

Upon completion of the teaching sequence, the learners are expected, through their interaction with the Arduino-controlled Newton's disc/discs, to have a measurable improvement in both their knowledge of optics and light in physics, as well as in their skills concerning operating educational robotic devices. Another expected outcome is gaining experience in programming (especially Scratch) and in algorithmic ways of thinking. Students may also achieve sensomotory, design and constructive skills. All the acquired knowledge and skills are linked to different cognitive subjects with the aim of promoting an interdisciplinary approach.

6 Conclusion

This submission is a scheduled project in progress for teaching light, using physical computing and educational robotics. With its future implementation in the classrooms and with the feedback from the users' communities, more measurable results and educational effects of the intervention are expected to rise. This teaching approach can be utilized not only in the third year of junior high school but also in high school classes as well as in students of pedagogical departments.

After completing the teaching intervention, students should be able to:

- relate the production of white colour to the angular velocity of the disc,
- describe the different optical result that the disc with 6 colours gives from the one with 7 colours,

- analyze problems into sub-problems, in order to deal with them more easily constructively and algorithmically,
- design, challenge, ensure and improve programs in the context of implementing algorithms or creative expression projects with simple variables, and.
- use a positive attitude and demonstrate perseverance, dedication and a sense of self-sufficiency towards understanding and solving problems.

References

Berns RS (2019) Billmeyer and Saltzman's principles of color technology Wiley, 4th edn, New York. https://doi.org/10.1002/9781119367314

Carvalho PS, Hahn M (2016) A simple experimental setup for teaching additive colors with Arduino. Phys Teach 54(4):244–245. https://doi.org/10.1119/1.4944370

Grover S, Pea R (2013) Computational thinking in K–12: a review of the state of the field. Educ Res 42(1):38–43. https://doi.org/10.3102/0013189X12463051

Houstoun RA (1917) Newton and the colours of the spectrum. Science Progress (1916–1919) 12(46):250–264

Juviler J (2021) What is GitHub? (and what is it used for?). https://blog.hubspot.com/website/what-is-github-used-for

Martini B, Tombolato M, D'Ugo R (2019) Primary colors as a source of possible misconceptions: an insight into teaching and learning about color. Cult E Sci Del Color-Color Cult Sci 11(02):25–33. https://doi.org/10.23738/CCSJ.110203

Newton I (1952) Opticks, or, a treatise of the reflections, refractions, inflections & colours of light. Courier Corporation

Nieh HW, Chen HY (2023) An Arduino-based experimental setup for teaching light color mixing, light intensity detection, and ambient temperature sensing. Phys Teach 61(2):133–137. https://doi.org/10.1119/5.0066060

Oprea M (2018) Arduino-based projects in physics education. In Conference proceedings of eLearning and Software for Education (eLSE) vol 14no. (2), Carol I National Defence University Publishing House pp 107–113. https://doi.org/10.12753/2066-026X-18-085

Organtini G (2018) Arduino as a tool for physics experiments. J Phys Conf Ser 1076(1):012026. https://doi.org/10.1088/1742-6596/1076/1/012026

Papert S, Harel I (1991) Situating constructionism. Constructionism 36(2):1–11

Petry A, Pacheco S, Lohmann D et al (2016) Project teaching beyond physics: integrating Arduino to the laboratory. Technol Appl Electron Teach (TAEE):1–6. https://doi.org/10.1109/TAEE.2016.7528376

Przybylla M, Romeike R (2014) Physical computing and its scope—towards a constructionist computer science curriculum with physical computing. Inform Educ 13(2):241–254

Saadi A, Chaouch Y, Belaidi H (2019) Propeller LED display using Arduino. In: Doctoral dissertation. https://dspace.univ-boumerdes.dz/handle/123456789/8258

Şahin Ç, İpek H, Ayas A (2008) Students' understanding of light concepts primary school: a cross-age study. In Asia-Pac Forum Sci Learn Teach 9:1–19

Sharma G, Bala R (eds) (2017) Digital color imaging handbook CRC Press, https://doi.org/10.1201/9781420041484

Williamson SJ, Cummins HZ (1983) Light and color in nature and art, vol 1. Wiley, New York. https://doi.org/10.1119/1.13928

Wing JM (2006) Computational thinking. Commun ACM 49(3):33–35. https://doi.org/10.1145/1118178.1118215

Wolfe WL (2007) Optics made clear: the nature of light and how we use it, vol 163. SPIE Press

Wyszecki G, Stiles WS (2000) Color science: concepts and methods, quantitative data and formulae, 2nd edn. Wiley Interscience

Building Physics Competences at the University Level

Enhancing Physics Education Through Dimensions of Learning

Claudio Fazio

Abstract One of the main aims of education research is to foster effective learning of new knowledge, concepts, and methods. In recent years, there has been growing interest in the educational community to specify and clarify the different aspects of learning processes that can help teachers and researchers offer students a multifaceted approach to constructing their knowledge, and physics education is no exception. By addressing cognitive, affective, and metacognitive aspects of learning, physics educators can create enriching experiences that empower students to become lifelong learners and proficient problem-solvers in physics. Frameworks developed by educational researchers emphasize that specifying the "dimensions" of learning enables educators to create dynamic learning environments that cater to diverse student needs and promote holistic understanding. Dimensions of learning, such as the development of attitudes and perceptions, learning styles, the acquisition and organization of knowledge, critical thinking, and metacognitive thinking, have been studied to improve students' engagement in study and exploration and their appropriation of subject matter. In this paper, I will discuss some relevant aspects of these frameworks and present a map, developed during research on learning physics related to surface phenomena in liquids with high school students, that makes explicit the dimensions of learning we wanted to study at that schooling level. Highlighting these dimensions can help teachers and researchers consciously design teaching activities and assess their efficacy, as well as select appropriate methods of data collection and analysis.

Keywords Physics education · Dimensions of learning · Teaching learning sequences

C. Fazio (✉)
Dipartimento di Fisica e Chimica—Emilio Segrè, Università degli Studi di Palermo, Palermo, Italy
e-mail: claudio.fazio@unipa.it

D. Sokołowska and E. McLoughlin (eds.), *Embracing Changes Together in Physics Education*, Challenges in Physics Education,
https://doi.org/10.1007/978-3-032-27385-7_6

1 Introduction

A primary objective of research in science education is to explore the challenges associated with teaching and empowering students' learning experiences.

Research has long shown that learning involves more than the simple memorization and recall of concepts and notions. Learning, as outlined by cognitive theories, involves transformations in mental processes and the arrangement of knowledge that stem from a learner's efforts to grasp the realities of the world (Eggen and Kauchak 2004). Researchers in cognitive science have outlined how individuals construct their own understanding of the world, coining the phrase 'mental model' and detailing its key features (Johnson-Laird 1983; Norman 1983).

Gilbert and Boulter (1998) emphasized that learning is inherently a personal experience, indicating that researchers need to rely on some "expressed forms" of mental models to infer what they are and their implications. This is largely achieved through external forms that illustrate an individual's reasoning, such as spoken words, nonverbal communication, written documents, or actions.

According to other authors (Ambrose et al. 2010; Mayer 2002), learning can be described as "a process that leads to change, which occurs as a result of experience and increases the potential for improved performance and future learning", means "acquiring knowledge and skills and having them readily available from memory so you can make sense of future problems and opportunities" (Brown et al. 2014), and "occurs through experience" (Schunk 2020).

So, following the literature, we can say that:

1. Learning is an evolving experience instead of a finished result. However, because this process takes place internally, in the learner's mind, we can only infer that it has occurred based on the work or performances exhibited by learners.
2. Learning involves a change in knowledge, beliefs, behaviors, or attitudes. This transformation occurs gradually and, if significant, it is enduring, significantly influencing students' thought processes and behaviors.
3. The process of learning is not a passive experience for students; it is something they actively undertake. It results from their interpretations and responses to their experiences, which can be both conscious and unconscious, and are rooted in their history and current situations.

Particularly, meaningful and lasting learning involves:

- Understanding and relating ideas
- Creating connections between previous and new knowledge
- The construction of independent and critical thinking
- The ability to transfer knowledge to new and different contexts.

and much more …

So, the teacher/researcher must keep in mind that there is much more to learning than merely acquiring content. It is necessary to pay attention not only to it but also to the cognitive processes related to learning and issues related to the emotional/

affective sphere (Cherukunnath and Singh 2022; Tyng et al. 2017; Pekrun et al. 2018).

Current insights reveal that tailoring educational experiences with foundational concepts from cognitive psychology can significantly enhance learning effectiveness by boosting students' drive, self-confidence, and cognitive advancement. Numerous proposed approaches for acquiring scientific knowledge, including the widely recognized Investigative Science Learning Environment (ISLE) model (Etkina et al. 2019), promote student involvement in instructional exercises that emulate the dynamics of science and collaboration, intending to cultivate substantial and transformative learning outcomes, "growth mindset" (Dweck 2016), and an overall sense of achievement in their studies.

According to American psychologist Carol Dweck, students' academic results are significantly influenced by their perceptions of whether their skills are fixed or can be developed over time. Students with a "growth mindset" are equipped to consciously and actively shape their educational experience through the projects their teachers assign. This inspires learners to engage in deliberate (Ericsson 2007) and context-sensitive practice (Scherr 2007), systematically addressing more challenging tasks, with the intention of moving beyond their "cognitive comfort zone" in their areas of proficiency.

While Dweck's mindset theory has faced criticism regarding its practical application in education, especially regarding teachers' perspectives (Canning et al. 2019; Muenks et al. 2020), studies indicate that mindset interventions can yield positive results across diverse settings (Yeager et al. 2019, 2021). A sense of confidence in their ability to develop their skills leads students to invest more energy in educational activities. As a result, dedicating time, energy, and support to managing progressively complex assignments allows learners to develop skills like those of experts, fostering intentional and continuous education, and supporting the growth of self-esteem and reflective thinking. A vital feature of deliberate practice is the sustained effort and progressive enhancement of skills, which supports students in effectively building these proficiencies (Mayer 2008). Through deliberate practice, learners can deepen their understanding of their skills and knowledge base. By understanding themselves better, learners can discern their strengths and weaknesses, prompting them to consider strategies to improve their academic performance. These strategies allow learners to introspect on their distinct learning/cognitive styles (Barbe et al. 1979; Fleming and Mills 1992; Fleming 1995; Felder and Silverman 1988) and determine their level of understanding in different environments. The aim is to help learners cultivate and enhance fresh abilities and mental processes, while also ensuring they do not linger excessively in their cognitive safe space, which, as highlighted, tends to be counterproductive to notable cognitive progress.

The process of considering how students learn, and their cognitive styles, is vital for educators and researchers in creating and implementing a teaching and learning framework or lesson plan. In fact, a meaningful educational journey for students must consider their preferred methods of gathering information from their surroundings (including the educational context) and elaborate on it to generate thorough descriptions and insights into real-life events and situations. In essence, having knowledge of theories about the possible learning styles of students, exemplified by the notable VARK model (Fleming and Mills 1992; Fleming 1995), a development of the earlier VAK learning styles framework (Barbe et al. 1979), can assist in the creation and application of educational strategies that successfully foster student learning, the key aim of any educational initiative.

2 Educational Pathways and Learning

Over the past few years, numerous theoretical models (Lijnse 1995; Kattmann et al. 1995; Duit et al. 2005, 2012) and pedagogical methods have been proposed and validated to enhance teaching and learning efficiency. On the basis of these models and methods, research has focused on developing and trialling educational pathways tailored to specific subjects, with the goal of assisting educators in organizing and tracking their teaching, ultimately improving student learning outcomes. Examples found in the literature include the so-called Teaching–Learning Sequences (TLSs) (e.g., Psillos and Kariotoglou 2016; Meheut and Psillos 2004) and Learning Progressions (LPs) (e.g., Schmidt 2005; Jackson et al. 2018; Kim and Care 2018).

A key feature of TLSs and LPs is their foundation in research-oriented evolutionary approaches that merge scientific understanding with learners' perspectives, describe and evaluate changes in students' thinking and skills, and optimize learning outcomes and their transferability (Psillos and Kariotoglou 2016). TLSs and LPs are usually composed of thoughtfully arranged learning segments that learners must complete to attain comprehensive educational targets. These segments frequently involve various sub-skills and essential knowledge frameworks. In many cases, TLSs and LPs are organized and used as research tools to investigate the effectiveness of proposed activities and evaluation techniques, and they also serve as innovative suggestions that show how specific learning-related issues can be effectively addressed.

One of the most frequently referenced and debated theoretical models aimed at enhancing the efficiency of teaching and learning processes is the "Educational Reconstruction" framework created by Kattmann and colleagues in 1995 (Kattmann et al. 1995; Duit et al. 2005, 2012). This framework looks carefully at the meaning of the term "learning", asserting that a well-structured learning process should stem from an in-depth investigation of the framework of scientific content and pertinent educational issues, such as familiarity with the context, student common sense ideas, learning knots and processes known from research, and social and ethical implications involved. To effectively "reconstruct" the material for instruction and design the educational journey, one must consider both the structural analysis of scientific concepts and the relevant educational factors. The model is based on an integrated constructionist perspective: Learning is understood as the learner's active engagement in a social and tangible context, with scientific knowledge being interpreted as a tentative human development (Meheut and Psillos 2004).

As a consequence of adopting this holistic constructionist framework, the primary educational strategies used to formulate and advance TLSs and LPs are fundamentally rooted in the concept of "Active Learning". Studies in Science Education indicate that shifting from traditional deductive teaching methods to strategies that engage students actively in their learning can enhance educational outcomes. This is especially evident in enhancing students' grasp of concepts, sustaining knowledge over time, and fostering critical thinking (e.g., Yew and Goh 2016; Duran and Dokme 2016; Oja 2011). Active learning involves learners taking a more hands-on (and minds-on) approach than merely attending a lecture. They engage in reading, writing, formulating, and discussing inquiries, sourcing data from multiple outlets, creating models, and tackling issues to foster the growth of knowledge, competencies, and perspectives.

The concepts that support active learning are evidently based on constructivist theory, which outlines how individuals can learn and gain knowledge effectively. The process of acquiring knowledge is dynamic and unfolds in multiple stages, during which learners adapt to their surroundings while actively formulating, evaluating, and modifying their beliefs about the world. Evidence from the literature on constructivist learning frameworks indicates that the most effective way to acquire practical knowledge is through participatory learning experiences (e.g., Herrington and Oliver 2000).

3 "Dimensions of Learning" Frameworks

In 1992, Marzano introduced the "Dimensions of Learning" framework (Marzano 1992), a comprehensive model designed to enhance teaching and learning by focusing on five critical aspects, or dimensions, of the learning process.

Grounded in studies of cognitive psychology and pedagogy, this model equips educators with an organized strategy for designing educational pathways and curricula, conducting and validating instruction, and assessing learners.

According to Marzano, the five dimensions are:

1. **Attitudes and Perceptions**: This dimension highlights how vital students' feelings and views about education are. It is essential to foster a supportive atmosphere where learners feel valued, secure, and at ease. When students have positive attitudes and perceptions, they are more likely to engage actively in the learning process.
2. **Acquire and Integrate Knowledge**: In this dimension, the priority is to support students in obtaining new knowledge and blending it with what they already know. This encompasses instructing students on both factual knowledge (data and details) and procedural knowledge (methods for executing tasks and processes). Successful teaching methods employ advanced organizers, analogies, and metaphors to connect new information to existing knowledge.

3. **Extend and Refine Knowledge**: Once learners have obtained new knowledge, it is essential for them to deepen and refine it by critically examining, assessing, and integrating various pieces of information. This dimension promotes the application of cognitive strategies such as comparison, categorization, abstraction, and both inductive and deductive reasoning. Engaging learners in activities that necessitate the use of these methods enhances their comprehension and fosters advanced cognitive skills.
4. **Use Knowledge Meaningfully**: This dimension focuses on the application of knowledge in real-world contexts. Learners are encouraged to apply their expertise to address challenges, reach conclusions, carry out research, and participate in creative activities. By providing opportunities for meaningful use of knowledge, educators help learners see the relevance of what they are learning and foster intrinsic motivation.
5. **Productive Habits of Mind**: This final dimension emphasizes the cultivation of effective cognitive practices that foster continuous education throughout life. These include critical thinking, creative thinking, and self-regulation. Promoting students' self-awareness of their thought processes, fostering receptiveness to new concepts, and encouraging perseverance in the face of difficulties enable them to become autonomous and proficient learners.

The comprehensive nature of Marzano's framework has led to its widespread implementation in educational contexts, as it effectively supports understanding and learning advancement. By addressing each of these dimensions, educators can create a balanced and effective instructional program that not only conveys information but also cultivates the abilities and mindsets essential for students to thrive as lifelong learners.

Research (e.g., Alrababah 2017; Kumar and Sethy 2025) has shown that applying the Dimensions of Learning model can improve student outcomes, positively impacting students' acquisition of scientific concepts, critical thinking, and problem-solving skills.

Building on Marzano's idea of the dimensions of learning, Tareq Rasheed introduced the Knowledge, Skill, Attitude, and Habit (KSAH) model of self-development in 2021 (Rasheed 2021). It is a framework designed to guide individuals' personal and professional development. The model highlights a step-by-step method for learning and personal growth, focusing on relevant aspects of learning, from gaining knowledge to establishing lasting habits.

1. **Knowledge**: This initial phase focuses on collecting and comprehending details about both personal identity and the surrounding environment. Self-recognition plays a key role here, as it allows individuals to assess their strengths, areas for improvement, core beliefs, and future aims. By gathering significant information, one prepares the way for further progress.
2. **Skills**: This phase emphasizes the hands-on use of acquired knowledge, organizing abilities into distinct categories:

    - Technical Skills: Specific to a particular field or profession.

- Interpersonal Skills: Also known as soft skills, these include communication, problem-solving, and teamwork abilities.
- Transferable Skills: Capabilities that span diverse domains, such as language skills or technological proficiency.

Mastering these skills enables individuals to effectively translate their knowledge into real-life applications.

3. **Attitude**: This component illustrates the way people interpret and tackle circumstances, shaped by their principles and convictions. A positive attitude (or mindset) enhances the use of expertise and abilities, promoting initiative and resilience when confronted with difficulties.
4. **Habits**: The last phase entails assimilating attitudes and actions into habitual practices. Cultivating beneficial habits promotes continuous growth both personally and professionally, as behaviors turn more reflexive and aligned with one's aspirations.

Advancing through these steps enables people to achieve ongoing personal growth and success. The KSAH model acts as a thorough resource for individuals aiming to improve their learning, focusing on self-awareness and effectiveness.

4 Planning a TLS on Surface Phenomena Using Dimensions of Learning

In this section, we discuss some aspects of the design of two TLSs on surface phenomena in liquids. We will not focus here on the specific content and phases of the TLSs, including their trialling and validation, that are described elsewhere (Fazio et al. 2023; Termini et al. 2025), but we will discuss some aspects of the use of the Dimensions of Learning framework during the planning phase of the TLSs.

The TLSs were designed to encourage high school students to delve into the intriguing realm of surface phenomena using two distinct perspectives: one focusing on macroscopic aspects and the other on mesoscopic ones. While educational discussions of surface phenomena, particularly in relation to modeling physical events, are rarely documented in the existing literature, a comprehensive understanding of surface interactions is essential for grasping concepts such as the buoyancy of lightweight objects. (insects, razor blades, …) on water, wetting, and capillarity, which hold considerable real-world relevance in fields such as materials science, biological studies, chemical research, and engineering disciplines. Because of its intricate details, this topic is commonly neglected in both school and introductory physics programs at the university level, or it is presented in overly simplistic ways that do not stimulate students' interest or curiosity (e.g., Marchand et al. 2011). Unfortunately, this approach hinders students' ability to examine and understand concepts related to fluid dynamics, surface chemistry, or the interactions of liquid interfaces in biological contexts. The usual methods of describing superficial phenomena, commonly utilized in educational institutions, might prevent students from fully understanding more sophisticated theories and make it difficult to connect theoretical knowledge with real-world phenomena. Concurrently, conventional educational techniques that rely on one-way communication, such as lectures with minimal student engagement, textbook reading, and limited laboratory time, are prevalent and do not contribute to producing significant learning.

Based on these insights, we sought out alternative explanations of surface phenomena that diverge from conventional methods, as well as teaching strategies that actively engage students in effectively grasping and comprehending this subject matter. To carry out these approaches, we designed and evaluated two TLSs focusing on hands-on experiments, engaging computer-based simulations, and collaborative activities that students can undertake with the support of multimedia resources (Fazio et al. 2023; Termini et al. 2025). Active student involvement in laboratory activities leads to better understanding and assimilation of the topics addressed. Interactive simulations that students can easily control enable them to observe in real time the changes in the characteristics and behavior of the physical systems being studied as they adjust the parameters of the corresponding model (Rutten et al. 2012; Kincaid et al. 2003). This helps students grasp the concepts being studied in a solid and relatable way. The use of computer-based simulations is a valuable approach for visualizing surface phenomena and facilitating the adjustment of parameters to help students describe them (e.g., Battaglia et al. 2019, 2010). This interactive technique links abstract theories to practical experiences, encouraging a richer grasp of the operational principles behind the physical phenomena under investigation.

The TLSs were designed to integrate hands-on laboratory experiments with computer simulations, enabling students to thoroughly investigate surface phenomena and effectively analyze their experimental results alongside simulation outputs to gain insights into the model's significance and how it is interpreted. All the activities proposed as part of the trialling are based on cooperative learning and teamwork. During lab and modelling sessions, students collaborate, share thoughts, tackle challenges, and collectively analyze outcomes. This fosters a nurturing, participatory learning environment where students exchange knowledge and gain insights from various perspectives.

Given that the model, i.e., the scale through which a given phenomenon can be described, significantly affects students' ability to learn and understand it, we chose to construct two separate TLSs that would facilitate experimentation with alternative approaches to the topic and help us identify which components would foster more successful and sustainable learning. Evidence from the literature indicates that macroscopic models are beneficial for teaching students about solid friction and fluid statics (Besson and Viennot 2004), yet they fall short in effectively conveying surface phenomena and promoting substantial student engagement in their educational experiences. Research suggests that using micro/mesoscopic models of various phenomena can greatly improve students' understanding of surface phenomena by helping them identify the causal links and mechanisms that drive the physical processes being studied (Battaglia et al. 2019; Besson and Viennot 2004; Katsoulakis and Vlachos 2004; Chittleborough and Treagust 2007; Corpuz and Rebello 2011).

Considering this, we formulated two different TLSs, distinguished by the scale of the modelling process (macroscopic and mesoscopic, respectively), to determine whether the selection of modelling scale, paired with specific educational approaches and teaching aids, can significantly affect the process of teaching and learning about complicated subjects like surface phenomena in liquids.

A preliminary study of the TLSs was conducted with upper secondary students, aiming to gather support for our hypothesis and to gain insights into how mesoscopic modelling functions within science education.

The literature does not frequently address surface phenomena from an educational viewpoint. The aim of evaluating two distinct methods regarding surface phenomena was not to establish which modeling technique yielded better educational results. Rather, the main aim was to investigate and identify the crucial features of each strategy that significantly promoted science learning.

However, the idea of "promotion of science learning" can be examined from various angles, making it challenging to fully understand its intricate nature. To gain clearer insight into this idea, we referred to the Dimensions of Learning framework and, following an extensive review of the existing literature, identified various aspects of learning that can be useful in Physics (or, more generally, Science) Education. Thus, we constructed a map that conveys the critical components of our framework (see Fig. 1). Drawing from existing studies, we have pinpointed three key aspects (or dimensions) of learning that pertain to our investigation: (1) Students' acquisition of conceptual knowledge; (2) Intellectual growth; and (3) Development of habits of mind suitable for learning science.

The process of acquiring conceptual knowledge entails students developing a deep grasp of the fundamental concepts and theories relevant to the subject matter they are exploring. It encompasses more than simply recalling information; it also involves the skill of using and relating ideas in a meaningful way, evolving common-sense conceptions into more scientific ones, and retaining knowledge. The process of intellectual growth encompasses the cultivation of analytical thinking, the capacity to tackle challenges, and the proficiency in executing scientific research. This entails the growth of higher-level cognitive functions, such as analysis, synthesis, and evaluation, alongside the development of creative and critical thinking skills and generalization skills. The development of habits of mind suited to learning science requires nurturing students' perspectives, convictions, and enthusiasm for studying science. This dimension is strictly linked to self-efficacy beliefs, metacognition in the learning process, growth mindset, and reflection on the nature of science.

So, every dimension contains multiple sub-dimensions. We used them as study variables in our research to examine the dimensions at a finer level of detail, focusing on the nuances of student learning. Moreover, analyzing the learning sub-dimensions from the data from the TLSs' trial enabled us to thoroughly grasp the impact of specific activities, teaching approaches, educational tools, or proposed model scales on enhancing or impeding student learning, as thoroughly explained elsewhere (Fazio et al. 2023; Termini et al. 2025).

While the three dimensions pertain to distinct aspects of education, they are intricately linked and mutually supportive. For instance, acquiring conceptual knowledge

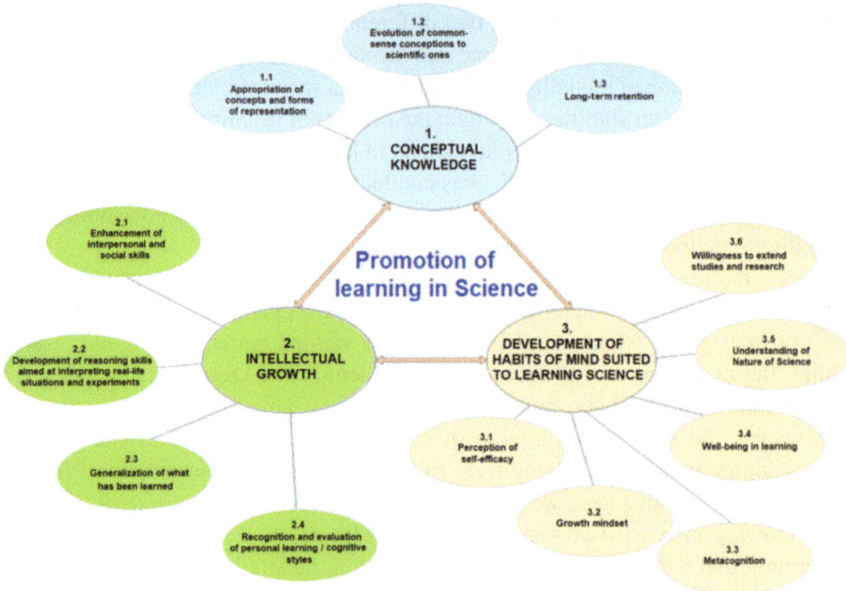

Fig. 1 Map showing the main research-informed dimensions and sub-dimensions of learning that we consider when we analyze the concept of "promotion of science learning"

can foster intellectual growth by providing a foundation for analytical thinking and problem-solving. Likewise, a favourable attitude towards learning science can boost students' motivation and participation, which, in turn, fosters a deeper understanding of concepts and promotes cognitive growth.

Overall, the map proposed in Fig. 1 provides a conceptual framework for studying the factors that play a key role in science learning, i.e., those that promote or hinder student learning within a pedagogical pathway.

Following, we give a concise description, drawn from the literature, of the various sub-dimensions shown in Fig. 1. Details on them can be found in the references reported after each description. They cannot cover the full literature on the topic, yet provide an overview of the key concepts related to promoting science learning, as we have addressed in our research.

4.1 Appropriation of Concepts and Forms of Representation

To achieve a deep and meaningful understanding of concepts, students must "own the content" and be able to represent it in different ways, modifying and transforming the reasoning patterns of their knowledge related to it. Appropriation of content implies a deep conceptual understanding, but it also involves a reflexive process of transforming scientific discourse in an authentic and personal way. On the other hand, the appropriation of forms of representing concepts is essentially related to the ability to use different representations and communication channels. Verbal, iconic, tabular, graphic, and analytical representations are among the most widely used in science (Levrini et al. 2015; Rogoff 1995; DiSessa and Sherin 1998).

4.2 Evolution of Common-Sense Conceptions to Scientific Ones

Students initiate the process of gaining and organizing knowledge by creating beliefs from their day-to-day experiences and cultural influences and organizing them into patterns and models of reality. When learning science, students should recognize the underlying assumptions and epistemological commitments that define their intuitive theories and acknowledge their differences and limitations in comparison to scientific theories. This means they must also become aware of their experience-based beliefs and develop new belief systems grounded in scientific reasoning and executive function skills (Ogborn 2006; Magnusson et al. 1997; Vosniadou 2019).

4.3 Long-Term Retention of Concepts

Memory is not static and unchangeable, but is an active, dynamic, continuously evolving process. The ability to store new information in one's long-term memory implies, in fact, the ability to effectively integrate and modify the network of one's previous knowledge. It is recognized that the value of education strongly depends on long-term retention of what has been learned (Kastrinos 1965; Arthur et al. 1998; Bahrick 1984; Persano-Adorno et al. 2018).

4.4 Enhancement of Interpersonal and Social Skills

The development of students' social and interpersonal skills must be a primary educational objective. Transversal skills such as knowing how to work in a team, cooperating, the ability to help and support those who are in difficulty, and recognizing and accepting differences represent the basis on which all the other competencies can be built and represent the key to effective intellectual growth (McFall 1982; Heckman and Kautz 2012).

4.5 Development of Reasoning Skills Aimed at Interpreting Real-Life Situations and Experiments

In Physics, an explanation of a system or of a situation/observation is different from a description, because it supposes that the system has properties which are not directly observable, but play a role in the observed regularities. The construction and validation of explanations require building several hypothesis typologies: empirical law hypothesis, synthesis of regularities (arising from phenomenological observations and condensed into rules), and hypothesis for the construction of explicative models introducing theoretical representations and often containing non-observable entities (Besson 2010; Fazio et al. 2013; Bao and Redish 2006; Gopnik 2000).

4.6 Generalization of What Has Been Learned

The generalization of what has been learned is often related to the systematization of knowledge. Indeed, combining knowledge of facts, phenomena, laws, and principles into a system allows students to discover new, previously unknown connections among concepts and their relationships. The generalization of knowledge can be considered a transition to a higher level of abstraction that highlights common signs (properties, relationships, etc.) of objects and phenomena. The generalization of knowledge leads to significant changes in the quality of knowledge and to its transfer (Gluck et al. 2016; Banich and Caccamise 2011; Shea and Bauer 2007).

4.7 Recognition and Evolution of Personal Learning/Cognitive Styles

Meaningful didactics aiming at promoting effective and personalized learning should be guided by the analysis of students' learning and cognitive styles. According to Keefe (1985; 1979), learning styles are "a composite of characteristic cognitive, affective, and physiological factors that serve as relatively stable indicators of how a learner perceives, interacts with, and responds to the learning environment." Cognitive styles indicate "Consistent individual differences in preferred ways of organising and processing information and experience" (Barbe et al. 1979; Fleming and Mills 1992; Fleming 1995; Felder and Silverman 1988; Riding and Rayner 2007; Reid 2002; Messick 1984; Sadler-Smith 2001).

4.8 *Self-Efficacy*

Self-efficacy is the belief in one's capacity to influence his/her own motivation, actions, and social environment. Self-assessments of cognition impact various aspects of human experiences, encompassing the objectives individuals aim for, the level of effort put into reaching those goals, and the probability of achieving specific levels of performance (Bandura 1993, 1997; Zimmerman et al. 1992; Zimmerman 2000; Tang and Zhang 2020; Cassidy 2015).

4.9 *Growth Mindset*

Learning is influenced by personal beliefs about one's abilities and predispositions. If a person believes that his/her abilities are not static but can evolve, he/she will approach learning activities more positively and actively, and will also accept criticism and seize growth opportunities more effectively. A Growth Mindset can help students free themselves from outdated mental patterns and take responsibility for directing their own learning. Moreover, it encourages students not to settle into a "cognitive comfort zone" that can be identified with skills that they believe they have mastered, and which makes further appreciable cognitive growth less easy (Dweck 2016; Ericsson 2007; Scherr 2007; Canning et al. 2019; Muenks et al. 2020; Yeager et al. 2019, 2021; Mayer 2008; Zhang et al. 2017).

4.10 *Metacognition*

Martinez defines it as "the monitoring and control of thought" (Martinez 2006). Components of metacognition are (Baker 2002; Harris et al. 2009; Pintrich 2002):

- Metamemory: How accurately can someone evaluate his/her own knowledge state
- Metacomprehension: To what extent can someone accurately judge his/her comprehension of written or spoken communication
- Problem solving: The pursuit of a goal when the route to it is unclear. It is done continuously, generating possibilities, weighing those options, exploring subsets of options, and evaluating the results
- Critical thinking: Appraising ideas for their quality, specifically analyzing whether or not they make sense

4.11 Well-Being in Learning

Well-being is crucial for the development of safe, dynamic, and effective learning settings. It is not an "add-on" that educators might address if they have extra time, as it plays a central role in the development of students' learning skills and in teachers' ability to offer top-notch teaching and learning environments. While it is essential for educators to create a supportive, inclusive, and emotionally safe environment, one that encourages openness, collaboration, and respect, well-being should not be seen solely as the responsibility of the teacher or the school context. Rather, it must also be cultivated by students themselves through active participation, self-reflection, and engagement in their educational journey (Noddings 2003; Konu and Rimpelä 2002; Sellström and Bremberg 2006; Boekaerts 1993).

4.12 Understanding of Nature of Science

Properly learning science requires students to understand what science is, what the procedures typical of science are, and modify the way they represent and explain the physical world, as well as their approaches to reasoning (Vosniadou 2019; Schwartz et al. 2004; Abd-El-Khalick 2012).

4.13 Willingness to Learn and Extend

It can be defined as "an impulse or desire within yourself to continue to search for information and understand various things in the context of self-development". This form of curiosity illustrates a dynamic engagement in the learning process, in which individuals possess an inherent drive to seek out new concepts, question existing beliefs, and deepen their understanding beyond mere goals or external rewards (McDevitt 1997; Freeman et al. 2014; Baker et al. 2002; Wößmann 2010).

5 Conclusion

In this paper, we discussed relevant aspects of frameworks developed by educational researchers to focus on specific "dimensions" of learning, with the aim of creating dynamic learning environments that cater to diverse student needs and promote holistic understanding. Understanding of different dimensions of learning, like the development of attitudes and perceptions, learning styles, acquisition and organization of knowledge, critical thinking, and metacognitive thinking, can help the teachers and the researchers to focus on different nuances of student learning, to improve student appropriation of subject matter, and to engage in study and exploration. Moreover, analyzing the learning sub-dimensions from trial data on pedagogical pathways can enable teachers/researchers to better understand the impact of specific activities, teaching approaches, and educational tools on student learning.

Moreover, we discussed a map, developed during research on learning the physics related to surface phenomena in liquids with high school students, that makes explicit the dimensions of learning that we wanted to study at that schooling level.

References

Abd-El-Khalick F (2012) Examining the sources for our understandings about science: Enduring conflations and critical issues in research on nature of science in science education. Int J Sci Educ 34(3):353–374. https://doi.org/10.1080/09500693.2011.629013

Alrababah FEA-R (2017) The effectiveness of Marzano's dimensions of learning model in the tenth-grade students' acquisition of scientific concepts in science and the development of critical thinking skills and their attitude towards science. British J Educ 5(1):76–84

Ambrose SA, Bridges MW, DiPietro M, Lovett MC, Norman MK (2010) How learning works. John Wiley & Sons Inc., San Francisco

Arthur W Jr, Bennett W Jr, Stanush PL, McNelly TL (1998) Factors that influence skill decay and retention: A quantitative review and analysis. Hum Perform 11(1):57–101

Bahrick HP (1984) Semantic memory content in permastore: fifty years of memory for Spanish learned in school. J Exper Psych.: General 113(1):1–29. https://doi.org/10.1037/0096-3445.113.1.1

Baker L (2002) Metacognition in comprehension instruction. In: Block CC, Pressley M (eds) Comprehension instruction: Research-based best practices. Guilford Press, NY, pp 77–95

Baker P, Goesling B, Letendre G (2002) Socioeconomic status, school quality, and national economic development: A cross-national analysis of the "heyneman-loxley effect" on mathematics and science achievement. Compar Educ Rev 46(3):291–312. https://doi.org/10.1086/341159

Bandura A (1993) Perceived self-efficacy in cognitive development and functioning. Educ Psychol 28(2):117–148. https://doi.org/10.1207/s15326985ep2802_3

Bandura A (1997) Self-efficacy: The exercise of control. WH Freeman and Co, NY

Banich MT, Caccamise D (2011) Generalization of knowledge: Multidisciplinary perspectives. Psychology Press, NY

Bao L and Redish EF (2006) Model analysis: Representing and assessing the dynamics of student learning. Phys Rev Special Topics—Phys Educ Res 2(1): 010103. https://doi.org/10.1103/PhysRevSTPER.2.010103

Barbe WB, Swassing RH and Milone MN (1979) Teaching through modality strengths: concepts practices. Zaner-Bloser, Columbus

Battaglia OR, Agliolo Gallitto A, Fazio C (2019) Computer simulations to approach surface tension by means of a simple mesoscopic mechanical model. Comp Appl Eng Educ 27(6):1333–1342. https://doi.org/10.1002/cae.22154

Battaglia OR, Fazio C, Guastella I, Sperandeo-Mineo RM (2010) An experiment on the velocity distribution of thermionic electrons. Am J Phys 78:1302–1308. https://doi.org/10.1119/1.3471937

Besson U (2010) Calculating and understanding: Formal models and causal explanations in science, common reasoning and physics teaching. Sci & Educ 19:225–257. https://doi.org/10.1007/s11191-009-9203-9

Besson U, Viennot L (2004) Using models at the mesoscopic scale in teaching physics: two experimental interventions in solid friction and fluid statics. Int J Sci Educ 26(9):1083–1110. https://doi.org/10.1080/0950069042000205396

Boekaerts M (1993) Being concerned with well-being and with learning. Educ Psychol 28(2):149–167. https://doi.org/10.1207/s15326985ep2802_4

Brown PC, Roediger HL, McDaniel MA (2014) Make It stick: The science of successful learning. The Belknap Press of Harvard University Press, Cambridge, Massachusetts, London, England

Canning EA, Muenks K, Green DJ and Murphy MC (2019) STEM faculty who believe ability is fixed have larger racial achievement gaps and inspire less student motivation in their classes. Sci Adv 5(2). https://doi.org/10.1126/sciadv.aau4734

Cassidy S (2015) Resilience building in students: The role of academic self-efficacy. Front Psychol 6(178):1–14. https://doi.org/10.3389/fpsyg.2015.01781

Cherukunnath D, Singh AP (2022) Exploring cognitive processes of knowledge acquisition to upgrade academic practice. Front. Psychol. 13(1). https://doi.org/10.3389/fpsyg.2022.682628

Chittleborough G, Treagust DF (2007) The modelling ability of non-major chemistry students and their understanding of the sub-microscopic level. Chem Educ Res Prac 8(3):274–292. https://doi.org/10.1039/B6RP90035F

Corpuz ED and Rebello NS (2011) Investigating students' mental models and knowledge construction of microscopic friction. Implications for curriculum design and development Phys. Rev. Special Topics—Phys Educ Res 7(2): 020102. https://doi.org/10.1103/PhysRevSTPER.7.020102

DiSessa AA, Sherin BL (1998) What changes in conceptual change? Int J Sci Educ 20(10):1155–1191. https://doi.org/10.1080/0950069980201002

Psillos D, Kariotoglou P (eds) (2016) Iterative design of teaching-learning sequences. Springer Science+Business Media, Dordrech

Duit R, Gropengießer H, Kattmann U, Komorek M, Parchmann I (2012) The model of educational reconstruction—a framework for improving teaching and learning science. In Jorde D, and J Dillon (eds) Science Education Research and Practice in Europe. Cultural Perspectives in Science Education vol 5. SensePublishers, Rotterdam

Duit R, Gropengießer H, Kattmann U (2005) Towards science education that is relevant for improving practice: The model of educational reconstruction. In: Fischer HE (ed) Developing standards in research on science Education. Taylor & Francis, London, pp 1–9

Duran M, Dokme I (2016) The Effect of the inquiry-based learning approach on student's critical thinking skills. Eurasia J Math, Sci Tech Educ 12(12):2887–2908. https://doi.org/10.12973/eurasia.2016.02311a

Dweck C (2016) Mindset: The new psychology of success updated edition. Random House, NY

Eggen PD, Kauchak DP (2004) Educational psychology: Windows on classrooms. Prentice Hall, Upper Saddle River, NJ

Ericsson KA (2007) Deliberate practice and the modifiability of body and mind: Toward a science of the structure and acquisition of expert and elite performance. Int J Sport Psych 38(1):4–34. https://psycnet.apa.org/record/2007-06716-002

Etkina E, Brookes DT, Planinšič G (2019) Investigative science learning environment. IOP Publishing, Bristol, UK, When learning physics mirrors doing physics

Fazio C, Battaglia OR and Di Paola B (2013) Investigating the quality of mental models deployed by undergraduate engineering students in creating explanations: The case of thermally activated phenomena. Phys Rev Special Topics—Phys Educ Res 9(2):020101. https://doi.org/10.1103/PhysRevSTPER.9.020101

Fazio C, Agliolo Gallitto A, Galiano G, Giarratano G, Grazia I, Termini G, Battaglia OR (2023) An approach to research-based design of teaching-learning sequences in the context of physics education: Theoretical frameworks, pedagogical methods, and examples of Data Analysis. Nuovo Cimento della Societa Italiana di Fisica C 46(6):199. https://doi.org/10.1393/ncc/i2023-23199-1

Felder RM, Silverman LK (1988) Learning and teaching styles in engineering education. Eng Edu 78(7):674–681

Fleming ND (1995) I'm different; not dumb. Modes of presentation (VARK) in the tertiary classroom. Research and development in higher education. In: Proceedings of the 1995 annual conference of the Higher Education and Research Development Society of Australasia (HERDSA), HERDSA, vol. 18, pp 308–313

Fleming ND, Mills C (1992) Not another inventory, Rather a catalyst for reflection. To Improve the Academy 11:137–155. https://doi.org/10.1002/j.2334-4822.1992.tb00213.x

Freeman S, Eddy SL, McDonough M, Smith MK, Okoroafor N, Jordt H, Wenderoth MP (2014) Active learning increases student performance in science, engineering, and mathematics. Proc Natl Acad Sci USA 111(23):8410–8415. https://doi.org/10.1073/pnas.1319030111

Gilbert JK, Boulter C (1998) Learning science through models and modelling. In: Fraser BJ, Tobin KG (eds) International handbook of science education. Kluwer Academic Publisher, Dordrecht, The Netherlands, p 53

Gluck MA, Mercado E, Myers CE (2016) Learning and memory: From brain to behavior, 3rd edn. Worth Publishers, Macmillan Learning, NY

Gopnik A (2000) Explanation as orgasm and the drive for causal knowledge: The function, evolution, and phenomenology of the theory formation system. In: Keil FC, Wilson RA (eds) Explanation and cognition. The MIT Press, Cambridge, Mass, pp 299–323

Harris KR, Graham S, Brindle M and Sandmel K (2009) Metacognition and children's writing. In Hacker DJ, Dunlosky J, Graesser AC (eds) Handbook of metacognition in education. Routledge, Abingdon, Oxon, pp 131–153

Heckman JJ, Kautz T (2012) Hard evidence on soft skills. Labour Econ 19(4):451–464. https://doi.org/10.1016/j.labeco.2012.05.014

Herrington J, Oliver R (2000) An instructional design framework for authentic learning environments. Educ Technol Res Dev 48(3):23–48. https://doi.org/10.1007/BF02319856

Jackson J, Adams R, Schwantner U, Waters C (2018) The role of learning progressions in global scales. https://www.brookings.edu/blog/education-plus-development/2018/05/29/the-role-of-learning-progressions-in-global-scales/

Johnson-Laird PN (1983) Mental models. Cambridge University Press, Cambridge, UK

Kastrinos W (1965) A study of the retention of biological facts by high school biology students. Sci Educ 49(5):487–491. https://doi.org/10.1002/sce.3730490513

Katsoulakis MA, Vlachos DG (2004) Mesoscopic modeling of surface processes. In: Abdallah NB, Arnold A, Degond P, Gamba IM, Glassey RT, Levermore CD, Ringhofer C (eds) Dispersive transport equations and multiscale models. Springer, NY, pp 179–198

Kattmann U, Duit R, Gropengießer H and Komorek M (1995) A model of educational reconstruction paper presented at the NARST annual meeting, San Francisco, CA

Keefe JW (1985) Assessment of learning style variables: The NASSP task force model. Theory Into Practice 24(2):138–144. https://doi.org/10.1080/00405848509543162

Keefe JW (1979) Learning style: An overview. In: Keefe JW, Kiernan OB (eds) Student learning styles: Diagnosing and prescribing programs, 1: 1–17. National Association of Secondary School Principles

Kim H, Care E (2018) Learning progressions: pathways for 21st century teaching and learning. https://www.brookings.edu/blog/education-plus-development/2018/03/27/learning-progressions-pathways-for-21st-century-teaching-and-learning/

Kincaid JP, Hamilton R, Tarr RW, Sangani H (2003) Simulation in education and training. In: Obaidat MS, Papadimitrioupp GI (eds) Applied system simulation: methodologies and applications. Springer, NY, pp 437–456

Konu A, Rimpelä M (2002) Well-being in schools: a conceptual model. Health Promot Int 17(1): 79–87. https://doi.org/10.1093/heapro/17.1.79

Kumar R, Sethy R (2025). Impact of Marzano's dimensions of learning model on achievement in mathematics, creative thinking and attitude towards mathematics among eighth standard

students. Sch Res J Humlty Sci & Engl Lang 13(71):54–64. https://doi.org/10.5281/zenodo.17590421
Levrini O, Fantini P, Tasquier G, Pecori B, Levin M (2015) Defining and operationalizing appropriation for science learning. J Learn Sci 24(1):93–136. https://doi.org/10.1080/10508406.2014.928215
Lijnse PL (1995) Developmental research as a way to an empirically based 'Didactical Structure' of Science. Sci Educ 79(2):189–199. https://doi.org/10.1002/sce.3730790205
Magnusson SJ, Templin M, Boyle RA (1997) Dynamic science assessment: A new approach for investigating conceptual change. J Learn Sci 6(1):91–142. https://www.jstor.org/stable/1466877
Marchand A, Weijs JH, Snoeijer JH, Andreotti B (2011) Why is surface tension a force parallel to the interface? Am J Phys 79(10):999–1008. https://doi.org/10.1119/1.3619866
Martinez ME (2006) What is metacognition? Phi Delta Kappan 87(9):696–699. https://doi.org/10.1177/003172170608700916
Marzano RJ (1992) A different kind of classroom: Teaching with dimensions of learning. Association for Supervision and Curriculum Development, Alexandria VA
Mayer RE (2002) The promise of educational psychology, volume 2: Teaching for meaningful learning. Merrill Prentice Hall, Upper Saddle River, NJ
Mayer RE (2008) Learning and instruction. Pearson Education Inc., Upper Saddle River
McDevitt B (1997) Learner autonomy and the need for learner training. Lang Learn J 16(1):34–39. https://doi.org/10.1080/09571739785200251
McFall RM (1982) A review and reformulation of the concept of social skills. Behav Assess 4(1):1–33. https://doi.org/10.1007/BF01321377
Meheut M, Psillos D (eds) (2004) Teaching-learning sequences: aims and tools for science education research. Int J Sci Educ 26(5):515–535. https://doi.org/10.1080/09500690310001614762
Messick S (1984) The nature of cognitive styles: problems and promises in educational research. Educ Psych 19:59–74. https://doi.org/10.1080/00461528409529283
Muenks K, Canning EA, LaCosse J, Green DJ, Zirkel S, Garcia JA and Murphy MC (2020) Does my professor think my ability can change? Students' perceptions of their STEM professors' mindset beliefs predict their psychological vulnerability, engagement, and performance in class. J Exp Psychol: Gen 149(11):2119–2144. https://doi.org/10.1037/xge0000763
Noddings N (2003) Happiness and education. Cambridge University Press, Cambridge
Norman DA (1983). Some observations on mental models. In Gentner D, Stevens A (eds) Mental models, L. Erlbaum Associates, Hillsdale, NJ
Ogborn J (2006) Science and commonsense. Revista Brasileira De Pesquisa Em Educação Em Ciências 6(1):1–21
Oja KJ (2011) Using problem-based learning in the clinical setting to improve nursing students' critical thinking: an evidence review. J Nurs Educ 50(3):145–151. https://doi.org/10.3928/01484834-20101230-10
Pekrun R, Muis KR, Frenzel AC, Götz T (2018) Emotions at school. Routledge, NY
Persano Adorno D, Pizzolato N, Fazio C (2018) Long term stability of learning outcomes in undergraduates after an open-inquiry instruction on thermal science. Phys Rev Phys Educ Res 14:010108. https://doi.org/10.1103/PhysRevPhysEducRes.14.010108
Pintrich PR (2002) The role of metacognitive knowledge in learning, teaching, and assessing. Theory Pract 41(4):219–225. https://doi.org/10.1207/s15430421tip4104_3
Rasheed T (2021) Self development sustainability model. J US-China Pub Admin 18(5):237–241
Reid JM (2002) Learning styles in the ESL/EFL classroom. Foreign Language Teaching and Research Press, Beijing
Riding R, Rayner S (2007) Cognitive styles and learning strategies: Understanding style differences in learning and behavior. Routledge, Abingdon, Oxon
Rogoff B (1995) Observing sociocultural activity on three planes: participatory appropriation, guided participation, and apprenticeship. In: Wertsch JV, del Rio P, Alvarez A (eds) Sociocultural studies of mind. Learning in doing: Social, cognitive and computational perspectives. Cambridge University Press, Cambridge, pp 139–164

Rutten N, Van Joolingen WR, Van Der Veen JT (2012) The learning effects of computer simulations in science education. Comput Educ 58(1):136–153. https://doi.org/10.1016/j.compedu.2011.07.017

Sadler-Smith E (2001) The relationship between learning style and cognitive style. Pers Individ Dif 30(4):609–616. https://doi.org/10.1016/S0191-8869(00)00059-3

Scherr R (2007) Modeling student thinking: an example in special relativity. Am J Phys 75(3):272–280. https://doi.org/10.1119/1.2410013

Schmidt WH, Wang HC and mcKnight CC (2005) Curriculum coherence: an examination of US mathematics and science content standards from an international perspective. J Curric Stud 37(5):525–559. https://doi.org/10.1080/0022027042000294682

Schunk DH (2020) Learning theories: An educational perspective, 8th edn. Pearson Education, New York, NY

Schwartz RS, Lederman NG, Crawford BA (2004) Developing views of nature of science in an authentic context: An explicit approach to bridging the gap between nature of science and scientific inquiry. Sci Educ 88(4):610–645. https://doi.org/10.1002/sce.10128

Sellström E, Bremberg S (2006) Is there a 'school effect' on pupil outcomes? A review of multi-level studies. J Epidemiol Community Health 60(2):149–155. https://doi.org/10.1136/jech.2005.036707

Shea TM and Bauer AM (2007) Behavior management: A practical approach for educators. Pearson Education, Harlow

Tang X, Zhang D (2020) How informal science learning experience influences students' science performance: A cross-cultural study based on PISA 2015. Int J Sci Educ 42(4):598–616

Termini G, Battaglia OR, Grazia I, Fazio C (2025) Gaining insights through thematic analysis about learning dimensions in teaching learning sequences on surface phenomena in liquids. Phys Rev Phys Educ Res 21:020143. https://doi.org/10.1103/4vsw-m71t

Tyng CM, Amin HU, Saad MNM and Malik AS (2017) The influences of emotion on learning and memory. Front Psychol 8(1). https://doi.org/10.3389/fpsyg.2017.01454

Vosniadou S (2019) The development of students' understanding of science. Front Edu 4 art 32. https://doi.org/10.3389/feduc.2019.00032

Wößmann L (2010) Families, schools and primary-school learning: Evidence for Argentina and Colombia in an international perspective. Appl Econ 42(21):2645–2665. https://doi.org/10.1080/00036840801964617

Yeager DS, Carrol JM, Buontempo J, Cimpian A, Woody S, Crosnow R et al (2021) Teacher mindsets help explain where a growth-mindset intervention does and doesn't work. Psychol Sci 33(1):18–32. https://doi.org/10.1177/09567976211028984

Yeager DS, Hanselman P, Walton GM et al (2019) A national experiment reveals where a growth mindset improves achievement. Nature 573:364–369. https://doi.org/10.1038/s41586-019-1466-y

Yew EHJ, Goh K (2016) Problem-based learning: An Overview of its process and impact on learning. Health Prof Educ 2(2):75–79. https://doi.org/10.1016/j.hpe.2016.01.004

Zhang J, Kuusisto E, Tirri K (2017) How teachers' and students' mindsets in learning have been studied: research findings on mindset and academic achievement. Psychology 8(9):1363–1377. https://doi.org/10.4236/psych.2017.89089

Zimmerman BJ (2000) Self-efficacy: An essential motive to learn. Cont Educ Psych 25(1):82–91. https://doi.org/10.1006/ceps.1999.1016

Zimmerman BJ, Bandura A, Martinez-Pons M (1992) Self-motivation for academic attainment: The role of self-efficacy beliefs and personal goal setting. Am Educ Res J 29(3):663–676. https://doi.org/10.3102/00028312029003663

Comparing Prior Knowledge of First-Semester Physics Students Between the Cohorts of 2013 and 2023

Dennys Gahrmann, Irene Neumann, and Andreas Borowski

Abstract German university instructors often report about an on-going decline of prior knowledge in incoming first-semester physics students. One reason for the decline might lie in the fact that, over the past decades, the primary goal of school science education shifted from focusing on training science specialists to focusing on scientific literacy for all, whereas tertiary education (notably universities) generally focusses on specializing in one field of expertise. This gap between school and university might lead to challenges for incoming physics students. The present study therefore investigated whether there is empirical evidence for the reported decline in physics students' prior knowledge. We refined a national study entrance test from 1978 to assess the physics prior knowledge of first-semester students in 2023 and compared two nationwide cohorts of first-semester physics students: 2013 ($N = 2251$) and 2023 ($N = 2017$). Overall, we found a small increase in students' prior physics knowledge from 2013 to 2023. In addition, we found that the 2023 students performed significantly better on more contemporary items (e.g. on scientific reasoning) compared to the more traditional items (e.g. on recalling factual knowledge).

Keywords Study entrance phase · Trend study · Freshmen physics student

D. Gahrmann (✉) · A. Borowski
Physics Education, Institute of Physics and Astronomy, University of Potsdam, Potsdam, Germany
e-mail: dennys.gahrmann@uni-potsdam.de

A. Borowski
e-mail: andreas.borowski@uni-potsdam.de

I. Neumann
Leibniz Institute for Science and Mathematics Education (IPN), Kiel, Germany
e-mail: ineumann@leibniz-ipn.de

D. Sokołowska and E. McLoughlin (eds.), *Embracing Changes Together in Physics Education*, Challenges in Physics Education,
https://doi.org/10.1007/978-3-032-27385-7_7

1 Introduction and Theoretical Background

Rooted in the ideals of *Humanistische Bildung*—a philosophy focused on individual self-cultivation and personal development—German education historically emphasized classical domains such as ancient languages, arts, and humanities (Neumann et al. 2010). These areas were seen as the foundation for nurturing well-rounded individuals who could engage with the world thoughtfully and critically. In contrast, natural sciences were often considered secondary, supporting a broader education rather than being valued as core disciplines in their own right. The goal was to cultivate *Bildung*, the holistic development of personality (Fischer et al. 2008).

This philosophical foundation of education was put into question in the mid-twentieth century, as global societal and political challenges demanded a more pragmatic approach. A key turning point came with the *Sputnik crisis* in 1957, when the Soviet Union's successful satellite launch sparked a global space race and an urgent need for expertise in science and technology (Picht 1964). The response, particularly pronounced in the United States, was a reorientation of educational priorities to address the growing demand for a scientifically literate and technically skilled workforce. Training specialists in the natural sciences became a critical objective, far removed from the humanistic ideals that prioritized individual formation over practical application (Herold 1974).

In Germany, this shift in perspective unfolded more gradually but came to the forefront with the *PISA crisis* in 2000. International large-scale assessments revealed that German students were underperforming in science compared to their international peers, with results significantly below the OECD average (Lewalter et al. 2023). These findings were a wake-up call, sparking debates about the efficacy of Germany's educational system and whether the traditional ideals of *humanistische Bildung* were still fit for purpose in a rapidly evolving world.

As a response, Germany started significant educational reforms. National Educational Standards (NES) were introduced to align the system with measurable outcomes rather than simply prescribed content (Kultusministerkonferenz 2020). This transition marked a fundamental shift from an input-oriented system—focused on what schools should teach—to an output-oriented system that prioritized what students should know and be able to do by the time they left school. For the sciences, this shift was embodied in the concept of *Naturwissenschaftliche Grundbildung*, which aimed to balance traditional notions of *Bildung* with the more utilitarian goals of scientific literacy. *Naturwissenschaftliche Grundbildung* emphasized the application of science to real-life situations, fostering conceptual understanding and inquiry-based learning over rote memorization and traditional declarative knowledge (Neumann et al. 2010).

While these reforms aimed to modernize the system and improve student outcomes, evidence suggests their impact has been mixed. Large-scale assessments like PISA have continued to report a decline in German students' science performance over the past two decades (Lewalter et al. 2023). In parallel, research on physics prior knowledge among first-semester university students shows a similar trend: A 2013 study by Buschhüter et al. revisited a physics entrance test originally conducted in 1978 and found a significant decline with medium effect size (Cohen's d = 0.53, p. 15) in the physics prior knowledge of the first-semester physics student (Buschhüter et al. 2017). In particular, the results indicated shifts in specific areas of knowledge, likely reflecting changes in educational priorities. However, the study primarily focused on declarative knowledge and did not address dispositions emphasized in contemporary curricula (e.g. competencies).

In line with Buschhüter et al.'s finding, university physics instructors anecdotally report a perceived decline in traditional declarative knowledge over time.

Against this background, the present study aims to investigate changes in physics prior knowledge by comparing the 2013 cohort of first-semester physics students with their counterparts in 2023. In this context, it is also important to consider potential disruptions to learning due to the COVID-19 pandemic, which may have affected students' preparation during their final years of schooling. Specifically, we seek to determine whether the decline in physics prior knowledge observed by Buschhüter et al. has continued or whether alternative trends have emerged over the past decade. By addressing this issue, we aim to examine whether the trends identified in large-scale school assessments are reflected in the prior knowledge of first-semester physics students.

2 Objective and Research Question

The German education system has undergone significant reforms since the early 2000s, following international results like the *PISA crisis*, which revealed deficits in student performance in Germany. As a response to the first PISA studies NES were introduced, shifting the focus from an input-oriented approach—emphasizing what content is taught—to an output-oriented approach that prioritizes competencies such as scientific inquiry and knowledge-based application (*Naturwissenschaftliche Grundbildung*).

In contrast, German universities have largely retained their discipline-specific model, emphasizing traditional declarative knowledge and specialization in one field over competencies (Konferenz der Fachbereiche der Physik (KFP) 2010). This misalignment might have led to perceptions among university instructors of a decline in first-semester physics students' prior knowledge, particularly in mathematical routines, traditional declarative physics knowledge, and foundational topics.

A 2013 study by Buschhüter et al. revisited a physics entrance test originally conducted in 1978 (Krause and Reiners-Logothetidou 1979) and found no significant decline in overall student performance (Buschhüter et al. 2017). Instead, the results indicated shifts in specific areas of knowledge, likely reflecting changes in educational priorities. However, the study primarily focused on traditional declarative knowledge domains and did not address newer competencies emphasized in contemporary curricula (Kultusministerkonferenz 2020).

Based on these developments and the continued decline in student performance reported in large-scale assessments like PISA (OECD 2023a), the present study aims to investigate changes in physics prior knowledge between the 2013 cohort and the 2023 cohort of first-semester physics students. With empirical evidence from the national comparison of physics prior knowledge in 2013 indicating that the average performance declined with educational reforms, the consistent decline in science performance in international assessments over the past two decades raises concerns. Specifically, it is unclear whether these broader educational school trends are reflected in the physics prior knowledge of students transitioning to university. By examining both, traditional knowledge domains and competencies aligned with contemporary educational standards, this study seeks to evaluate whether and how these shifts impacted students' preparation for university physics programs. In particular, we investigated the following two research questions:

- How does the physics prior knowledge of current (2023) first-semester physics students differ from that of the previous cohort from 2013?
- How does the performance of the 2023 first-semester physics students compare on traditional items focusing on classical declarative knowledge, and how do they perform on items assessing more contemporary competencies?

3 Method

3.1 Content Validity of the 2023 National Study Entrance Test

To investigate these research questions, the 2013 study entrance test (originally developed in 1978) was carefully revised. After a first inspection and review of remarks by Buschhüter et al., it became apparent that a revision was necessary as content and curricular validity, particularly with respect to school curricula, were no longer sufficient (Buschhüter et al. 2016, 2017). Given the shifts in the educational system, it was crucial to update the test in order to accurately reflect the current state of students' prior knowledge and educational background. To ensure a direct comparison of items between both tests, however, the 2013 study entrance test was used as the basis for the revision.

Items for the revised study entrance test (SET 23) were drawn from two sources: anchor items retained from the original SET 78 and new items adapted from established instruments, widely recognized for their validity in assessing physics competencies at the study entrance level (e.g. Maries and Singh 2013). All items underwent curricular validation against the German NES (Kultusministerkonferenz 2020) to ensure alignment with current school curricula. Subsequently, a survey involving 22 university physics instructors from across Germany was conducted to further refine the selection of items. In this online survey, the instructors reviewed the curricular-validated items and identified those most suitable for inclusion in SET

23, ensuring the test's curricular validity (school perspective) as well as the content validity (university perspective).

The revision of SET 23 was undertaken to ensure the instrument's validity while addressing the changes in NES (curricular validity). Content and curricular validity were ensured by aligning the test items with the current NES and incorporating the results from a nationwide survey of university physics instructors (content validity). This process ensured that the items reflected the essential knowledge and competencies required at the start of a physics program.

3.2 *The 2023 National Study Entrance Test*

The final version of the study entrance test (SET 23) was designed to assess the prior knowledge and competencies of first-semester physics students, with a particular focus on physics-related skills. The test consists of four parts: (1) Demographic information (e.g. school GPA and participation in preparatory courses), (2) affective scales (e.g. interest), (3) mathematics items and (4) physics items. For the present analysis, we will focus specifically on parts 1 (demographic information) and 3 (physics items).

SET 23 included 32 physics items in total, 17 of which were anchor items from the original SET 78. These 17 items with an open format covered fundamental topics in physics (10 mechanics, 1 thermodynamics, 4 electricity, 2 optics) and primarily assessed traditional declarative knowledge (e.g. "What is the formula for the kinetic energy of a point mass?"; "Briefly describe the process of electromagnetic induction").

In addition, SET 23 included 15 more contemporary items, designed to assess other aspects of physics prior knowledge.

These items focused on scientific reasoning (Bao et al. 2009), diagram literacy (Lachmayer 2008), and problem-solving skills (Binder et al. 2019), and competencies that are increasingly emphasized in German schools after the educational reforms. New items were adapted from established instruments, including the TUG-K (Beichner 1994), the Lawson Classroom Test for scientific reasoning (Zavala et al. 2017), and the Test about measurement uncertainties (Schulz 2022), and are widely recognized for their validity in evaluating physics competencies at the study entrance phase in physics. These new items incorporated a multiple-choice format with three to five options, and their placement in the SET 23 followed the anchor items to maintain logical sequencing and cognitive consistency with the SET 78.

3.3 *Testing Procedure in 2023*

The updated version of the test was administered nationwide during the first week of the autumn semester 2023/24, following the testing procedure employed in 2013 (and in 1978) (Buschhüter et al. 2017; Krause and Reiners-Logothetidou 1979). The test had a duration of approximately 90 min and was administered as a paper–pencil test in 34 German universities. Upon completion, local staff at the participating institutions scored the responses based on a scoring manual. The scoring categories included "correct response," "incorrect response," and "omitted response."

After the completed tests were returned to the authors, a second scoring was conducted on 10% (or at least 10 tests) of each university's submissions to calculate the interrater reliability and ensure scoring accuracy. As in 2013, the scorers in both rounds (initial and secondary) were considered pseudo-raters, as different individuals performed the first and second scorings. To compare the scorings, a weighted Kappa (κ^*) was calculated following the procedure of Wirtz and Casper, as employed by Buschhüter et al. (2016; Wirtz and Caspar 2022). This approach accounts for the number of students at each university, making it more appropriate than an unweighted Kappa.

The interrater reliability for 2023 ($M\,(\kappa^*_{2023}) \approx 0.81$; $Mdn(\kappa^*_{2023}) \approx 0.84$) can be classified as "almost perfect" (Landis and Koch 1977), consistent with the reliability in 2013 after excluding items with low weighted interrater reliability ($M\,(\kappa^*_{2013}) \approx 0.91$; $Mdn(\kappa^*_{2013}) \approx 0.93$). This indicates that scoring in both years was highly reliable, with no items showing significant discrepancies between the initial and secondary scorings in either 2013 or 2023. Following the scoring process, the data was subjected to statistical analysis.

3.4 The Samples of 2013 and 2023

The cohort of 2013 (2023) consists of N = 2251 (2007) students from N = 27 (34) German universities, with all students being enrolled in introductory physics courses. Table 1 provides an overview of the demographic characteristics. The total number of students is slightly lower in 2023 (N = 2017) compared to 2013 (N = 2251), with the majority in both cohorts enrolled as physics majors rather than physics teacher candidates. The gender distribution shifted slightly over the decade, with the proportion of female students increasing from 21% in 2013 to 28% in 2023. The mean school grade point average (GPA) did not change statistically significant from 2.0 (0.6) in 2013 to 1.8 (0.6) in 2023, reflecting a similar school performance among incoming physics students. The percentage of students who participated in preparatory courses before starting their studies remained relatively stable, with 70% in 2013 and 67% in 2023. The demographic data indicate that the 2013 and 2023 cohorts are comparable in terms of key characteristics such as age, gender distribution, and educational background, ensuring a valid basis for comparing their physics prior knowledge.

Table 1 Demographic information of 2013 and 2023 cohort

Demographic characteristic	2013	2023
$N_{students}$	2251	2007
$Percentage_{Male}$|$Percentage_{Female}$	78%|21%	72%|28%
School GPA M (SD)	2.0 (0.6)	1.8 (0.6)
Participation rate in preparatory courses before university	70%	67%

Table 2 Correlations of the sum scores of both SETs with physics exam grades at the end of the first semester

	N	Kendall's tau B	p	Lower 95% CI	Upper 95% CI	Effect size (Fisher's z)	SE effect size
2023	629	–0.45	< 0.001	–0.49	–0.41	–0.49	0.02
2013	268	–0.40	< 0.001	–0.47	–0.32	–0.42	0.03

3.5 Prognostic Validity of the Instrument

Before exploring the above research questions, we employed these data to examine the revised SET's prognostic validity in comparison to the 2013 test. Therefore, the physics sum scores from the entrance test and first-semester physics exam grades were correlated. For both the 2013 and 2023 cohorts, a strong correlation was found, demonstrating the test's ability to predict academic success in physics (see Table 2).

Kendall's tau B was used, as scores of both cohorts are not normally distributed (Shapiro–Wilk Test for Multivariate Normality $p < 0.001$ and indication by Q-Q-plot).

4 Results

4.1 How Does the Physics Prior Knowledge of Current (2023) First-Semester Physics Students Differ from that of the Previous Cohort from 2013?

We investigated this question from two perspectives. First, we compared the distribution of students achieving different sum scores between the cohorts of 2013 and 2023. This comparison provides an overview of performance trends across the two groups. Second, we examined in more detail the percentage of students who correctly solved each item in the entrance tests, administered in 2013 and in 2023.

It was tested whether the two distributions of anchor item sum scores were the same. The Kolmogorov–Smirnov test showed a statistically significant difference ($D = 0.080$, $p < 0.001$). The descriptive statistics in Table 3 further highlight differences between the two distributions. For the 2023 cohort, the mean score (M =

6.45) is slightly higher than that of the 2013 cohort (M = 5.64), while both distributions exhibit similar standard deviations and skewness, indicating a slight positive asymmetry. Notably, neither distribution adheres to normality, as evidenced by the Shapiro–Wilk test ($p < 0.001$ for both cohorts).

The Wilcoxon rank sum test confirmed that the difference in raw scores between the two cohorts is statistically significant (W = 2,059,546, p < 0.001). However, the effect size is small (Cohen's d = 0.21, 95% CI [0.15, 0.27]) according to Cohen's classification (Cohen 1988). Still, Yeager et al. note that an effect of this size is roughly equivalent to the improvement typically seen after a year of classroom learning for ninth-grade students, as measured by standardized tests (Yeager et al. 2019) (Fig. 1).

To further examine the physics prior knowledge of the 2013 and 2023 cohorts, we focus on the percentage of students who solved each item correctly, plotted against an arbitrary item index, as shown in Fig. 2. The items are categorized into three groups: "old items," exclusively included in SET 78; "anchor items," common to both SET 78 and SET 23; and "new items," solely in SET 23.

In the "anchor items" category, used to directly compare the two cohorts, minor variations in performance are observable; however, these differences are not substantial, further supporting the conclusion that both cohorts have similar physics prior knowledge (see Fig. 1 and Table 3).

While the overall pattern for the anchor items appears similar across both cohorts, two items show notable differences (more than 10% difference in correct solutions):

Table 3 Statistical information of both distributions

	N	M	Mdn	SD	Skewness	Kurtosis	p-value Shapiro–Wilk Normality test
2023	2007	6.45	6	3.97	0.35	–0.67	< 0.001
2013	2251	5.64	5	3.87	0.45	–0.60	< 0.001

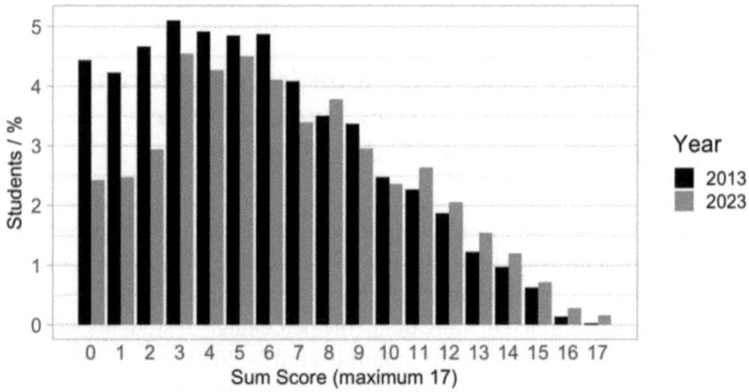

Fig. 1 Comparison of sum score distributions (maximum score: 17) for two cohorts of students in 2013 (black/left bars) and 2023 (gray/right bars). The x-axis represents the sum score, while the y-axis indicates the percentage of students achieving each score

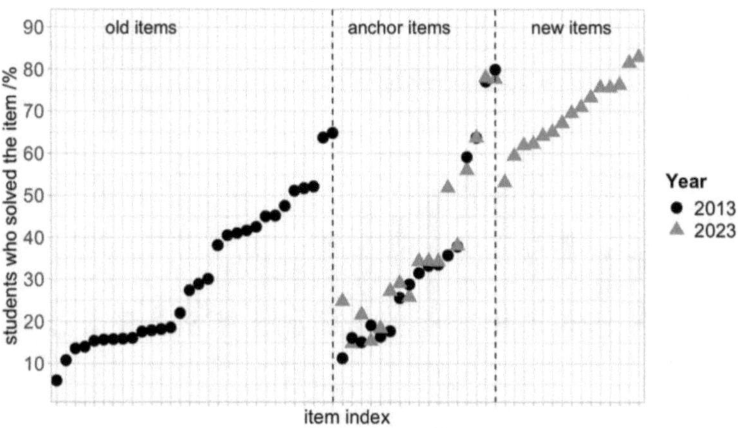

Fig. 2 Item-solving percentages for the 2013 (black circles) and 2023 (gray triangles) students, categorized by old (left), anchor (middle), and new items (right)

These items were solved by more students in 2023 (sketching an electrical circuit: 51% in 2023, 36% in 2013 and describing the principle of induction: 25% in 2023, 11% in 2013).

These findings indicate that both cohorts performed similarly on the anchor items. From this perspective, the overall analysis suggests that while a statistically significant change in physics prior knowledge is detectable between the two cohorts, the effect is negligible in practical terms.

4.2 *How Does the Performance of the 2023 First-Semester Physics Students Compare on Traditional Items Focusing on Classical Declarative Knowledge, and How Do They Perform on items Assessing Newer Competencies?*

To answer the second research question, the performance of the 2023 cohort was analyzed for anchor items (representing more traditional physics knowledge) and new items (focusing on competencies such as scientific reasoning). Figure 3 illustrates the distribution of sum scores for these two item groups. The anchor item scores range from 0 to 17, while the new item scores range from 0 to 15. The maximum values are indicated by dotted vertical lines.

The distribution of scores for the anchor items (black bars) is relatively broad, with the largest proportion of students scoring in the mid-range. This reflects a more balanced spread of student performance across different sum scores. The mode of the anchor item distribution is less pronounced, indicating that students' results are more evenly distributed over a wider range of scores.

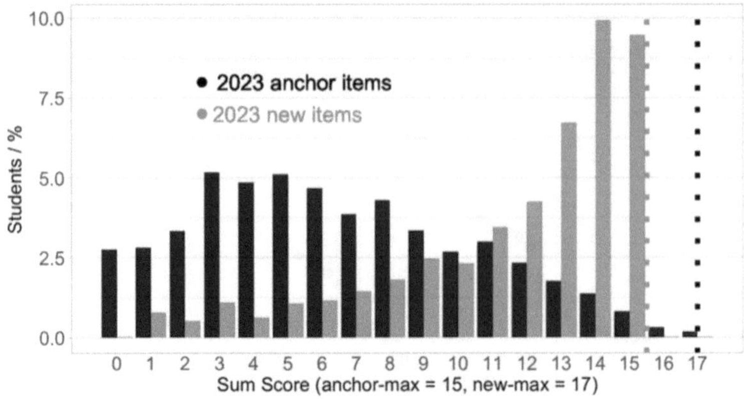

Fig. 3 Distribution of sum scores for 2023 anchor (black/left bars) and new items (gray/right bars), with maximum scores of 15 and 17, respectively

In contrast, the distribution of scores for the new items (gray bars) is heavily skewed toward the upper end of the scale. The highest proportion of students achieved scores clustered at a mode of 14. This suggests that the new items are easier for students to solve or align more closely with their competencies, resulting in a higher overall performance. The gray bars also show a narrower distribution, with relatively fewer students scoring in the mid or lower ranges compared to the anchor items.

A statistical comparison of the two distributions confirms these observations. The Kolmogorov–Smirnov test reveals a significant difference between the two score distributions ($D = 0.53$, $p < 0.001$). Furthermore, a Wilcoxon rank sum test with continuity correction indicates a statistically significant difference between the two item groups ($W = 615{,}072$, $p < 0.001$) with a large effect size (Cohen's $d = 1.37$, 95% CI [1.29, 1.44]).

Again, Fig. 2 supports these findings: A notable trend emerges in the "new items" category, where the 2023 cohort generally achieves higher percentages of correct solutions compared to the performance on the "old items" by the 2013 cohort. This suggests that the 2023 cohort may have been better prepared to solve these competency-based items, than the 2013 cohort was prepared to solve traditional declarative knowledge. These findings suggest that the 2023 students performed significantly better on the new items, which assess competencies such as scientific reasoning, compared to the anchor items, which primarily focus on traditional declarative knowledge. This performance gap could indicate that the current cohort is prepared differently for tasks requiring problem-solving skills, reflecting changes in physics education practices in schools. On the other hand, the relatively lower performance of the 2023 cohort on the anchor items may point to a decline or shift in focus away from traditional physics content in favor of broader competency development.

Overall, the results highlight a significant distinction in student performance between traditional and competency-based physics items. The new items seem to align better with students' current preparation during high school education.

5 Discussion

Large-scale studies like PISA have consistently reported declining student performance in science over the past two decades, raising concerns about the effectiveness of educational reforms aimed at improving scientific literacy. Additionally, prior research by Buschhüter et al. found that physics prior knowledge among first-semester physics students declined significantly between 1978 and 2013, suggesting that these broader trends may extend to students transitioning to university-level physics (Buschhüter et al. 2017). This study sought to examine whether this decline persisted into the next decade, comparing cohorts from 2013 and 2023 against a background of ongoing decline in PISA, competency-focused teaching, and eventual disruptions of the COVID-19 pandemic (OECD 2023a, 2023b).

The results indicate that overall physics prior knowledge has remained stable between the 2013 and 2023 cohorts, with a statistically significant yet negligible difference favoring the 2023 cohort. These findings extend the work of Buschhüter et al., who reported a decline in physics prior knowledge between 1978 and 2013. While the current results do not contradict Buschhüter et al.'s findings, they provide additional insights by demonstrating continuity in students' general preparedness for university physics over the subsequent decade. This further comparison with a new cohort highlights the stability of physics prior knowledge in recent years. Moreover, the 2023 cohort demonstrated significantly stronger performance on tasks emphasizing competencies and scientific reasoning compared to traditional declarative knowledge, reflecting the shift in secondary education toward problem-solving and scientific literacy.

However, this stability in overall performance contrasts sharply with university instructors' widespread perception of declining student preparedness (Käß et al. 2022). This misalignment represents a key finding of this study, and the question remains why university instructors perceive a decline in prior knowledge even though it cannot be statistically found. The stronger performance of the 2023 cohort on competency-based items may reflect the success of educational reforms that followed the *PISA crisis*, which emphasized competencies and scientific literacy. Yet, these reforms may inadvertently de-emphasize traditional declarative knowledge, which is perceived to be essential for mastering the beginning of a physics study (e.g. Buschhüter and Borowski 2014).

6 Conclusion

A key point of the study is that despite the overall stability in performance, university instructors have expressed concerns that students are less prepared for university-level physics, which points to a possible misalignment of the university instructors' perceptions: These perceptions may stem from instructors' expectations or a greater emphasis on certain types of knowledge they consider most important, rather than reflecting an actual decline in students' abilities. It would be valuable to gather a representative opinion to determine whether this perceived decline by university instructors is recognized by the majority or if it is primarily driven by a few vocal voices.

While trend studies like PISA, which assess student performance at the school level, have consistently reported a decline in overall student knowledge, our findings of stable physics prior knowledge between the 2013 and 2023 cohorts provide an interesting contrast. The stability in physics prior knowledge at the university entrance level suggests that while broader educational trends may indicate a decline, students entering university physics courses have maintained a solid foundation in subject-specific content.

It is crucial to reflect on the structure of the study entrance phase in physics, and potentially other scientific disciplines, especially as universities focus on specialization. While a strong foundation in traditional declarative knowledge remains essential for success in foundational courses, modern educational practices place growing importance on competencies such as problem-solving and scientific reasoning. These competencies complement traditional knowledge and are crucial for tackling complex scientific challenges. However, transitions between educational stages often involve discontinuities such as in the necessary knowledge types, and it is not the responsibility of one institution alone to address these gaps. Effective preparation for higher education requires collaboration between secondary and tertiary institutions, ensuring a seamless integration of both types of knowledge.

7 Competing Interests

The authors have no conflicts of interest to declare that are relevant to the content of this chapter.

8 Ethics Approval

Approval was granted by the Ethics Committee of the University of Potsdam (27.07.2023/ No. 52/2023). Informed consent to participate and to publish was obtained from all individual participants involved in the study.

References

Bao L, Cai T, Koenig K, Fang K, Han J, Wang J, Liu Q, Ding L, Cui L, Luo Y, Wang Y, Li L, Wu N (2009) Learning and scientific reasoning. Science 323(5914):586–587. https://doi.org/10.1126/science.1167740

Beichner RJ (1994) Testing student interpretation of kinematics graphs. Am J Phys 62(8):750–762. https://doi.org/10.1119/1.17449

Kultusministerkonferenz (eds) (2020) Beschlüsse der Kultusministerkonferenz: Bildungsstandards im Fach Physik für die Allgemeine Hochschulreife. Beschluss vom 18.06.2020. [Resolutions of the Standing Conference of the Ministers of Education and Cultural Affairs: Educational Standards in Physics for the General Higher Education Entrance Qualification (Resolution of 18 June 2020)]. https://www.kmk.org/fileadmin/Dateien/veroeffentlichungen_beschluesse/2020/2020_06_18-BildungsstandardsAHR_Physik.pdf

Binder T, Sandmann A, Sures B, Friege G, Theyssen H, Schmiemann P (2019) Assessing prior knowledge types as predictors of academic achievement in the introductory phase of biology and physics study programmes using logistic regression. Int J STEM Educ 6(1):33. https://doi.org/10.1186/s40594-019-0189-9

Buschhüter D, Borowski A (2014) Modellierung von Eingangsanforderungen für das Studienfach Physik. [Modeling Entry Requirements for University Studies in Physics]. In Bernholt S (ed) Naturwissenschaftliche Bildung zwischen Science- und Fachunterricht, pp 540–542

Buschhüter D, Spoden C, Borowski A (2016) Mathematische Kenntnisse und Fähigkeiten von Physikstudierenden zu Studienbeginn. [Mathematical Knowledge and Skills of Freshmen Physics Students]. Z Für Didakt Naturwissenschaften 22(1):61–75. https://doi.org/10.1007/s40573-016-0041-4

Buschhüter D, Spoden C, Borowski A (2017) Physics knowledge of first semester physics students in Germany: A comparison of 1978 and 2013 cohorts. Int J Sci Educ 39(9):1109–1132. https://doi.org/10.1080/09500693.2017.1318457

Cohen J (1988) Statistical power analysis for the behavioral sciences. Routledge

Fischer HE, Kauertz A, Neumann K (2008) Standards of science education. In Four decades of research in science education-from curriculum development to quality improvement, pp 29–42. Waxmann

Herold J (1974) Sputnik in american education a history and reappraisal. McGill J Educ/Rev Sci L'Éducation McGill 9(002). https://mje.mcgill.ca/article/view/6971

Käß H, Berger T, Boin M, Fujan K, Güßmann M, Hardy EH, Karsten F, Nandi G, Nawrodt R, Raudzis C (2022) Mindestanforderungskatalog Physik. [Catalogue of Minimum Requirements in Physics]. PhyDid B—Didaktik Physik—Beiträge Zur DPG-Frühjahrstagung, 1. https://ojs.dpg-physik.de/index.php/phydid-b/article/view/1284

Konferenz der Fachbereiche der Physik (KFP) (2010) Zur Konzeption von Bachelor- und Master-Studiengängen in der Physik. [Concept for the Design of Bachelor's and Master's Degree Programmes in Physics]. https://www.kfp-physik.de/dokument/KFP_Handreichung_Konzeption-Studiengaenge-Physik-101108.pdf

Krause F, Reiners-Logothetidou A (1979) Der bundesweite Studieneingangstest Physik 1978. [The National Physics Entrance Test 1978]. Physik J 35(11):495–510. https://doi.org/10.1002/phbl.19790351103

Lachmayer S (2008) Entwicklung und Überprüfung eines Strukturmodells der Diagrammkompetenz für den Biologieunterricht. [Development and Validation of a Structural Model of Diagram Competence for Biology Education]. https://macau.uni-kiel.de/servlets/MCRFileNodeServlet/dissertation_derivate_00002471/Diss_Lachmayer.pdf

Landis JR, Koch GG (1977) The measurement of observer agreement for categorical data. Biometrics 33(1):159. https://doi.org/10.2307/2529310

Lewalter D, Diedrich J, Goldhammer F, Köller O, Reiss K (eds) (2023) PISA 2022: Analyse der Bildungsergebnisse in Deutschland. [PISA 2022: Analysis of Educational Outcomes in Germany] Waxmann Verlag GmbH. https://doi.org/10.31244/9783830998488

Maries A, Singh C (2013) Exploring one aspect of pedagogical content knowledge of teaching assistants using the test of understanding graphs in kinematics. Phys Rev Spec Top Phys Educ Res 9(2):020120. https://doi.org/10.1103/PhysRevSTPER.9.020120

Neumann K, Fischer HE, Kauertz A (2010) From PISA to educational standards: The impact of large-scale assessment on science education in Germany. Int J Sci Math Educ 8(3):545–563. https://doi.org/10.1007/s10763-010-9206-7

OECD (2023a) PISA 2022 results (Volume I): The state of learning and equity in education. OECD. https://doi.org/10.1787/53f23881-en

OECD (2023b) PISA 2022 Results (Volume II): Learning during—and from—disruption. OECD. https://doi.org/10.1787/a97db61c-en

Picht G (1964) Die deutsche Bildungskatastrophe: Analyse und Dokumentation [The German Education Catastrophe: Analysis and Documentation], vol 349. Deutscher Taschenbuch Verlag München

Schulz J (2022) Entwicklung eines Testinstrumentes zur Erfassung von Kompetenzen im Umgang mit Messunsicherheiten. [Development of a test instrument for assessing competence in dealing with measurement uncertainties] https://doi.org/10.18452/23957

Wirtz MA, Caspar F (2022) *Beurteilerübereinstimmung und Beurteilerreliabilität: Methoden zur Bestimmung und Verbesserung der Zuverlässigkeit von Einschätzungen mittels Kategoriensystemen und Ratingskalen.* [Interrater Agreement and Reliability: Methods for Determining and Improving the Consistency of Judgments Using Category Systems and Rating Scales] Hogrefe, Verlag für Psychologie

Yeager DS, Hanselman P, Walton GM, Murray JS, Crosnoe R, Muller C, Tipton E, Schneider B, Hulleman CS, Hinojosa CP, Paunesku D, Romero C, Flint K, Roberts A, Trott J, Iachan R, Buontempo J, Yang SM, Carvalho CM, Dweck CS (2019) A national experiment reveals where a growth mindset improves achievement. Nature 573(7774):364–369. https://doi.org/10.1038/s41586-019-1466-y

Zavala G, Tejeda S, Barniol P, Beichner RJ (2017) Modifying the test of understanding graphs in kinematics. Phys Rev Phys Educ Res 13(2):020111. https://doi.org/10.1103/PhysRevPhysEducRes.13.020111

Teaching a Quantum Physics Class Using a Student-Centered Collaborative Learning Approach at the University Level

Gerald Feldman, Marcel Vreeswijk, Auke Pieter Colijn, Natasa Brouwer, and André Heck

Abstract We report on the implementation of the SCALE-UP collaborative pedagogical approach in a first-year undergraduate quantum physics class at the University of Amsterdam. The performance of the students on two quizzes, a mid-term exam and a cumulative final exam was measured, as well as results from a conceptual pre- and post-test assessment using the Quantum Mechanics Conceptual Survey (QMCS), in order to gauge the effectiveness of this approach. Surveys were also given to get feedback on how students were reacting to this innovative style of pedagogy.

Keywords Collaborative group learning · Active learning · SCALE-UP · Studio physics · Quantum physics

1 Introduction

Studies of undergraduate STEM education have demonstrated that students need to be actively engaged in the learning process in order for it to be effective. A passive lecture environment ("teaching by telling") has been shown to be largely ineffective in developing students' skills in critical thinking and problem solving (McDermott 2001; Freeman et al. 2014). Active-learning strategies are becoming more common in physics classes, although most institutions around the world still rely on the standard lecture approach in combination with tutorial or exercise sessions. One active-learning technique that has gained popularity in the United States is the SCALE-UP (or studio physics) group-learning approach that was developed at North Carolina State University for large physics classes (Beichner et al. 2007;

G. Feldman (✉)
George Washington University, Washington, DC, USA
e-mail: feldman@gwu.edu

G. Feldman · M. Vreeswijk · A. P. Colijn · N. Brouwer · A. Heck
University of Amsterdam, Amsterdam, The Netherlands

D. Sokołowska and E. McLoughlin (eds.), *Embracing Changes Together in Physics Education*, Challenges in Physics Education,
https://doi.org/10.1007/978-3-032-27385-7_8

Beichner 2008) and has been adopted by over 200 institutions around the United States (including George Washington University). In this classroom format, students work together in small groups on guided conceptual and numerical activities, and the instructor serves as more of a facilitator or "coach" instead of a lecturer. When one of us (G. Feldman) had a sabbatical semester in Spring 2024, we embarked on a pilot implementation of SCALE-UP in a quantum physics class for first-year students at the University of Amsterdam (UvA).

2 Background on Quantum Physics Education

While the literature on teaching upper-level courses is not as extensive as the work on introductory courses, there is still a reasonable body of research on the subject. In the area of quantum physics, we can broadly generalize that the main pedagogical themes cover three basic categories: (1) specific topics to teach in the course, (2) students' comprehension of quantum concepts, and (3) instructional methods for teaching the class. In the present work, we have focused exclusively on point #3, and we have employed the active-learning SCALE-UP approach as an innovative pedagogical method in a university environment for the Quantum Physics 1 class at UvA.

A very recent Focused Collection of review papers (Lane et al. 2025) provides an excellent overview of current work on quantum education across various levels. In that regard, we try to make a distinction between quantum physics in a university setting and secondary school. For example, in the paper by Donhauser et al. (2024), the authors reach the following important conclusion:

> *"Looking at the structure of teaching strategies, it is noticeable that approaches based on innovative social structures of activation strategies are exclusively aimed at schools. This automatically raises the question of why social structures and their effects on learning are not taken into account in university courses or further education in the context of quantum science. When the learning of complex content works for school students within innovative social learning environments, this can also have positive effects on university students or professionals. It could therefore be interesting for educational research to replicate the social innovations of these studies and test them in the context of university courses or further education."*

Thus, the investigation of active-engagement methods for teaching quantum physics in a university setting is strongly encouraged, and this is specifically the objective of our study.

In a paper by Buzzell et al. (2025) on instruction in modern physics classes (a precursor to upper-level quantum physics), the authors reported that only about 7% of the 167 syllabi sampled in their study employed active-learning, studio-based, or flipped-classroom approaches. Collaborative group work was also highlighted in the work of Brundage et al. (2023) in a quantum mechanics class for 3rd or 4th-year university students and in a study by Deslauriers and Wieman (2011) in a modern physics class for 2nd-year university students. In both cases, student performance was evaluated using research-validated conceptual diagnostic surveys. Further insights were obtained in a recent study by Körhasan and Pospiech (2024), in which specific factors relative to collaborative group learning were investigated. Finally, Porter and Heckler (2020) demonstrated the advantages of guided group work sessions in graduate-level courses, although they "*have observed that there are some important differences in how to run these sessions compared to sessions with novice undergraduate students*". For completeness, we note that there have also been studies of group work in secondary school (e.g., Bungum et al. 2018), but those studies have concentrated on very specific isolated concepts in quantum physics (in that case, student understanding of wave-particle duality and the Schrödinger's cat gedanken experiment) which do not overlap with our present study in terms of the student population.

3 Course Setup and Classroom Facilities

The Quantum Physics 1 course was delivered during a period of 8 weeks in the second half of the spring semester of 2024, following the block-scheduling system employed for classes at UvA. The course is not taught at an introductory level, and it is expected that students know the relevant mathematics and are already familiar with basic concepts in quantum physics. Approximately 130 students participated, mainly physics majors, but also double-degree bachelor students and a few students in other study programs at UvA. The course was taught in previous years based on a structure of weekly lectures given by the UvA lecturers, followed by exercise sessions led by teaching assistants (TAs) who were older bachelor, master, or PhD students.

Courses at the UvA Faculty of Science are typically taught as traditional lectures for large groups (100-200 students), with separate tutorials for small groups of about 25 students. The classroom facilities consist of large lecture theaters where there is no possibility to organize group work which can be monitored by instructors in the classroom. For our project, we utilized the largest flat classroom on campus (accommodating up to 70 students) where we reconfigured the classroom by joining pairs of rectangular tables to form 11 composite shapes to organize 66 students per class session into 22 groups of three (two groups per table), as shown by the schematic in Fig. 1.

In this setup, there was sufficient space for the instructor and the TAs to circulate throughout the classroom to observe the students working and to easily reach any groups that might have questions. The blue dots represent the positions of the students at one table, indicating where the students were sitting for group work. Small portable whiteboards (45 cm × 60 cm) were distributed to each group for working together on numerical and conceptual questions in class. We used the electronic response system Wooclap (www.wooclap.com) to deliver conceptual questions periodically to the class, following the Peer Instruction method pioneered by Mazur (1997). Several photos of the classroom itself are depicted in Figs. 2 and 3.

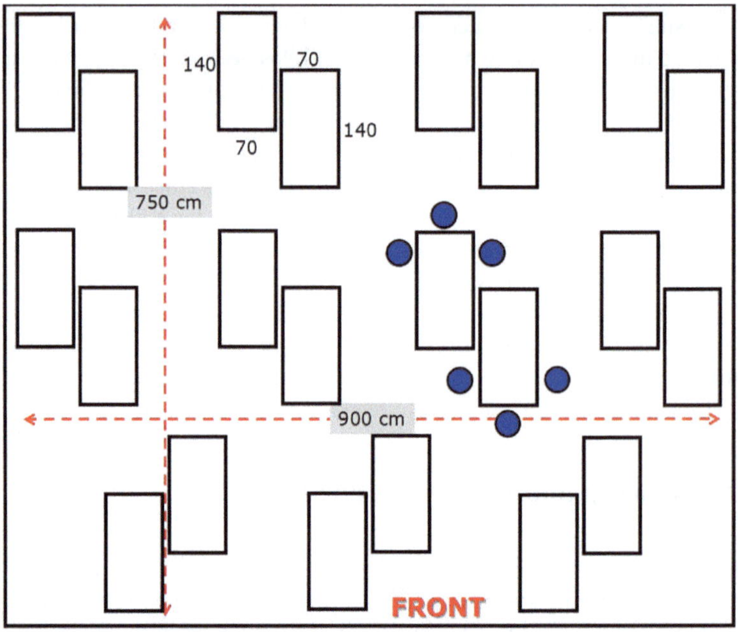

Fig. 1 The classroom configuration with 11 composite table setups for SCALE-UP, also showing the seating pattern for students (two groups of three) at one of the stations. Not depicted in this figure: a table at the front for the lecturer to connect a laptop to a beamer and a large screen for displaying slides

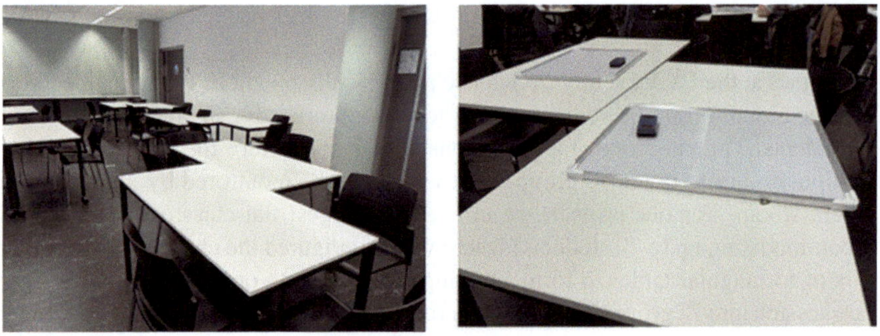

Fig. 2 Photos of the SCALE-UP classroom configuration, showing the formation of the tables (left) and the small whiteboards used by the student groups (right)

The instructional schedule for the class consisted of four sessions each week, with each session divided into two 45-minute blocks, with a 15-minute break:

1. Monday: a conventional lecture for all 130 students in a large lecture hall, delivered by one of the two UvA lecturers.

Fig. 3 Photos of the SCALE-UP class in action, showing the students working together on their whiteboards, assisted by the instructor and the TAs

2. Tuesday: two separate exercise sessions for 66 students, each run by one of the two UvA lecturers who gave the lecture on Monday and 3 teaching assistants in the format of the traditional exercise session in the flat classroom (Fig. 2). The class sessions were divided in this manner because a flat classroom large enough to accommodate all 130 students at the same time could not be secured, so the group activities had to be done in two consecutive shifts.
3. Wednesday and Thursday: two separate SCALE-UP sessions for 66 students in the flat classroom (Fig. 3), each in sequential sessions, following the collaborative group-learning model and given by the visiting lecturer who specialized in the SCALE-UP approach, assisted by 3 teaching assistants.

In addition to the three primary lecturers, the course was staffed by 8 graduate TAs. Each class session on Tuesday, Wednesday and Thursday had three TAs in the room to monitor the progress of the students and to help with questions as they arose. The TAs were also responsible for grading the weekly homework assignments and the quizzes. They participated in the grading of the exams as well, along with the instructors. The SCALE-UP sessions were observed by two didactic experts, who were involved in the research planning and data analysis.

The physics topics covered were based on the first two chapters of the textbook *Introduction to Quantum Mechanics* (3rd edition) by Griffiths and Schroeter (2018). This covered wave functions, statistics and probability, the uncertainty principle, and numerous solutions to the time-independent Schrödinger equation, including square well, harmonic oscillator and delta function potentials, as well as solutions for a free particle. While students received an overview of the material for each week in the Monday lecture, they were expected to read the textbook and come to class on the other days prepared to work on the group activities together, which included conceptual questions that they would discuss and numerical problems that they would solve.

4 Pedagogical Methodology

There were quite a few novel features in the delivery of this class that were very different from the conventional lecture-tutorial environment to which the students were accustomed. One day per week was reserved for a conventional lecture presentation by one of the local UvA instructors, which was recorded for students' later use and in which students could also ask questions. The lecture was, however, mostly a passive environment. As opposed to the Monday lecture, the focus of the other class sessions during the week was primarily on the students, allowing them to wrestle with the quantum physics material themselves in their groups (supported by real-time guidance from the instructors and TAs, as needed)—it was truly a "student-centered" classroom.

The Tuesday exercise session utilized a traditional collaborative group-learning approach, essentially unstructured, where students were doing exercises in the way they preferred, in groups or individually at their own pace, and where they could ask questions individually. There were no interruptions to report solutions or to discuss the problems with the entire class.

For the SCALE-UP in-class activities, exercises were generally taken from the end-of-chapter problems in the textbook or other problems contained in the instructor's course script. Many of these problems were lengthy multi-part questions that were broken into smaller pieces for the students to solve collectively in their groups using their whiteboards. Solutions to these questions and problems were reported by the students themselves in the classroom, displaying their group's work on their portable whiteboards to the rest of the class and explaining their steps in the solution. After one part was finished, the lecturer moved the groups ahead to the next part, continuing in this manner until the entire problem had been solved and discussed. In addition to the numerical problems, conceptual multiple-choice questions (Mazur 1997) were periodically posed to the students, who would respond after their group discussions using the Wooclap response system (www.wooclap.com), with the results being displayed to the class for further discussion and resolution. Students had no experience with such an approach in previous courses in their studies. Outside of class, aside from reading the textbook and preparing for the week's group activities, the students had homework problems assigned in the online learning management system used at UvA. The weekly problem sets consisted of about 4-5 numerical problems that were not formally graded, but they were checked by the TAs and were required to be completed by the students in order to qualify for participation in the final exam for the course.

5 Research Methodology

From a research point of view, we wanted to explore how the students responded to the new features of the course, as well as other considerations that applied uniquely to the SCALE-UP approach:

- Reduced lecture hours – less formal "transmission" of information than usual.
- Necessity of coming to class prepared to work on exercises.
- Reactions to new class elements (group work, electronic response system, whiteboards, etc.).
- Willingness to speak up and report solutions to the entire class.
- Different (and dynamic!) classroom environment compared to lecture and exercise sessions.

To address our research interests and to make some determination about the effectiveness of the novel SCALE-UP pedagogy, student data were collected (see Table 1) to provide both a quantitative evaluation and some qualitative feedback on the general impact of our intervention. For summative assessments, a mid-term exam at the middle of the semester (end of week #4) and a comprehensive final exam at the conclusion of the course (end of week #8) were given. In order to provide intermediate measurements of the students' progress during the semester and to provide feedback on their learning, two short quizzes were included as low-stakes formative assessments in the course schedule halfway to the mid-term exam (end of week #2) and then halfway to the final exam (end of week #6). These assessments were all required and counted as graded elements in the Quantum Physics 1 course, and the relative point values in the overall grading scheme are included in Table 1.

For a diagnostic conceptual assessment, we utilized the Quantum Mechanics Conceptual Survey (QMCS) which was developed by McKagan et al. (2010) at the University of Colorado and is available at the PhysPort website (www.physport.org). The focus of the QMCS is on standard topics covered in most second-year modern physics courses (e.g. wave functions and probability, wave-particle duality, Schrödinger equation, quantization of states, uncertainty principle, and tunneling).

Table 1 Summary of data collected for the Quantum Physics 1 class. The percentages shown for the quizzes and exams indicate the relative weighting of those course components in the final course grades

Summative assessments	Mid-term exam at end of week #4 (30%) Final exam at end of week #8 (60%)
Formative assessments	First quiz at end of week #2 (5%) Second quiz at end of week #6 (5%)
Diagnostic conceptual assessment	Quantum mechanics conceptual survey (QMCS) as pre-test and post-test
Feedback and comments	Mid-semester questionnaire Standard UvA course evaluation

Using the QMCS as a pre-test (on the first course day) made sense because first-year physics students at UvA had already been taught some quantum physics at a conceptual level in the compulsory "Quantum World" domain of the Dutch secondary school physics examination program for nationwide exams (Staderman 2022). The post-test was given in the last week of the course. Students participated in both assessments voluntarily, and the QMCS was not part of the graded components of the course.

Finally, a short mid-semester questionnaire for intermediate feedback and a longer standard UvA end-of-semester course evaluation (for all courses across the institution) were used to elicit student reactions and get their impressions of the pedagogical innovations.

6 Results

The distribution of scores on the mid-term and final exams is shown in Fig. 4, with scores binned in intervals of 10 points and bars at the middle of each bin. The average score for the mid-term exam was 60.0% for $N = 111$ students and the average score for the final exam was 65.3% for $N = 130$ students. While the distributions for both exams are rather broad, the class average did improve slightly for the final exam at the end of the semester. This was also the heaviest graded element in the entire course, contributing 60% of the points to the final course grade.

The dotted black line in Fig. 4 indicates the threshold for an exam score of 60%. A closer look at the students who scored over 60% in these two exams shows that only 58/111 = 52% of the students who took the mid-term exam exceeded that level, whereas 82/130 = 63% of the students who took the final exam successfully passed that threshold. This is a notable increase in the number of students who passed the second exam, which indicates that some progress had been made in the course.

The results of the two quizzes given during the semester are less clear. The distribution of quiz scores is shown in Fig. 5, and it is apparent that the students performed better on the first quiz. This could be an effect of pressure to perform well in the early stages of the course, which might tend to diminish as the semester goes on and students get fatigued. The second quiz scores are spread out quite widely, and also fewer students took the second quiz. Since each quiz was only worth 5% of the final grade in the course, it was easy for the students to not take these quizzes seriously, even though they were designed to provide useful and informative feedback on each student's progress.

The distribution of scores on the pre- and post-test, which consisted of the 12 conceptual multiple-choice questions of the QMCS (McKagan et al. 2010), is presented in Fig. 6. The results of both tests were statistically analyzed for item difficulty, discrimination of items, and reliability indices such as Cronbach's alpha and item-total correlations. The overall conclusion of the psychometric analysis was that the reliability of the QMCS as a pre-test was fine, but somewhat questionable as a post-test in the Quantum Physics 1 course.

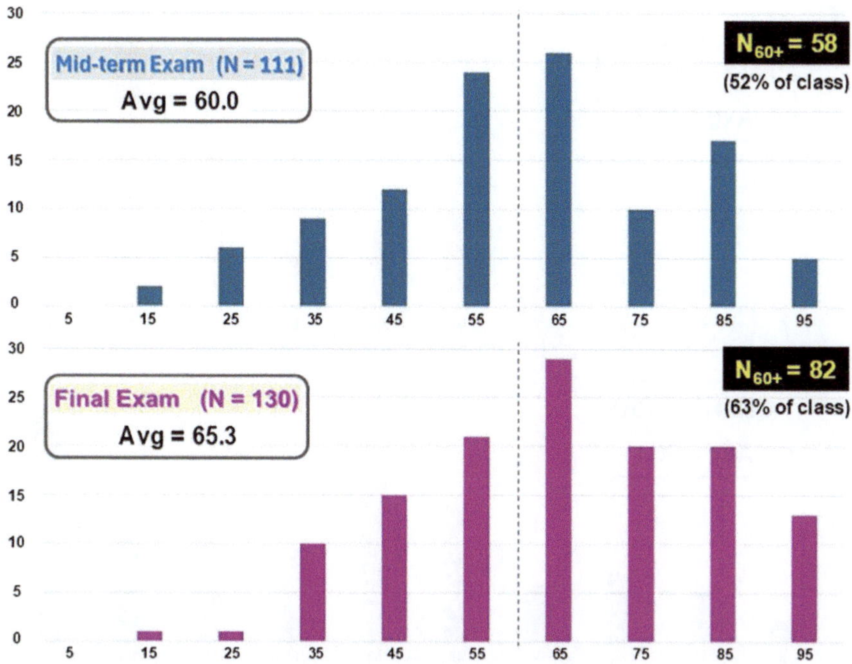

Fig. 4 Distribution of scores on the mid-term exam (upper graph) and the final exam (lower graph) in the Quantum Physics 1 course. Data are binned in grade intervals of 10 points, with bars at the middle of each bin

The average post-test score was slightly higher than the pre-test score, but of course this could be expected at the end of the course. However, a relevant indicator of student performance is the normalized gain of averages, given by $\langle g \rangle = (\langle \text{post} \rangle - \langle \text{pre} \rangle)/(12 - \langle \text{pre} \rangle)$, where brackets indicate class averages. This characterizes the fraction of missed points on the pre-test that were recovered in the post-test. The computed normalized gain of averages $\langle g \rangle = 0.10$ was low. When the normalized gain was computed for each of the students who completed both tests and the average of this group of 66 students was computed, the averaged normalized gain was $\langle g \rangle_{\text{avg}} = 0.02$. It is useful to put this into context by comparing our current result with some other QMCS results summarized in the original paper (McKagan et al. 2010), as shown in Table 2.

We have also added the QMCS results from Deslauriers and Wieman (2011) taken from two modern physics classes at the University of British Columbia (UBC). The UvA students had a relatively high pre-test score (6.70/12 = 55.8%) compared to other student populations. However, while the post-test score (7.21/12 = 60.1%) was generally consistent with the other populations, the normalized gain of averages and the averaged normalized gain were quite low due to the high pre-test score. In other words, the UvA students started at a high level, and from that point, they only advanced very little after the course was completed.

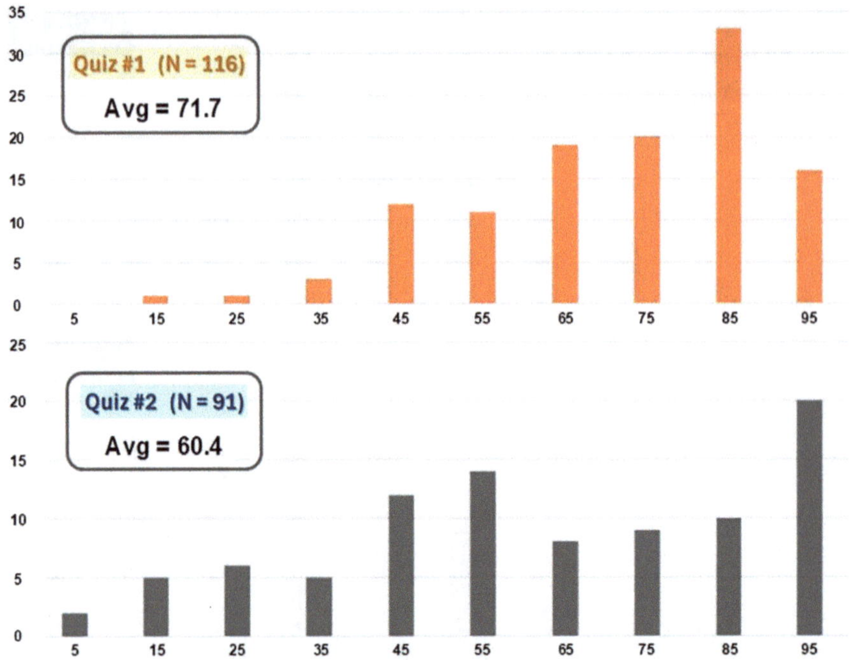

Fig. 5 Distribution of scores on quiz #1 (upper graph) and quiz #2 (lower graph) in the Quantum Physics 1 course. Data are binned in grade intervals of 10 points, with bars at the middle of each bin

Since students participated voluntarily in the QMCS, it is possible that students scoring high in the pre-test opted out of the post-test, which lowers the normalized gain. However, another indicator might be more suitable: Marx and Cummings (2007) suggested calculating the "normalized change", which removes students who score 0 or 100% on both the pre- and post-test, and it also uses a different normalization for students with negative gains, namely (post − pre)/pre. The averaged normalized change for the group of 66 students who completed both tests was $c_{avg} = 0.11$. When the group was split into low achievers on the pre-test (score ≤ 6) and high achievers on the pre-test (score ≥ 7), then the averaged normalized change was $c_{avg} = 0.20$ for the 35 low achievers and $c_{avg} = 0.01$ for the 31 high achievers. Thus, there was a moderate average normalized change for the low-achieving group, but there was no change for the high-achieving group on the QMCS.

We also wanted to examine the interplay between the pre- and post-tests and the course performance variables quiz 1, quiz 2, mid-term exam, and final exam. In order to establish relationships between these variables, Spearman's rank correlation coefficients were computed, based on 51 students for whom all data were available. Table 3 lists the correlation coefficients and their significance.

The significant correlation coefficients in boldface in Table 3 indicate the following: (i) students' scores on the pre-test and quiz 1 could already have provided

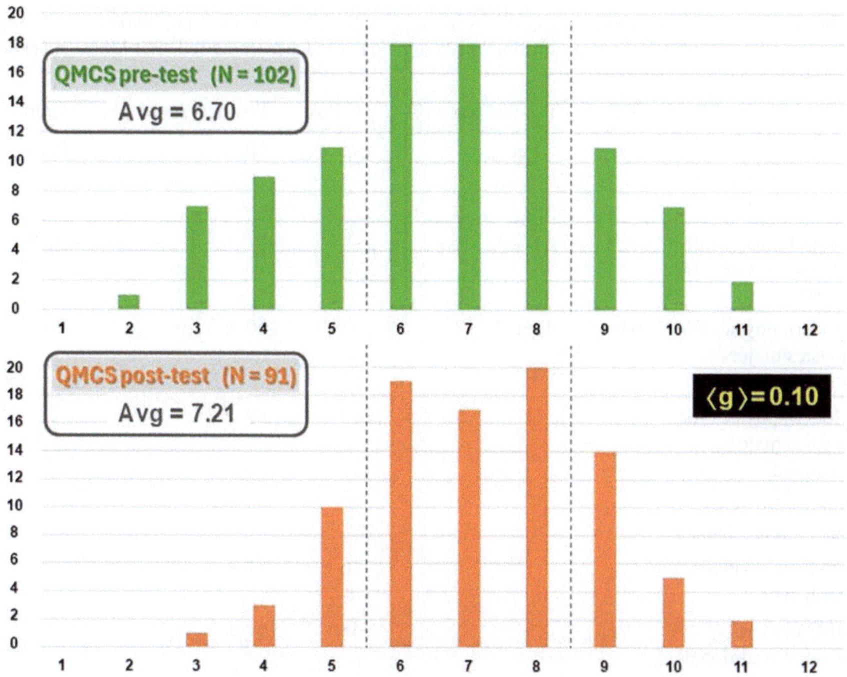

Fig. 6 Distribution of scores on the QMCS pre-test and post-test in the Quantum Physics 1 course. Each correct answer in the 12-item multiple-choice assessment has a point value of 1. The averages of the pre- and post-test scores and the normalized gain of averages $\langle g \rangle$ are shown in the figure

good information about the expected student success on the mid-term exam, and (ii) students' scores on the pre-test, quiz 1 and mid-term exam could already have been useful predictive indicators of success on the final exam. In other words, these course components could potentially have provided an early warning to the students at risk.

Student feedback was collected in the middle of the semester to get a sense of how students were reacting to the novel pedagogical environment of the SCALE-UP class and to have the opportunity to make any adjustments. This survey included several 5-point Likert-scale questions, as well as one open-ended question where students could offer personal comments. Two of the Likert-scale questions were:

1. "How effective is learning for you in the following settings?"
   a) Tutorial for 30 students with one TA
   b) Tutorial for 70 students with one instructor and several TAs
   c) SCALE-UP approach for 70 students with one instructor and several TAs
2. "I am learning better when . . ."
   a) I collaborate in my small group
   b) I work individually without any group interaction at any time

Table 2 Summary of results for the QMCS when used with different student populations, including a comparison of traditional versus reformed pedagogical approaches (modified after McKagan et al. 2010)

Class	N Pre	Pre-test score (%)	N Post	Post-test score (%)	Gain
Modern physics for engineering majors, traditional	60	34 ± 2	43	60 ± 3	0.39
Modern physics for engineering majors, reformed	246	35 ± 1	202	67 ± 2	0.49
Modern physics for physics majors, traditional	63	41 ± 2	57	50 ± 2	0.15
Modern physics for physics majors, reformed			69	69 ± 2	
Junior (3rd year) quantum mechanics			22	79 ± 4	
Graduate quantum mechanics			16	81 ± 6	
UBC modern physics, traditional			48	67 ± 3	
UBC modern physics, reformed			62	85 ± 2	
University of Amsterdam Quantum Physics 1	102	55.8	91	60.1	0.10

Table 3 Correlation coefficients between pre-test and post-test results and the course performance variables quiz 1, quiz 2, mid-term exam, and final exam. (** Correlation significant at 0.01 level, 2-tailed)

	quiz 1	*mid-term exam*	*quiz 2*	*post-test*	*final exam*
pre-test	.39**	**.45****	.22	.65**	**.36****
quiz 1		**.40****	.36**	.27	**.51****
mid-term exam			.49**	.54**	**.67****
quiz 2				.24	.48**
post-test					.37**

The distribution of answers to the first and second questions is shown in Figs. 7 and 8, respectively. In Fig. 7, the SCALE-UP approach scored the highest average rating of 3.7 out of 5, shown by the green histogram. The majority of students gave this approach rating values of 4 or 5, as opposed to the tutorial choices that predominantly had ratings of 3 or 4. In Fig. 8, the group work scored a higher rating of 3.9 out of 5 relative to the individual work, shown by comparing the blue vs. red histograms. In

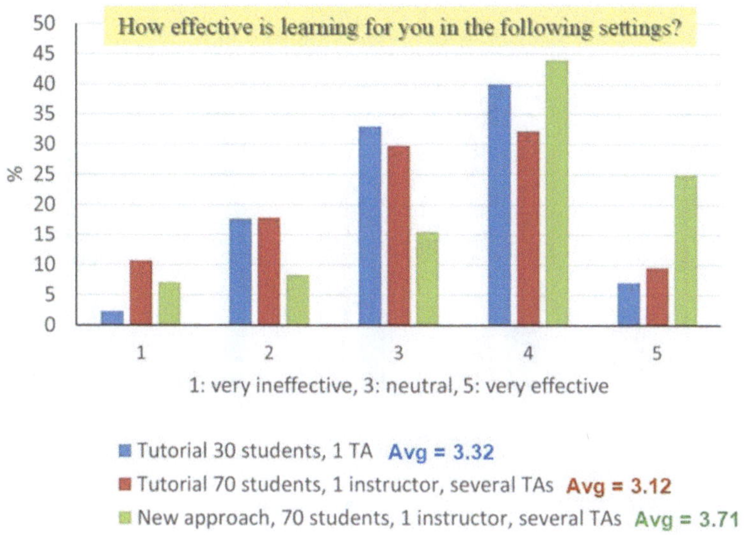

Fig. 7 Summary of the results for the student feedback on the Likert-scale question #1. Answer options a, b, c are colored blue, red, and green, respectively

this case, the students mostly gave the group work rating values of 4 or 5, while the individual work predominantly had ratings of 2 or 3. The combination of these two questions constitutes a compelling endorsement by the students of the collaborative group-learning approach offered by the SCALE-UP methodology after exposure to the novel pedagogical approach during the first half of the semester.

Some representative open-ended comments from the students are presented below (translated):

- I find it much better than the traditional way of teaching because you can discuss the material with your fellow students a lot, and even if you don't know something, or even if your whole group doesn't know, it's easier to find the right answer with multiple people.
- It's more effective because you learn by helping other people in your group, and you also get help from them.
- You're constantly encouraged to work, which keeps you attentive throughout the entire lecture. Because you're so engaged with the subject matter, you also have to do less work at home, which saves a lot of time. Overall, I experience the new approach very positively.
- It motivates me to actually be engaged during the tutorial. That's something I don't experience during the other tutorials, which sometimes leads me to not work on the material and fall behind.

It is noteworthy that the students recognized the value of the collaborative group discussions and the benefit of sharing various insights among the group members to help each other understand the material. They also stated that they felt more attentive

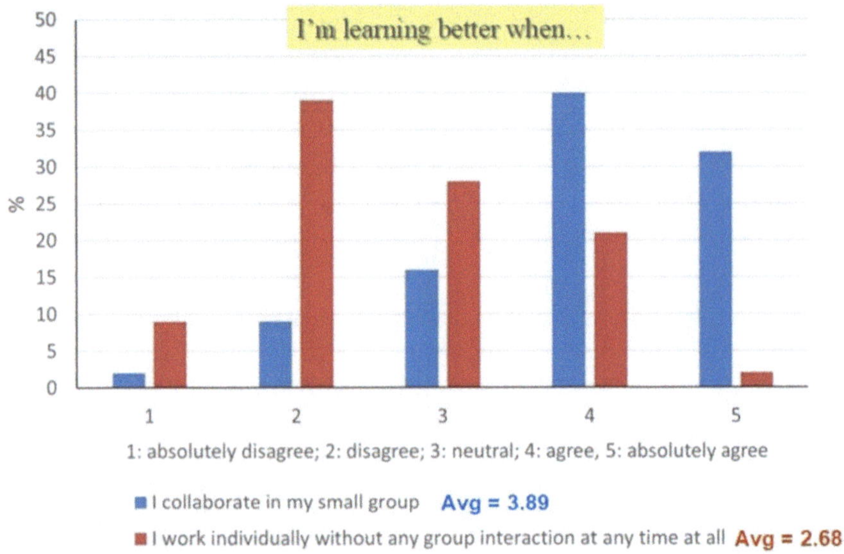

Fig. 8 Summary of the results for the student feedback on the Likert-scale question #2. Answer options a and b are colored blue and red, respectively

and engaged in the active-learning setting, which is often not the case in a more conventional (passive) lecture classroom setting.

However, there were also some negative comments, which are worth discussing:

- The SCALE-UP lessons are too chaotic. It's nice that answers are discussed right away, but it often happens too quickly, and there's not enough time to work on the assignments before they're discussed. This leads to falling behind very quickly. Regular tutorials with 20 students and one TA are better.
- To be honest, the pace in the tutorial is very slow, and I could learn two/three times faster individually.
- I feel like we don't have enough time to ask questions, and sometimes we rush through things too quickly. I think having one tutorial in the new format would be good, or maybe half of the classes, so we have enough time to ask questions properly.

In this case, the students seemed to complain about the structured nature of the SCALE-UP classroom activities. It is true that the collaborative group work is periodically interrupted so that one group can present the solution and the entire class can all review the problem together. In that regard, not all groups will have finished their work, and so the interruption can potentially be a bit discouraging. However, it is curious that two of the comments above contradict each other—one student felt that the SCALE-UP pace was too slow, but another student indicated that "we rush through things too quickly." As you can see, it is difficult to accommodate the pace of learning for all of the students at the same time!

Student feedback was also collected at the end of the course through the regular UvA course evaluations, and the current results could be compared with feedback from the academic year 2021-22. Both cases were very positive and had similar quantitative outcomes, but the reformed Quantum Physics 1 course received higher positive scores on the items "activating teaching and learning", "collaboration" and "interest in the subject". Generally, students in the SCALE-UP class perceived less work stress, greater stimulus to actively participate in class, and larger improvement in collaboration skills.

Some representative comments from students about the SCALE-UP approach in response to the neutral question "what they had appreciated most in the Quantum Physics 1 course" were (translated):

- I think the SCALE-UP sessions work well because you get to see all course subjects once and you could finish at home. Because questions were answered quickly, you had more time to engage with the lesson material.
- I really enjoyed working with the SCALE-UP workgroups. The whiteboard and the collaboration that came with it helped me understand the material. It was also great that the assignments were discussed and evaluated right away.
- For a student like me who struggles with self-motivation to do homework, the SCALE-UP classes are fantastic as they actually make me do assignments that I probably wouldn't have started on until it was too late; so definitely keep doing them.

After the course was finished, most students apparently better understood and greatly appreciated the SCALE-UP approach. But like in the mid-term survey, not all students were positive. In response to the neutral question "what do they suggest for improving the course", some student comments were made about the SCALE-UP sessions:

- During the SCALE-UP session, it was maybe a bit too noisy in the classroom.
- For some, collaboration was difficult when there were differences in the levels of students' understanding of the material. This made it difficult for some to join in.
- The new type of teaching was very boring, there were few opportunities to do more difficult assignments if you already understood it.
- The SCALE-UP lessons went very fast. As a result, I did not always understand what the correct answer was and why. Asking questions later is also not possible, because we were working in class the whole time. I think it would be useful to leave part of the lesson(s) for self-study, so that questions can be asked.

Here again, we see a variety of different reactions to the SCALE-UP collaborative environment. Actually, noisy classrooms are inevitable in SCALE-UP, but as one student remarked: "it may be a matter to point out to students talking during the lecturer's explanation that this hinders others who want to listen." The observers noticed that some students who were seemingly unwilling to collaborate at first, changed later to become real participants in a group, whereas a minority kept working alone. The lecturer and the TAs stimulated collaboration but did not enforce it.

7 Conclusions

In summary, the students were highly engaged during the group sessions, and they seemed to fully appreciate the opportunity to work with their partners during the classroom activities. In the SCALE-UP sessions, students were not at all hesitant to speak up during class—group members freely reported their results to the entire class, and follow-up questions flowed naturally from the students during the class period. In that regard, it was quite valuable to have alternating periods of intense group work followed by whole class discussions of the results for each particular problem. This enabled all class members to see the solution of each problem, even if they did not have quite enough time to complete the task themselves.

In the major assessments given to the class, a small improvement in the class average on the final exam (65.3% for $N = 130$) was found compared to the mid-term exam (60.0% for $N = 111$). While intermediate quizzes may have helped students prepare for these high-stakes exams, a separate analysis of the quiz scores themselves did not yield any pertinent conclusions.

In order to quantify possible learning gains in the class, the QMCS (McKagan et al. 2010) was administered as a pre- and post-test combination to evaluate the students' conceptual knowledge about quantum physics. The average pre-test score at UvA was rather high compared to other similar student populations, but the average post-test score was largely consistent with those other cases. As a result, the normalized gain for the UvA class was rather small. But a deeper analysis showed that the applied instructional approach certainly raised the post-test level of performance for students who had low scores on the pre-test. However, the interpretation of the QMCS results for gauging student learning in the Quantum Physics 1 course was somewhat uncertain. The QMCS was primarily designed to be used in a typical modern physics class in the United States. While the UvA course was aimed at first-year students in the Netherlands, the level of the course was somewhat higher than a modern physics class (although somewhat lower than an upper-level quantum mechanics class, which could also use Griffiths as the textbook). In that regard, a careful evaluation of the questions in the QMCS revealed that questions 1-4 were really covering background information probably found in Dutch high schools and had no relevance to the material in the UvA Quantum Physics 1 course.

The most significant piece of evaluative evidence in our UvA class might come from the student feedback itself, collected in a mid-semester questionnaire and the end-of-semester UvA course evaluation. Likert-scale questions in our survey indicated that students preferred the SCALE-UP approach with its more structured set of collaborative activities, and that they overwhelmingly favored the collaboration with their partners as opposed to working alone. While student comments in an open-ended question were predominantly positive, some critiques of the collaborative group approach that seemed to stifle the progress of some students were observed. While it could be true that regulating the pace in the group setting might tend to "slow some students down" for the very talented individuals, the main point is to offer a pedagogical setting that can better serve the overall student population as a whole.

In that regard, we think that we were able to provide such a setting for a majority of the students.

8 Outlook

In this project, the students had a clearly positive response for including the collaborative group-learning approach in the Quantum Physics 1 course. However, there was some "tension" expressed in the student comments about the difference between the unstructured group work on Tuesday and the more organized SCALE-UP format on Wednesday and Thursday. The former offered a freer and more open setup, where students worked together but at their own pace (which varied widely between groups). The latter provided a framework that alternated between group work and whole class discussion periods which kept the entire class moving together at the same rate. It seems logical to take the message from the students that both formats are valuable and that both should be incorporated into the class in a sensible manner, but perhaps in a different order. A reasonable suggestion to achieve the appropriate combination might be to begin each week with a lecture, followed by two SCALE-UP sessions, and finishing with the open, unstructured group work, with TAs monitoring the group activities. In this manner, the pedagogical approach flows from an initial instructor-centered format (the class lecture), to structured group-learning activities where real-time guidance is provided, then finally leading to a more free-form approach to "let the students loose" to engage in groups by themselves, move forward at their own speed, and if necessary have further discussions with TAs to help them grasp the material. Further improvements were suggested to leverage the group work in a more advantageous way and also to temper the students' expectations of the pedagogical environment in which they find themselves. Hopefully this innovative course reform will continue at UvA, and will also extend to other courses in Physics, as well as other STEM disciplines.

Acknowledgements One of us (G.F.) would like to thank the Faculty of Science at the University of Amsterdam for their generous support of the visit during the Spring 2024 semester that made this project possible. My co-authors were very gracious hosts, dedicated educators, and extremely pleasant colleagues with whom I was fortunate to spend a remarkable five months. I am also grateful to the students and TAs in the UvA Quantum Physics 1 class for their pioneering spirit and willingness to engage in this novel pedagogical endeavor.

References

Beichner R et al (2007) The student-centered activities for large enrollment undergraduate programs (SCALE-UP) project. In Research-based reform of University Physics, Reviews in PER Vol. 1, edited by EF Redish and PJ Cooney (AAPT, College Park, MD, 2007). https://doi.org/10.1119/RevPERv1.1.4

Beichner RJ (2008) The SCALE-UP Project: a student-centered active learning environment for undergraduate programs. Natl Acad White Pap. https://www.per-central.org/items/detail.cfm?ID=14699. Accessed 23 Feb 2026.
Brundage MJ, Malespina A, Singh C (2023) Peer interaction facilitates co-construction of knowledge in quantum mechanics. Phys Rev Phys Educ Res 19:020133. https://doi.org/10.1103/PhysRevPhysEducRes.19.020133
Bungum B, Bøe MV, Henriksen EK (2018) Quantum talk: how small-group discussions may enhance students' understanding in quantum physics. Sci Educ 102:856–877. https://doi.org/10.1002/sce.21447
Buzzell A, Atherton TJ, Barthelemy R (2025) Quantum mechanics curriculum in the U.S.: quantifying the instructional time, content taught, and paradigms used. Phys Rev Phys Educ Res 21:010139. https://doi.org/10.1103/PhysRevPhysEducRes.21.010102
Deslauriers L, Wieman C (2011) Learning and retention of quantum concepts with different teaching methods. Phys Rev ST Phys Educ Res 7:010101. https://doi.org/10.1103/PhysRevSTPER.7.010101
Donhauser A et al (2024) Empirical insights into the effects of research-based teaching strategies in quantum education. Phys Rev Phys Educ Res 20:020601. https://doi.org/10.1103/PhysRevPhysEducRes.20.020601
Freeman S et al (2014) Active learning increases student performance in science, engineering, and mathematics. Proc Natl Acad Sci 111:8410–8415. https://doi.org/10.1073/pnas.1319030111
Griffiths DJ, Schroeter DF (2018) Introduction to quantum mechanics, 3rd edn. Cambridge University Press, New York, NY. https://doi.org/10.1017/9781316995433
Körhasan ND, Pospiech G (2024) Examination of students' quantum physics knowledge construction with peers and metacognitive evaluations. Phys Rev Phys Educ Res 20:020153. https://doi.org/10.1103/PhysRevPhysEducRes.20.020153
Lane WB, Michelini M, Singh C (2025) Editorial: focused collection: investigating and improving quantum education through research. Phys Rev Phys Educ Res 21:010002. https://doi.org/10.1103/PhysRevPhysEducRes.21.010002
Marx JD, Cummings K (2007) Normalized change. Am J Phys 75:87–91. https://doi.org/10.1119/1.2372468
Mazur E (1997) Peer instruction: a user's manual. Prentice-Hall, Upper Saddle River, NJ
McDermott LC (2001) Oersted medal lecture 2001: "physics education research—the key to student learning." Am J Phys 69:1127–1137. https://doi.org/10.1119/1.1389280.
McKagan SB, Perkins KK, Wieman CE (2010) Design and validation of the quantum mechanics conceptual survey. Phys Rev ST Phys Educ Res 6:020121. https://doi.org/10.1103/PhysRevSTPER.6.020121
PhysPort website provides access to the QMCS at: https://www.physport.org/assessments/assessment.cfm?I=2&A=QMCS. Accessed 23 Feb 2026
Porter CD, Heckler AF (2020) Effectiveness of guided group work in graduate level quantum mechanics. Phys Rev Phys Educ Res 16:020127. https://doi.org/10.1103/PhysRevPhysEducRes.16.020127
Staderman K (2022) Connecting secondary school quantum physics and nature of science: possibilities and challenges in curriculum design, teaching, and learning. (Doctoral dissertation, University of Groningen), chap 1. https://doi.org/10.33612/diss.195942915
Wooclap website is located at: https://www.wooclap.com. Accessed 23 Feb 2026

Non-constant Acceleration Kinematics: University Students' Difficulties Related to Graphical Representation Systems

Any Urrutia, Cristian Merino, Kristina Zuza, and Jenaro Guisasola

Abstract In physics courses, teachers use graphical and algebraic representations, and in university courses kinematic situations with non-constant acceleration are included. Despite the instruction, students face difficulties both in the treatment within the representations and in the conversion between them, which affects their understanding of the physical phenomenon. In this paper, we present the results obtained in the research that has been carried out in relation to the treatment of information within the graphical representation. The theoretical framework of semiotic representations and a phenomenographic methodological approach are applied to analyse how 239 engineering students interpret and extract information in this type of representation. The results reveal that in graphical representations, conceptual confusions and problems in recognising and processing information are observed. The findings suggest the need to develop teaching materials that promote a deeper understanding of kinematic concepts of non-constant acceleration.

Keywords Non-constant acceleration · Difficulties · Graphical representation · Semiotic representation

---

A. Urrutia (✉)
Department of Physics and Astronomy, Faculty of Exact Sciences, Universidad Andres Bello, Viña del Mar, Chile
e-mail: any.urrutia@unab.cl

C. Merino
Institute of Chemistry, CIDSTEM, Pontificia Universidad Católica de Valparaíso, Valparaíso, Chile

K. Zuza
Department of Applied Physics, University of the Basque Country (UPV/EHU),IKASGARAIA, San Sebastián, Spain

J. Guisasola
IMH Campus, Azkue Auzoa, Elgoibar, Spain

D. Sokołowska and E. McLoughlin (eds.), *Embracing Changes Together in Physics Education*, Challenges in Physics Education,
https://doi.org/10.1007/978-3-032-27385-7_9

1 Introduction

Research in physics teaching has provided teachers and researchers with insights into students' conceptions, allowing them to anticipate their reasoning and plan instruction effectively (Viennot, 2001). Previous studies of kinematics in introductory physics courses, beginning in the 1980s, have explored students' ideas and patterns of reasoning about fundamental concepts in this area (Brasell and Rowe, 1993; McDermott et al. 1987).

There are studies that indicate that students face difficulties with both qualitative and quantitative definitions of velocity and acceleration, as well as their interrelationship (Shaffer and McDermott, 2005). However, despite advances in this area, conceptual errors persist, such as not recognising that instantaneous velocity is tangent to the trajectory, assuming that acceleration is zero when the module of the velocity is constant or not understanding that a change in the direction of velocity in two dimensions implies a non-zero acceleration (Shodiqin and Taqwa, 2021; Viennot, 2001).

In addition, students have been found to have problems solving physics problems. Students tend to have greater difficulty interpreting the physical context and, although they use an algorithmic approach, they find it difficult to transfer mathematical knowledge to physical situations (Guisasola and Zuza, 2024). Despite advances in research, few studies analyse in depth the difficulties faced by university students when dealing with the treatment of graphical representations in the study of kinematics with non-constant acceleration, and although teachers use algebraic and graphical representations to facilitate understanding in this topic, students often have difficulty conceptualising this non-uniform change, further complicating the physical interpretation of mathematical solutions (Pospiech, 2019).

This study distinguishes itself from other research on students' ideas of kinematics in physics courses by focusing specifically on the analysis of motion situations with non-constant acceleration and by taking into account the difficulties students face in interpreting the graphical representations involved in the analysis of these motions. Similar studies in other subjects have inspired the analysis of the difficulties associated with handling semiotic representations in this type of context (Campos et al. 2022). The aim of this paper is to investigate students' difficulties in identifying the physical meaning and characteristics of motion with non-constant acceleration in its semiotic graphical representation. We intend to relate their difficulties of recognition to those of conceptual understanding of physics at university level.

Thus, our research question is as follows:

What are the obstacles and reasoning patterns that emerge when second-year engineering students are confronted with kinematics topics with variable acceleration using graphical semiotic representations as their main tool?

The work presented below is part of a broader investigation that analyses students' reasoning when faced with different types of graphical and algebraic semiotic representations, their treatment within each of the representations and the conversions between them. In this case, we focus on students' difficulties when faced with the treatment within graphical representations for the study of the kinematics of rectilinear motion with non-constant acceleration in advanced mechanics courses in science and engineering degrees.

2 Background

2.1 Semiotic Representation in Physics and Mathematics

In this study, we analyse the difficulties students face in understanding graphical representations of motion with non-constant acceleration, using the theory of semiotic representational registers (Duval, 2006). This theory considers graphical and algebraic representations as symbols that use a set of rules and associations to represent various motions. This allows us to investigate how these representations can function as individual beliefs and complex signs, influencing the acquisition of knowledge about representational systems. Using this theoretical framework enables us to explore the cognitive characteristics of different representational systems used in university physics teaching, such as position, velocity and acceleration versus time graphs describing the kinematics of motion with constant acceleration, as we agree with Duval (Duval, 2006) that mathematical objects are not accessible through direct perception, which can hinder the understanding of physics and hinder learning.

Students' difficulties may arise when they have to recognise mathematical objects in different representations (Duval, 2017). This is particularly evident in the relationships between position, velocity and acceleration functions with their respective graphical representations of a body in motion with non-constant acceleration that will be the subject of study in future work. Since students must be able to move within the same representation in order to understand physical phenomena of kinematics in this type of movement, it requires not only the handling of physical concepts specific to this type of movement, but also mathematical skills to respond to them.

Difficulties in understanding the treatment of information within a single representation and in making conversions between different representations are due to the need to develop two cognitive skills, according to Duval (Duval, 2006, 2017): extracting information from the content of the representation of the object it represents and recognising the same object across different representations. This phenomenon, evidenced in mathematics (Duval, 1996), can also manifest itself in the physical sciences (Campos et al. 2023). While the use of multiple representations can enrich conceptual understanding, it can also create challenges. In addition, equations of motion with varying acceleration, which are often non-linear, demand advanced mastery of differential and integral calculus, which may contribute to additional difficulties in student learning in this context (Duval, 2006, 2017).

2.2 Graphical Representations in University Physics

Early studies on graph comprehension in physics by McDermott, Rosenquist and van Zee (McDermott et al. 1987) identified students' difficulties in connecting graphs to physical concepts, focusing on the interpretation and construction of graphs, especially in kinematics with constant acceleration, which has prompted further research in this field (Brasell and Rowe, 1993; Wemyss and Van Kampen, 2013).

Later research, such as that of Beichner (Beichner, 1994) related to the kinematics of motion with constant acceleration, confirmed students' difficulties in understanding graphs in these subjects, especially in the confusion between position, velocity and acceleration, and the relationship between slope and area under the curve, using the Test of Understanding Graphs in Kinematics (TUG-K), which has been fundamental for evaluating educational interventions in physics and, inspired by it, the TUG-C was also developed to evaluate mathematical concepts (Zavala et al. 2017), where they highlight that the context in which concepts related to graphs of derivatives and antiderivatives are taught has an impact on students' understanding and performance (Barniol et al. 2024).

Recent studies (Ivanjek et al. 2016; Wemyss and Van Kampen, 2013) have shown that students still face significant difficulties in interpreting graphs of motion with constant acceleration, especially with regard to the direction of motion, calculating slope and interpreting the area under the curve, as well as frequently resorting to incorrect formulae to solve kinematic problems. In the complex relationship between physics and mathematics in university contexts, it is essential that students are able to interpret each mathematical symbol with its corresponding physical meaning, ensuring that mathematical operations are consistent with the relevant physical theory. This will contribute to strengthening students' learning and perception of physics in these contexts.

Despite advances in research in the areas of physics and mathematics, few studies comprehensively address the difficulties faced by university students in interpreting motion graphs with non-constant acceleration, a topic that remains a challenge due to the complexity of the concept of acceleration. For example, (Shodiqin and Taqwa, 2021) conducted a study on the understanding of the concept of acceleration in kinematics based on various representations (mathematical, verbal, pictorial and graphical), revealing that students still have significant difficulties in understanding this concept.

Finally, in university physics courses, especially in mechanics courses, teachers use various graphical and algebraic representations, where the most common mathematical representations in these contexts are usually manifested as algebraic expressions (such as formulas, equations or functions). In addition, graphical representations play a fundamental role, as they combine iconic elements with abstract and symbolic features acting as a crucial bridge between iconic and algebraic representations (Pospiech, 2019). However, students often face difficulties in conceptualising non-uniform changes, especially in the context of kinematics with non-constant acceleration. In this process of mathematical modelling, students must be able to move from a physical logic to a mathematical logic, which implies moving different semantic registers.

3 Methodology

The aim of this paper is to investigate students' difficulties in identifying the physical meaning and characteristics of motion with non-constant acceleration in a graphical semiotic representation. We aim to relate their difficulties in recognising and processing this type of representation to those of conceptual understanding at university level. To achieve the objective of this study, an open-ended questionnaire was designed with space to justify the students' answers in written form. Three of these questions allowed the student to respond by processing within the graphical representation.

The three questions were designed on the basis of the key ideas extracted from epistemological analysis. After identifying the key ideas and confronting these ideas with the corresponding curriculum (second year physics for engineering), the learning objectives for the course were defined (see Table 1). The questions were aligned with these objectives. Each question consisted of a statement accompanied by an initial graphical representation, followed by a space where students had to answer using the corresponding final representation.

The analysis of the data obtained was carried out using a phenomenographic approach (Marton, 1981), which considers the various ways in which people interpret and understand reality. According to this approach, each structure, concept and form of reasoning constitutes a descriptive category of reality that can be collectively appreciated. Thus, these representations reflect a type of collective intellect, where a particular category is manifested through different reasoning. In this framework, data collection and analysis focused on students' explanations from which a bottom-up study was conducted (Guisasola et al. 2023).

This method allowed the creation of descriptive categories and an outcome space illustrating the relationships between them. Each category focuses on critical aspects of the understanding of a specific phenomenon, interrelated in a hierarchical and inclusive manner (Guisasola et al. 2023). This enables the development of more advanced and complex categories, ensuring their internal consistency. In this study, we identified descriptive categories that explain students' abilities to

Table 1 Relation of questions, learning objectives and initial and final representations

Question	Learning objective	Representation
Q1 Q2 Q3	Identifies, relates, compares and analyses the movement of bodies of variable acceleration in graphic representation and carries out an adequate treatment of the information within the same representation	Initial: graphical Final: graphical

Below is a position versus time graph that describes the motion of a body moving in a straight line with variable acceleration.
With the available information, answer:

a) What is the <u>velocity v/s time</u> graph that best represents the movement described above?

b) Justify how did you arrived at that choice?

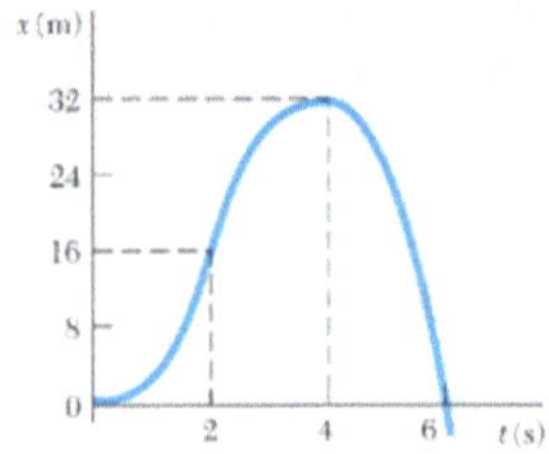

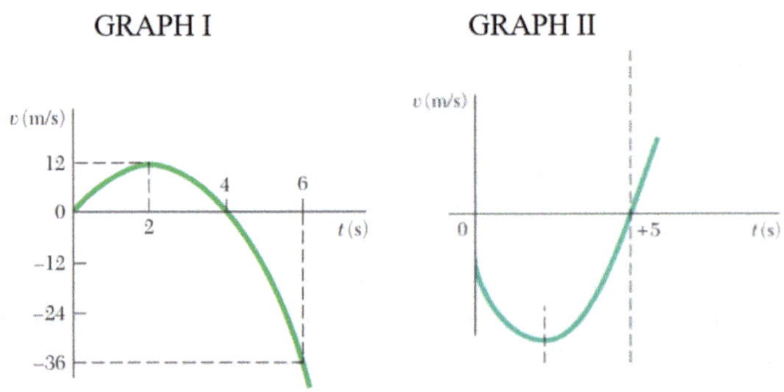

Fig. 1 *Task used in the study.* **a** Position–time graph representing the motion of a body moving along a straight line with non-constant acceleration. **b** Two candidate velocity–time graphs (Graph I and Graph II) presented to students, who were asked to determine which graph correctly represents the motion and to justify their reasoning

recognise graphical representations for the kinematics of motion with non-constant acceleration.

To gain a more detailed understanding of the students' reasoning in answering the open-ended questionnaire, interviews were conducted using the 'Thinking Aloud' technique (Van Someren et al. 1994). This methodology allowed us to explore their thought processes as they reviewed and commented on their responses, which were audio-recorded. The transcription of these recordings facilitated a deeper analysis of the cognitive strategies employed. The students selected for the interviews were intentionally chosen by the researchers because they offered more complete explanations that provided a clear picture of their reasoning.

The combined use of open-ended questionnaires and interviews provided a more complete picture of the students' difficulties in working with semiotic representations in university physics. While the questionnaires offered a first approximation of the problem areas, the interviews allowed a deeper insight into the underlying cognitive processes (Van Someren et al. 1994). Together, these methods facilitated the identification of reasoning patterns and enriched the analysis of students' justifications. This mixed approach also allowed for a triangulation of data, which improved the validity of the findings and provided a more holistic understanding of cognitive difficulties.

3.1 Context

This study involved 239 students in a calculus-based mechanical physics course offered at a private Chilean university with three campuses in different regions of the country. The course corresponded to the third semester of the engineering degree and was taken by students who had already completed two physics courses (Introduction to Physics and Experimental Physics) and two mathematics courses (Introduction to Mathematics and Differential Calculus). The participants were distributed in 18 parallel classes, with between 35 and 45 students per group. Although all classes shared the same timetable and pedagogical approach, each was taught by a different lecturer. All classes have followed the usual traditional format of university teaching with lectures and problem-solving classes, all of them with non-active methodologies in which the teacher has been the centre of teaching.

The first chapter of the mechanics course covered motion in one, two and three dimensions, including concepts such as vector displacement, velocity and acceleration, both constant and non-constant. This chapter lasted 4 weeks, with three lectures per week and an hour and a half devoted to practical workshops. Teaching took place in a traditional transmissive instructional context, complemented by active learning strategies during the workshops. Each week, formative assessments were conducted online, with immediate feedback. At the end of the chapter, students took a final exam, composed of problems similar to those at the end of the textbook chapters. Data collection was conducted during a face-to-face class at the end of the 4 weeks of instruction and after the final exam.

3.2 Instruments

To address the research question, a questionnaire of open-ended questions was designed: three on graphical semiotic representations of motion with non-constant acceleration. The aim was for students to provide detailed explanations on the recognition and processing of these representations, aligned with the curriculum objectives, which include interpreting, discriminating and extracting information from algebraic and graphical representations of position, velocity and non-constant acceleration as a function of time. Students had to demonstrate their ability to understand the characteristics of motion with non-constant acceleration, justifying their answers in writing (see Table 1).

The questionnaire was validated by two physics experts and two specialists in physics education research, who had access to the open-ended questionnaire and completed a specification table evaluating the alignment between the learning objective and each question. Following this review process, the handwritten questionnaire was improved, addressing aspects such as robustness, usefulness, readability and alignment between objectives and graphical semiotic representations (see Table 2).

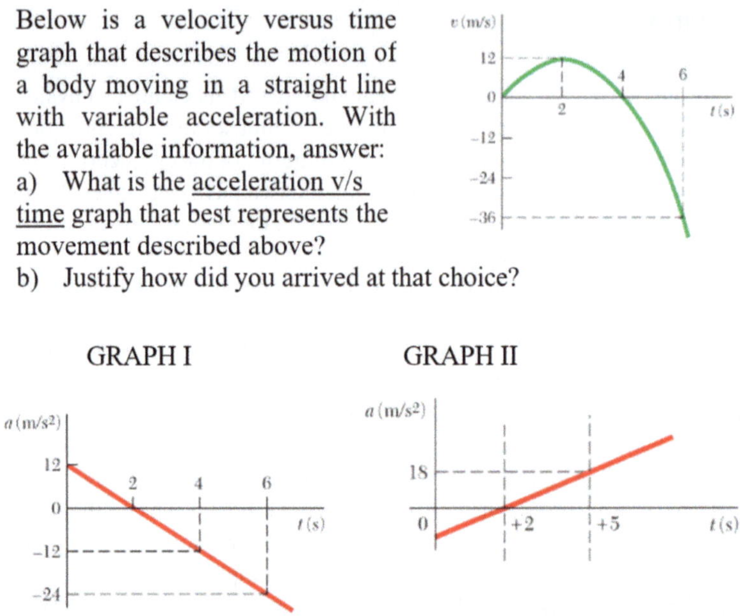

Fig. 2 *Task used in the study.* **a** Velocity–time graph describing the motion of a body moving along a straight line with variable acceleration. **b** Two candidate acceleration–time graphs (Graph I and Graph II) presented to students, who were asked to determine which graph correctly represents the motion and to justify their reasoning

Before administering the questionnaire, students signed an informed consent form approved by the Bioethics and Biosafety Committee of the Pontifical Catholic University of Valparaíso (code BIOEPUCV-H 647–2023). They were provided with a detailed description of the study, including the measures taken to protect their privacy and personal data. Additionally, they were assured that incorrect answers would not result in any penalties and were asked to complete the questionnaire in its entirety to gain a comprehensive understanding of their reasoning and difficulties.

After administering the open-ended questionnaire, 12 interviews were conducted with 4 students selected from each region of the country. The interviews focused on the questionnaire questions, aiming to explore students' reasoning in greater depth (Van Someren et al. 1994). During the interviews, students reviewed their answers while verbalising their thoughts, strategies, and reasoning, which were recorded and transcribed for analysis. This approach allowed for a direct exploration of students' cognitive processes, the identification of conceptual difficulties and misunderstandings, and a deeper understanding of the cognitive barriers in handling representations of motion with non-constant acceleration.

Table 2 Questions on non-constant acceleration kinematics. Initial/final graphical representation

Question 1	
See Fig. 1	*The x/t graph in the statement shows that the motion of the particle can be analysed through the slopes:* ● *From the origin (x = 0 m) and with zero velocity, the particle experiences a positive acceleration that allows it to gain speed and move in the positive direction between t = 0 s and t = 2 s.* ● *At t = 2 s, the velocity reaches its maximum value. Between t = 2 s and t = 4 s, the particle continues moving in the positive direction but at a decreasing speed.* ● *At t = 4 s, the velocity is zero, and the particle reaches its maximum position. From that moment, the velocity becomes negative, and the particle begins to move in the negative direction, accelerating.* ● *Between t = 4 s and t = 6 s, the particle continues moving rapidly in the negative direction.*
Question 2	
See Fig. 2	*The v/t graph in the statement shows that the motion of the particle can be divided into three stages through slope analysis:* ● *The body starts its motion from rest with positive acceleration, which decreases over time.* ● *At t = 2 s, the acceleration is zero, and the velocity reaches its maximum value of 12 m/s.* ● *From t = 2 s onward, the slope is negative, meaning the acceleration is negative, indicating that the body accelerates but moves in the negative direction with increasing speed.*
Question 3	
See Fig. 3	*The a/t graph in the statement shows that the motion of the particle can be divided into three stages through slope analysis:* ● *The body starts its motion with positive acceleration, which decreases over time.* ● *At t = 2 s, the acceleration is zero, and the velocity reaches its maximum value of 12 m/s, as calculated from the area under the curve of the a/t graph.* ● *From t = 2 s onward, the acceleration is negative, meaning that the body accelerates but moves in the negative direction with increasing speed.*

4 Data Analysis

The data analysis was conducted following the methodology described in the previous section. In the phenomenographic analysis, the required steps were followed (Guisasola et al. 2023), and a validation was performed to measure inter-rater reliability using Cohen's kappa coefficient, obtaining an average value of 0.95.

In the statement of each of the questions (see Table 2), a graphical semiotic representation is initially presented. In question Q1, a position versus time graph is shown, along with two possible velocity versus time graphs. In question Q2, a velocity versus time graph is provided, accompanied by two acceleration versus time graphs. Finally, in question Q3, an acceleration versus time graph is presented, along with two velocity versus time graphs to choose from (see Table 2).

To answer, students must select one of the two graphical representations that correctly correspond to the initial graph, allowing for an assessment of their ability to

Below is an acceleration versus time graph that describes the motion of a body moving in a straight line.

With the available information, answer:

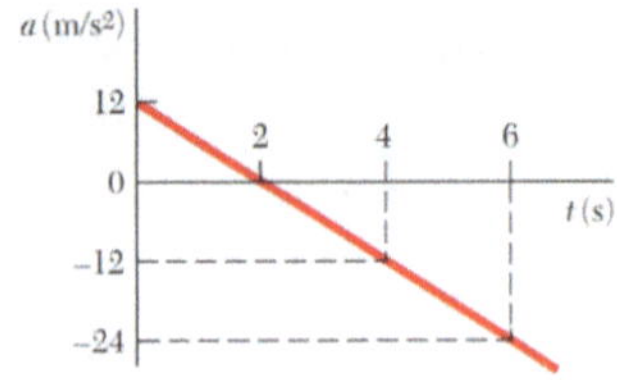

a) What is the velocity vs. time graph that best represents the movement described above?

b) Justify how did you arrived at that decision?

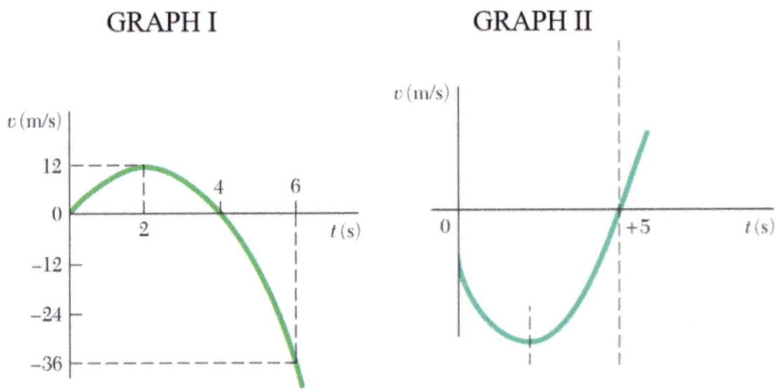

Fig. 3 *Task used in the study.* **a** Acceleration–time graph describing the motion of a body moving along a straight line. **b** Two candidate velocity–time graphs (Graph I and Graph II) presented to students, who were asked to determine which graph correctly represents the motion and to justify their reasoning

work within the same graphical representation. The common objective of these three questions is to ensure that students can interpret, differentiate and extract information from position, velocity and acceleration graphs as functions of time, specifically in rectilinear motion with variable acceleration. Additionally, these questions evaluate students' ability to correctly apply differentiation and integration techniques based on graphical representations, ensuring they interpret the results within the physical context of motion with variable acceleration.

Table 3 shows the results of the descriptive categories from the phenomenographic analysis for the treatment of graphical representations in the open-ended questionnaire. Both the initial representation (provided in the statement) and the final semiotic representation (in the student's choice) are graphical representations.

For question Q1, students had to choose between two velocity versus time graphs with similar characteristics. Table 3 shows that, in this case, only 25% of students correctly justified their choice between the two graph options, achieving the objective of interpreting and extracting information from a graphical representation related to motion with constant acceleration. 21% of students used mathematical tools without

Table 3 Students answer categories and percentages per question

Category	Description of the category	Results (%)		
		Q1	Q2	Q3
A	Correctly interpret, differentiate, and extract information	25	35	37
B	Use mathematical tools without physical meaning	21	23	18
C	Misunderstanding of physical meaning	37	32	29
F	No answer	17	10	16

physical meaning when choosing between the two graphs presented. This is reflected in the interview analysis when a student says, for category B:

'By differentiating a function, you get velocity, and then you replace the obtained value'.

Another student states:

'The velocity versus time graph is the derivative of the position versus time graph... so, we apply the integral to the velocity versus time graph, and it gives us the position versus time graph'.

In both examples, it can be seen that students 'apply' mathematical resources in a rote manner without giving them any physical meaning. In the same question, the largest percentage of students who answered the questionnaire (37%) misunderstood the physical meaning of the graph. This can be evidenced when, in interviews:

'I think I would choose graph one… since its speed increases and then decreases. I reached this conclusion by observing the meters it advances over time'.

Another student says:

'Because in the position versus time graph, the acceleration is seen to decrease, so the velocity also decreases, as observed in graph one... that's why I chose it'.

It is interesting to note that 17% of the students faced with question Q1 could not provide a justification supporting their choice, or they did not answer the question.

In question Q2, students had to respond to a question that presented a velocity versus time graph with variable acceleration. It was found that 35% of them correctly understood the relationship between velocity and acceleration. They identified that the slope of the velocity versus time graph is zero at $t = 2$ s and recognised that the body decelerates before accelerating in the negative direction (Category A).

Meanwhile, in Category B, 23% of students used mathematical tools without physical meaning. For example, one student stated:

'I choose graph 1 because in graph 2, it goes from positive to negative, whereas graph 1 goes from.

negative to positive'.

Similar to question Q1, question Q2 showed the highest percentage in category B (32%) where students misinterpret physical concepts and confuse them. For instance, during an interview, one student explained:

'Since acceleration is variable with respect to time, for velocity to decrease, acceleration must become smaller with each passing second'.

Furthermore, 10% of students could not answer the question or did not know how to justify their selection.

Finally, question Q3 presented a statement with the graphical representation of a body's acceleration as a function of time. Students had to choose between two velocity versus time graphs and provide written justifications for their choice in the questionnaire.

For this question, only 37% of students achieved the objective, while 18% applied mathematical tools without physical meaning. For example, one student stated:

'One parabola is positive, and the other parabola is negative'.

Another student explained:

'I choose graph one… because acceleration is a decreasing straight line, so the other graph should.

also be decreasing'.

A significant percentage of students (29%) confused the concepts of acceleration and deceleration with those of positive and negative acceleration. This confusion is evident in the interviews. For example, one student explained:

'It is because the acceleration is becoming increasingly negative; therefore, the velocity will continue decreasing into the negatives and at a faster rate'.

Another student, demonstrating the same misunderstanding, stated:

'I reached this conclusion because the time in the example graph is decreasing, just like the time in graph 1… and also because the time in graph 2 is increasing, which would be the opposite of what is.

shown in the reference image'.

Similar to questions Q1 and Q2, in question Q3, 16% of students, despite instruction and having completed the chapter's evaluation, were unable to provide an answer.

These results indicate that students have important difficulties when working within the same graphical representation, despite the fact that the initial and final graphs were provided in the questionnaire, and they only had to choose one and justify their selection. It is reasonable to assume that the difficulties would be even greater if they had to construct the graphs instead of selecting them. However, challenges persist in the physical interpretation of the graphs, especially regarding the relationship between the different representations of position, velocity, and acceleration versus time, and their physical meaning. Although some students improve in the application of mathematical tools, many continue to use these tools without fully understanding the physical context. It is advisable to strengthen the teaching of the relationship between graphical representations and the underlying physical phenomena to enhance the understanding of motion with variable acceleration and its physical representation through semiotic graphical representations.

5 Conclusions

In this study, we have integrated the theoretical framework of semiotic representation with a phenomenographic approach to explore students' difficulties in recognising graphical representations and their relationship with the conceptual understanding of the concepts of position, velocity and acceleration in kinematics with non-constant acceleration. The applied phenomenographic methodology (Guisasola t al. 2023; Marton, 1981) has allowed us to identify difficulties in recognising graphical representations by linking them to problems of conceptual understanding, supported by the theory of semiotic representation registers (Campos et al. 2022).

Through the phenomenographic analysis of the questions, we have concluded that slightly less than a third of the students answer the proposed questions correctly with the required justifications. However, around a quarter of the students apply the mathematical tools they have acquired by rote learning but are not able to make physical sense of the process or the results obtained. What is most striking is that the rest of the students, around 40%, have even more severe difficulties, as they confuse the physical meaning of the concepts presented in the questions.

In addition, the analysis of the interviews reinforced the idea that many students misunderstood the relationship between the graphical representations, especially with regard to acceleration and its impact on velocity. In several responses, students confused fundamental concepts such as positive and negative acceleration with those of deceleration, leading to an incorrect choice of graphs. These results suggest that, although students have access to graphical representations during their learning, they still lack a deep understanding of how variations in one physical quantity are reflected in the others, which hinders the understanding of fundamental concepts of kinematics. This disconnect was also evident in the application of mathematical tools without considering their physical meaning, indicating that, even at advanced levels of mechanical physics, a separation persists between mathematical skills and their physical understanding.

These results are in line with those obtained in previous similar studies for algebraic representation (Urrutia et al. 2024); the extraction and processing of information in graphic representation are more complicated for students, since the percentages of correct answers, which are around 50% for algebraic representations, drop to around 30% for graphic representations. Furthermore, these findings are in agreement with existing literature on students' difficulties in understanding kinematic concepts, especially in the context of constant-acceleration movements (Reif and Allen, 1992; Shaffer and McDermott, 2005). However, what our research contributes is the identification of a critical dissociation between algebraic and graphical notation and its physical interpretation, extending the understanding of previous documented difficulties.

To complete this study about difficulties, once the difficulties in the processing of information by students in graphic and algebraic representations have been detected, it will be necessary to continue with studies analysing the difficulties that students may have in questions related to the conversion between graphic and algebraic representations and vice versa.

Despite the limitations of this study in terms of sample, size and context, the results underline the urgent need to review pedagogical strategies in the teaching of kinematics at university level. The difficulties observed in this context where the teaching methodology was traditional, indicate that teachers should not assume that students automatically understand graphical representations and their physical meaning. An active teaching approach, which encourages discussion of the physical meaning of mathematical notation, is crucial to improve students' understanding. In future research, we plan to design Teaching Learning Sequences that facilitate understanding of the concepts of motion with non-constant acceleration through different representations, as well as to explore teaching approaches that more explicitly connect mathematics and physics in this and other educational contexts.

References

Barniol P, Zavala G, Dominguez A (2024) The effect of calculus and kinematics contexts on students' understanding of graphs. Eurasia J Math, Sci Technol Educ 20(11):em2533. https://doi.org/10.29333/EJMSTE/15581

Beichner RJ (1994) Testing student interpretation of kinematics graphs. Am J Phys 62(8):750–762. https://doi.org/10.1119/1.17449

Brasell HM, Rowe MB (1993) Graphing skills among high school physics students. Sch Sci Math 93(2):62–70. https://doi.org/10.1111/j.1949-8594.1993.tb12196.x

Campos E, Zavala G, Zuza K, Guisasola J (2022) El papel de la utilización de representaciones de los conceptos en las dificultades de su comprensión: el caso del campo eléctrico. Revista Brasileira de Ensino de Física 44:e20220146. https://doi.org/10.1590/1806-9126-RBEF-2022-0146

Campos E, Zuza K, Guisasola J, Zavala G (2023) Recognition and conversion of electric field representations: the case of electric field lines. Phys Rev Phys Educ Res 19(2):020117. https://doi.org/10.1103/PhysRevPhysEducRes.19.020117

Duval R (1996) Quel cognitif retenir en didactique des mathématiques? Recherches En Didactique Des Mathématiques 16(3) https://revue-rdm.com/1996/quel-cognitif-retenir-en/

Duval R (2006) Un tema crucial en la educación matemática: La habilidad para cambiar el registro de representación. La Gaceta de La Real Sociedad Matemática Española 9:143–168. https://doi.org/10.17398/2444-8646.09.1.143

Duval R (2017) Understanding the mathematical way of thinking—the registers of semiotic representations. Underst Math Way Thinking—Regist Semiot Represent 1–117. https://doi.org/10.1007/978-3-319-56910-9

Guisasola, J, Campos, E, Zuza, K, Zavala, G. (2023) Phenomenographic approach to understanding students' learning in physics education. In Physical review physics education research (Vol. 19, issue 2, p. 020602). American Physical Society. doi:https://doi.org/10.1103/PhysRevPhysEducRes.19.020602

Guisasola, J., & Zuza, K. (2024). A physics curriculum for the modern world. Nat Phys, 20(3), 342–344. doi: https://doi.org/10.1038/S41567-023-02370-3;SUBJMETA=25,639,766;KWRD=APPLIED+PHYSICS,PHYSICS

Ivanjek L, Susac A, Planinic M, Andrasevic A, Milin-Sipus Z (2016) Student reasoning about graphs in different contexts. Phys Rev Phys Educ Res 12(1):010106. https://doi.org/10.1103/PhysRevPhysEducRes.12.010106

Marton F (1981) Phenomenography—describing conceptions of the world around us. Instr Sci 10(2):177–200. https://doi.org/10.1007/BF00132516

McDermott LC, Rosenquist ML, van Zee EH (1987) Student difficulties in connecting graphs and physics: examples from kinematics. Am J Phys 55(6):503–513. https://doi.org/10.1119/1.15104
Pospiech G (2019) Framework of mathematization in physics from a teaching perspective. Math Phys Educ 1–33. https://doi.org/10.1007/978-3-030-04627-9_1
Reif F, Allen S (1992) Cognition for interpreting scientific concepts: a study of acceleration. Cogn Instr 9(1):1–44. https://doi.org/10.1207/S1532690XCI0901_1
Shaffer PS, McDermott LC (2005) A research-based approach to improving student understanding of the vector nature of kinematical concepts. Am J Phys 73(10):921–931. https://doi.org/10.1119/1.2000976
Shodiqin MI, Taqwa MRA (2021) Identification of student difficulties in understanding kinematics: focus of study on the topic of acceleration. J Phys Conf Ser 1918(2):022016. https://doi.org/10.1088/1742-6596/1918/2/022016
Urrutia A, Merino C, Zuza K, Guisasola J (2024) Dificultades de estudiantes universitarios chilenos relacionadas con representaciones semióticas algebraicas en cinemática con aceleración no constante. Revista Brasileira de Ensino de Física 46:e20240280. https://doi.org/10.1590/1806-9126-RBEF-2024-0280
Van Someren, MW, Barnard, Y. F., & Sandberg, JAC (1994) The think aloud method: a practical approach to modelling cognitive. Vol. 11, pp. 29–41. Academic press, London, 1994. https://hdl.handle.net/11245/1.103289
Viennot L (2001) Reason Phys Part Common Sense:230. https://books.google.com/books/about/Reasoning_in_Physics.html?hl=es&id=h0X0T3sYxE0C
Wemyss T, Van Kampen P (2013) Categorization of first-year university students' interpretations of numerical linear distance-time graphs. Phys Rev Spec Top Phys Educ Res 9(1):010107. https://doi.org/10.1103/PHYSREVSTPER.9.010107/FIGURES/5/MEDIUM
Zavala G, Tejeda S, Barniol P, Beichner RJ (2017) Modifying the test of understanding graphs in kinematics. Phys Rev Phys Educ Res 13(2):020111. https://doi.org/10.1103/PHYSREVPHYSEDUCRES.13.020111/FIGURES/2/MEDIUM

Advancing Physics Education for a Modern Society

How Thematic Analysis can Inform Designing, Planning and Evaluation of Teaching/Learning Sequences

Giulia Termini, Onofrio Rosario Battaglia, Giusy Giarratano, Ilaria Grazia, and Claudio Fazio

Abstract This paper explores how Thematic Analysis (TA) can be applied in Physics Education Research (PER), specifically in relation to improving didactic practices through Teaching and Learning Sequences (TLSs). TA is a qualitative research method that identifying patterns or themes within qualitative data can enable a deeper understanding of students' and teachers' perspectives. In the context of TLSs, which are structured educational frameworks designed to guide the sequence of lessons and activities in teaching, TA provides valuable insights into the effectiveness of these sequences in promoting student understanding. By analysing various types of qualitative data such as classroom observations, student interviews and teacher reflections, TA can highlight the underlying mechanisms that contribute to or hinder learning. The paper presents the key stages of thematic analysis in our research, its possible applications in PER and illustrates how this method can inform the refinement of teaching strategies to better align with the cognitive and emotional students' needs.

Keywords Thematic analysis · Physics education research · Teaching and learning sequences · Qualitative educational research

1 Teaching and Learning Sequences (TLSs): Some General Aspects

Physics Education Research (PER) has evolved significantly over the past few decades, incorporating a broad range of methodologies to address the challenges faced in physics teaching and learning (Bao and Koening 2019). One of the primary

G. Termini (✉) · O. R. Battaglia · I. Grazia · C. Fazio
Department of Physics and Chemistry Emilio Segrè, University of Palermo, Palermo, Italy
e-mail: giulia.termini01@unipa.it

G. Giarratano
Department of Psychological, Pedagogical, Physical Exercise and Training Sciences, University of Palermo, Palermo, Italy

D. Sokołowska and E. McLoughlin (eds.), *Embracing Changes Together in Physics Education*, Challenges in Physics Education,
https://doi.org/10.1007/978-3-032-27385-7_10

goals of PER is to improve teaching practices and student outcomes by better understanding the ways students engage with complex physical concepts and how different didactic approaches can facilitate or hinder learning (Elby 2001).

TLSs are structured pathways designed to guide the teaching of specific concepts in a way that promotes progressive learning (Méheut and Psillos 2004; Psillos and Kariotoglou 2016). TLSs are an integral part of effective curriculum design and teaching practice, as they aim to promote knowledge by organizing content, pedagogy and assessment into a coherent, logically ordered progression. These sequences typically build on students' prior knowledge and experiences, helping them make connections between new and existing information. They also serve as a means of addressing common misconceptions and ensuring that new concepts are introduced in a manner that encourages deep, conceptual understanding. One of the defining features of TLSs is their cyclical nature. Unlike traditional lesson plans that may be linear and one-time-only in their approach, TLSs are designed to be dynamic, where teaching strategies are continuously adjusted based on ongoing student feedback and assessments. This iterative process ensures that teaching is responsive to the students' evolving needs, enhancing engagement and deepening understanding over time. Teachers are encouraged to assess students' comprehension regularly, using both formal and informal methods, and adjust the pace, content or didactic strategies as needed. This adaptive model of teaching fosters a learning environment where students are empowered to take ownership of their learning, receiving ongoing guidance and support as they progress through the sequence. TLSs' learning goals are usually framed in terms of conceptual understanding, skills development or application of knowledge. These pathways provide a clear sense of direction for both teachers and students, ensuring that all activities within the TLS are aligned with the desired outcomes.

The methods and activities used to engage students and facilitate learning have to be planned accurately. Didactic strategies include a wide range of approaches, such as problem-solving tasks, collaborative group work, inquiry-based learning and hands-on activities.

TLSs can also perform a monitoring and evaluation function (Méheut 2005; Duschl et al. 2011). As assessment tools, they help monitor students' progress throughout the sequence. These tools can include quizzes, formative assessments, classroom observations and even informal interactions between teachers and students. Feedback is an essential part of this process, as it provides students with specific, actionable insights into their learning, guiding them towards improvement. Formative assessments allow teachers to gauge students' understanding in real-time and adjust their teaching methods, accordingly, ensuring that the didactic sequence remains responsive to student needs.

The evaluation of TLS outcomes is a critical aspect, making it essential to identify suitable tools for assessing their effectiveness in fostering learning. While quantitative analysis methods are essential for measuring student outcomes, qualitative methods can offer more nuanced insights into how students experience and interact with the content presented in a TLS. Among the qualitative methods that can effectively inform teaching practices in PER, there is Thematic Analysis (TA) (Braun and Clarke 2006; Clarke and Braun 2017).

2 Thematic Analysis in Educational Research

Thematic Analysis (TA) is a qualitative research method widely employed in social sciences used to identify and analyze patterns within qualitative data (Braun and Clarke 2006; Clarke and Braun 2017). As a flexible and adaptable method, TA allows researchers to explore a wide range of research questions, making it particularly useful in fields such as education, where complex, context-dependent phenomena require a deep understanding. Unlike other qualitative methodologies such as grounded theory or ethnography, thematic analysis does not require the researcher to commit to a specific theoretical framework at the outset. This open-ended approach makes TA a highly versatile tool that can be applied to a variety of research designs, from exploratory studies to more targeted investigations into specific aspects of educational practice (Braun and Clarke 2006; Clarke and Braun 2017).

TA is particularly valuable in educational research as it allows for the identification of underlying patterns in the ways that students and teachers experience and interact with educational processes (Peel 2020). These patterns can shed light on how educational interventions or teaching strategies impact student engagement, how classroom dynamics influence learning outcomes and how teachers reflect on and adapt their didactic practices. By identifying recurring themes, researchers can gain insights into the shared experiences, perceptions and behaviours of individuals within the educational context. TA provides a nuanced way of examining qualitative data, offering a rich and detailed understanding of complex educational phenomena (Peel 2020). One of its key advantages is its flexibility; it can be applied to a variety of data sources such as interviews, focus groups, classroom observations and written reflections. This flexibility makes it particularly useful for exploring diverse aspects of education.

TA can be used to explore how students interact with didactic materials, engage with peer interactions or navigate learning environments. By analyzing students' perspectives, researchers can uncover the strategies that promote or hinder engagement and identify factors that influence motivation and active participation. It can help researchers understand how various elements of the classroom environment, including teacher-student interactions, group work and class discussions, affect student learning outcomes. Similarly, it is useful for examining the impact of different didactic strategies within TLSs. By identifying themes related to how students respond to specific teaching approaches, such as inquiry-based learning, problem-solving tasks or collaborative activities, researchers can assess which practices are most effective in fostering conceptual understanding. TA can also be employed to investigate how teachers reflect on their own practices. Through the examination of interviews or reflective journals, researchers can explore how teachers perceive the effectiveness of their teaching strategies, what challenges they encounter and how they adapt their approaches based on student feedback or classroom dynamics.

This method, though not yet widely used in PER, can allow a deeper understanding of learning aspects in PER that may be challenging to explore through quantitative methods or to gather more details on learning aspects already studied through quantitative analysis methods. This can help the researchers to carefully design, plan and evaluate Physics Education didactic practices and to improve TLSs or other pedagogical pathways.

2.1 Thematic Analysis in Our Research

In this paper, we show how we applied thematic analysis to qualitative data collected during the trailling and validation of two TLSs on surface phenomena (Fazio et al. 2023; Termini 2023). The qualitative data collected in our research consist of students' worksheets, audio recordings of students' group discussions at the end of each activity proposed in the TLS, students' contributions during the brainstorming phase on the final day and feedback on the didactic experience.

Through specific examples, we show how the results obtained can inform the design of the following educational pathways. The research presented builds on the conceptual framework of the Dimensions of Learning Map, introduced in previous works (Fazio et al. 2023), which serves as a guiding structure for understanding and analyzing student learning trajectories within the proposed TLSs. The map outlines, on the basis of thorough literature research, key dimensions of learning (Marzano 1992; Marzano and Pickering 2011), offering a comprehensive lens through which to plan and assess the effectiveness of didactic strategies. As reported in (Marzano 1992; Marzano and Pickering 2011), these dimensions of learning are: *Conceptual knowledge, Intellectual growth, Development of mind suited to learning science. Conceptual knowledge* refers to the appropriation and long-term retention of scientific concepts and representational forms (including graphs, language, models). The consolidation of the conceptual knowledge involves the progressive transformation of everyday or common-sense conceptions into coherent scientific understanding and encompasses the ability to use appropriate representations to interpret phenomena, explain mechanisms and make predictions.

Intellectual growth refers to the development of higher-order thinking skills that enable students to interpret complex situations and experimental results, make connections across different contexts and transfer what they have learned to new problems. Additionally, intellectual growth involves the enhancement of interpersonal and social skills which are essential for participating in co-constructing understanding and reasoning within group settings. *Development of habits of mind suited to learning science* is related to the cultivation of personal dispositions and attitudes that supports engagement in scientific learning. It encompasses students' perception of self-efficacy as well as the adoption of a growth mindset, which encourages persistence in the face of challenges. This dimension includes metacognitive awareness, allowing students to monitor and regulate their own learning processes, and a sense of well-being and confidence within educational contexts. It also involves developing an informed understanding of the nature of science (i.e. of its methods, limits and epistemic values) and a willingness to extend studies through further inquiry and exploration.

Each dimension has been more clearly defined in terms of learning subdimensions, which we have considered as guiding variables for our research (Fazio et al. 2023; Termini 2023). By leveraging this framework, the present study situates its findings within a broader research context, connecting the codes identified through thematic analysis to specific aspects of the Dimensions of Learning Map. This connection allows for a deeper exploration of how innovative tools, methodologies or strategies contribute to promote aspects of learning linked to these dimensions, while providing actionable insights for refining future TLSs design. In this sense, this paper specifically aims to show how thematic analysis can be applied within the context of PER to provide insights that can guide the design and planning of didactic pathways.

The main steps followed in conducting the analysis of our data are inspired by the literature regarding thematic analysis methods (e.g. Braun and Clarke 2006; Clarke and Braun 2017) and are outlined in Table 1.

Our analysis followed an inductive approach (Braun and Clarke 2006; Clarke and Braun 2017; Williams and Moser 2019) where codes were developed based on the data rather than a predetermined codebook. Through repeated readings of the data, initial codes (see Table 2) were identified based on recurring words and sentences. Then, these codes were refined iteratively, leading to the final 'overarching codes' A-F (see Table 2), obtained by eventually merging the initial codes when they conveyed similar information.

One of the tools that most aided us in the analysis process is the codes-variable correspondences table showing the correspondences among the codes and the variables (subdimensions of learning) of our study (see Table 3). This table helps us to detect relationships between codes and learning variables, i.e. allows us to easily identify the type of information conveyed by the data labelled with a specific code. While it provides an overview of all the identified code-variable correspondences across the qualitative databases collected, not all correspondences will appear in every dataset. Thus, the table can be customized for more precise analysis based on the specific data under consideration.

To ensure reliability, three researchers independently reviewed the data and collaborated through multiple rounds of discussion to resolve discrepancies and enhance the accuracy of the coding (Belotto 2018). It is important to emphasize that the decision to 'merge' two or more codes as well as the correspondence of a given code with one or more variables was carefully negotiated and agreed upon by the researchers involved in the analysis. This collaborative approach ensures that the final coding scheme (see Tables 2 and 3) reflects the collective expertise and consensus of the

Table 1 Main steps followed to conduct the thematic analysis of the data

Step 1	Familiarization with the data	We reviewed all the collected data to become familiar with their content and context.
Step 2	Selection of the data	We recognized inside the data text segments suitable for achieving the research purposes
Step 3	Raw data coding	We identified some preliminary codes (also named initial codes), i.e. keywords that helped us to synthesize the information provided by the data (see Table 2)
Step 4	Advanced data coding	After repeated readings, we decided to introduce some overarching codes embedding all the information carried by more than one initial code (see Table 2)
Step 5	Data labelling	We labelled text segments of analytic interest recognized in step 2 first through the codes identified in step 3 and then in step 4
Step 6	Code-variable correspondence table	We built a table containing code-variable correspondences (see Table 3). This correspondence clearly shows what kind of information is carried by a given segment of text labelled with a given code (or overarching code)
Step 7	Information processing	At this stage, we start to study the identified text segments in relation to specific sub-dimensions of learning based on the code-variable correspondences reported in Table 3

Table 2 Summary of the codes identified during the thematic analysis of the data. The letters A-F indicate the overarching codes obtained by merging the initial codes reported in the first column

Initial codes	Overarching codes
Procedure-methodology understanding/tools-skills/theory versus practice/ traditional lecture versus innovative lecture	A
Content understanding	B
Debate/perspective/role	C
Language	D
Reflection/acknowledgement	E
Engagement-interest/comfort/Proactiveness	F

researchers, contributing to an accurate interpretation of the data. By reducing subjective bias and ensuring consistency, this process enhances the validity and reliability of the findings.

For each overarching code, we provide a description that clarifies the aspects of the data it emphasizes. These descriptions are based on both the manifest content

Table 3 Table showing code-variables correspondences that help us in tracing the information on specific aspects of learning carried by a given segment of text labelled with a given code

	Variables												
	Conceptual knowledge			Intellectual growth				Development of habits of mind suited to learning science					
Overarching codes	1.1	1.2	1.3	2.1	2.2	2.3	2.4	3.1	3.2	3.3	3.4	3.5	3.6
A					X	X			X	X		X	X
B	X	X	X		X	X							
C		X		X	X		X	X	X	X	X	X	X
D	X	X		X		X				X		X	
E					X	X	X	X	X	X			X
F									X		X		X

of the data, such as verbatim student quotations, and our interpretation when we identified underlying, though less explicit, patterns.

As an example, we provide the description of the overarching code E, which was used to label the data that will be discussed in the next paragraph.

Overarching Code E (Reflection/Acknowledgement): Student reflection is a crucial part of the learning process, enabling them to assess their progress, recognize strengths and weaknesses and deepen their understanding of the knowledge and skills acquired. Through critical evaluation of activities, students identify areas for growth, learn from mistakes and set future goals. Students acknowledge their self-awareness and demonstrate autonomy by reflecting on their actions, identifying areas for improvement and setting goals for the future. They show a deeper understanding of themselves and how they can apply their knowledge and skills in real-life situations.

3 An Application of Thematic Analysis in our Research

In this paragraph, we present an application of thematic analysis to some qualitative data collected during the experimentation of the TLSs on surface phenomena described in detail in (Fazio et al. 2023; Termini 2023). The data consist of the students' comments on the activities carried out during the TLSs. In a satisfaction questionnaire (Termini 2023) administered at the end of the trial, students were asked to provide feedback on their learning experience and suggest improvements for the activities conducted within the TLSs. Each response was assigned a specific code, as detailed in (Termini 2023), which provides insights into aspects of learning linked to the dimensions of learning considered in our research.

All the following extracts, offering valuable information regarding the metacognitive dimensions of the learning process, have been labelled with the overarching code E.

Textual Segment n. 1: *'After participating in all these activities, I thought about how my vision of the studio has changed. I have always been good at school, but honestly if I went back, I would study differently. I would not waste time memorising formulas and definitions, but I would work to understand more deeply what I study because only this allows me to remember in the long term'.*

The student acknowledges that his/her approach to studying has changed. This indicates that the activities provided a shift from traditional methods of learning, such as memorizing formulas, to a deeper, more conceptual approach. This is a positive indicator that the activities promoted critical thinking and understanding. The emphasis on deep understanding as the key to long-term retention suggests that the activities might have involved real-world applications or deeper explorations of physics concepts, which helped the student realize the importance of understanding concepts fully. Although the student recognizes the importance of deeper understanding, they also imply that their past study methods were not as effective. This could be used as feedback to further emphasize the importance of concept-driven learning across more segments of the curriculum.

Textual Segment n. 2: *'The use of tools different from those I was used to, such as computer simulations, and in general, the possibility of being involved in experiments of increasing difficulty made me think about what I feel most prepared and comfortable in and what I should do instead still working to reach a satisfactory level'.*

The student highlights the benefits of using new tools, like computer simulations, and engaging in progressively difficult experiments. This suggests that the activities encouraged exploration and hands-on learning, which helped the student evaluate their strengths and weaknesses. Student's reflection on feeling both comfortable and challenged is a sign that the activities successfully pushed them beyond their comfort zone. He/she is now aware of areas where he/she needs further development. The mention of areas where the student needs to improve shows that there could be additional support or guidance provided, especially in more challenging areas. Perhaps scaffolding or additional practice could be introduced to support students as they reach for higher levels of proficiency.

Textual Segment n. 3: *'Reflecting on my experience with these activities, I realize that my approach to learning has evolved. I used to focus mainly on getting the right answers, but now I understand the importance of exploring concepts and asking questions. In the past, I might have just memorized facts, but now I see that deep understanding is the key to retaining knowledge and solving problems effectively'.*

The student's shift from focusing solely on correct answers to exploring concepts and asking questions reflects the effectiveness of activities that promote inquiry-based learning. It suggests that the didactic approach encouraged deeper engagement and understanding of physics. The student now recognizes that understanding is essential for solving problems, indicating that the activities might have included problem-solving tasks or scenarios that required the application of conceptual knowledge. The feedback underscores the importance of focusing on conceptual understanding over memorization. This insight can help educators further refine their approach to teaching, ensuring that students are consistently encouraged to explore and ask questions about the material.

Textual Segment n. 4: *"Using new methods like interactive learning tools and hands-on activities has made me reconsider my strengths and weaknesses in physics. I've realized that while I feel confident with theoretical aspects, I need to improve my practical skills. The more challenging experiments pushed me to identify areas where I can grow and have shown me how much I still need to practice feeling truly competent'.*

This student's reflection on feeling confident with theoretical knowledge but acknowledging the need to improve practical skills shows that the didactic experience offered a balance between theory and application. The inclusion of interactive tools and hands-on activities seems to have been effective in helping the student realize the importance of practical skills. The mention of challenging experiments encourages the idea that the activities were appropriately designed to push students beyond their comfort zone, which is critical for deep learning. The student's recognition of areas for growth suggests that the curriculum is stimulating self-assessment and encouraging personal development. The student's feedback indicates that while practical skills need further development, they are also necessary for becoming competent in physics. Future courses might consider incorporating more hands-on activities or practical challenges to help students improve these skills.

4 Discussion

The reflections provided by students offer valuable insights into the effectiveness of the TLSs, shedding light on the transformative impact of innovative didactic strategies. A recurring theme across the feedback is the shift in students' approach to learning, moving from traditional methods such as memorization to more concept-driven, inquiry-based approaches. This transformation underscores the success of the TLSs in fostering deeper understanding, critical thinking, and long-term retention of knowledge. The incorporation of innovative tools, particularly computer simulations and hands-on activities, played a pivotal role in this process. Students appreciated these methods for their ability to engage them actively and help them explore scientific concepts in a dynamic and interactive manner. The use of progressively challenging experiments further encouraged self-assessment and growth, enabling students to identify both their strengths and areas requiring improvement. This heightened metacognitive awareness reflects a significant step towards self-directed learning, a critical skill for lifelong education.

Another key observation is the balance achieved between theoretical and practical learning. While some students expressed confidence in their theoretical knowledge, they also recognized the need to strengthen their practical skills. This highlights the importance of designing TLSs that challenge students beyond their comfort zones while offering adequate support and scaffolding to address areas of difficulty. Students' reflections reveal the value of inquiry-based strategies in promoting curiosity and problem-solving skills. The emphasis on asking questions and exploring concepts rather than focusing solely on correct answers suggests that the TLSs were successful in creating an environment conducive to active learning and intellectual engagement.

These insights underscore the importance of integrating tools like computer simulations, interactive methods and hands-on experiments into TLSs to achieve a well-rounded learning experience. Future implementations could further refine these strategies, ensuring a seamless alignment of theoretical and practical components while providing targeted support to foster deeper learning and greater student confidence in all aspects of science education.

5 Conclusions

The advantages of TA have been discussed in the previous sections, and we emphasize that, in our opinion, it represents a very useful tool for informing and evaluating didactic design. Despite its strengths, we are fully aware of the several challenges associated with applying thematic analysis, particularly when using it to evaluate the efficacy of a TLS in PER.

One of the primary challenges is that the process of coding and analyzing qualitative data can be time-consuming. In contrast to quantitative analysis, which can be automated or simplified with statistical tools, thematic analysis requires careful, manual coding and data review, which can be labour-intensive, especially when dealing with large datasets or complex students' interactions (Braun and Clarke 2006; Clarke and Braun 2017; Belotto 2018). Researchers must carefully sift through data to identify meaningful codes and themes, which can require substantial time and attention to detail.

Researchers can potentially rely on the support of software for data analysis, such as MAXQDA (Schmieder and Rogers 2024), which offers advanced tools for managing, coding and interpreting qualitative data. This type of software allows for the systematic organization of large amounts of information, identification of recurring patterns, creation of interactive visualizations and generation of detailed reports, facilitating thorough and rigorous analysis.

Beyond the methods or tools used to conduct TA, one challenge is represented by subjectivity (Braun and Clarke 2006; Clarke and Braun 2017). The process of interpreting qualitative data is often influenced by the researcher's perspectives, beliefs and experiences. This subjectivity can introduce potential biases in the identification and interpretation of themes. For example, a researcher's prior knowledge of physics or educational theory may influence how they code data or perceive the significance of certain patterns. To mitigate this, it is essential to maintain a high level of reliability and consistency throughout the analysis process.

Finally, generalizability can be a limitation in thematic analysis, particularly when conducting studies within a specific classroom context. Findings from a single study or a small sample may not always be applicable to other settings or populations. This is particularly relevant in educational research, where teaching practices, student backgrounds and classroom environments can vary widely. Therefore, when interpreting the results of TA, it is crucial to consider the specificity of the context in which the research was conducted, acknowledging that findings may be more applicable to similar educational settings rather than being universally applicable.

References

Bao L, Koenig K (2019) Physics education research for 21st century learning. Discipl Interdiscip Sci Educ Res 1:2. https://doi.org/10.1186/s43031-019-0007-8

Belotto MJ (2018) Data analysis methods for qualitative research: managing the challenges of coding, interrater reliability, and thematic analysis. Qual Rep 23:2622–2633. https://doi.org/10.46743/2160-3715/2018.3492

Braun V, Clarke V (2006) Using thematic analysis in psychology. Qual Res Psychol 3:77–101. https://doi.org/10.1191/1478088706qp063oa

Clarke V, Braun V (2017) Thematic analysis. J Posit Psychol 12:297–298. https://doi.org/10.1080/17439760.2016.1262613

Duschl R, Maeng S, Sezen A (2011) Learning progressions and teaching sequences: a review and analysis. Stud Sci Educ 47:123–182. https://doi.org/10.1080/03057267.2011.604476

Elby A (2001) Helping physics students learn how to learn. Am J Phys 69:S54–S64. https://doi.org/10.1119/1.1377283

Fazio C et al (2023) An approach to research-based design of teaching-learning sequences in the context of physics education: theoretical frameworks, pedagogical methods, and examples of data analysis. Nuovo Cimento Soc Ital Fis C 46. https://doi.org/10.1393/ncc/i2023-23199-1

Marzano RJ (1992) A different kind of classroom: teaching with dimensions of learning. ERIC, Washington, DC

Marzano RJ, Pickering DJ (2011) Dimensions of learning teacher's manual. ASCD, Alexandria, VA

Méheut M (2005) Teaching-learning sequences: tools for learning and/or research. In: Boersma K et al (eds) Research and the quality of science education. Springer, Dordrecht, pp 195–207. https://doi.org/10.1007/1-4020-3673-6_16

Méheut M, Psillos D (2004) Teaching–learning sequences: aims and tools for science education research. Int J Sci Educ 26:515–535. https://doi.org/10.1080/09500690310001614762

Peel KL (2020) A beginner's guide to applied educational research using thematic analysis. Pract Assess Res Eval 25. https://doi.org/10.7275/RYR5-K983

Psillos D, Kariotoglou P (2016) Iterative design of teaching-learning sequences. Springer. https://doi.org/10.1007/978-94-007-7808-5

Schmieder C, Rogers A (2024) Using MAXQDA for a large qualitative dataset: identifying relevant data and analyzing it with thematic analysis. In: Schmieder C, Rogers A (eds) The practice of qualitative data analysis: research examples using MAXQDA, vol 2. Routledge, New York, p 127

Termini G (2023) Pedagogical approaches to surface phenomena in liquids: investigation-based laboratory and modelling activities to improve students' learning. PhD thesis

Williams M, Moser T (2019) The art of coding and thematic exploration in qualitative research. Int Manag Rev 15:45–55

Defining and Assessing Learning Goals in Physics Lab Courses

Paul Logman, Ian Bearden, Micol Alemani, Sergej Faletič, Sebastian Schellhammer, and K. Dunnett

Abstract Many physics lab courses are designed as places where students demonstrate physics concepts, often by completing highly instructed 'cookbook' activities. This is in contrast to studies that demonstrate that lab courses centred on learning goals related to experimental competencies result in improved student learning and motivation. However, reforming lab courses is often challenging, requiring time and resources as well as didactic expertise that may not be available locally. This manuscript reflects on the current state of research activities and teaching experiences on open questions of learning goals, assessment and the interplay between the two in physics lab courses at all levels of education. With this reflection, we provide guidance for the development of learning-goal-centred physics lab courses and identify open research questions.

Keywords Learning goals · Assessment · Rubrics · Alignment · Physics lab · Higher education · Open research question

P. Logman (✉)
Leiden University, Leiden Institute of Physics, Leiden, The Netherlands
e-mail: logman@physics.leidenuniv.nl

I. Bearden
University of Copenhagen, Niels Bohr Institute, Copenhagen, Denmark

M. Alemani
University of Potsdam, Physics and Astronomy Institute, Potsdam, Germany

S. Faletič
University of Ljubljana, Department of Physics, Ljubljana, Slovenia

S. Schellhammer
Technische Universität Dresden, Institute for Applied Physics, Dresden, Germany

K. Dunnett
University of Oslo, Department of Geosciences, Oslo, Norway

D. Sokołowska and E. McLoughlin (eds.), *Embracing Changes Together in Physics Education*, Challenges in Physics Education,
https://doi.org/10.1007/978-3-032-27385-7_11

1 Introduction

Physics is inherently an experimental science and experimental education (hereafter lab work) is a cornerstone of physics curricula (Trumper 2003). For today's students to counter urgent challenges such as the climate crisis and conspiracy theories, they need a wide range of skills, including critical thinking, data analysis, problem-solving and communication. The practical nature of lab work means such courses have significant potential to equip students with these crucial skills (Hofstein and Lunetta 2004). Despite the opportunities presented by lab work, lab courses appear to receive less attention in the physics education discourse than the traditional teaching form of lectures on theoretical material followed by exercise classes to learn to solve problems mathematically.

This disconnect between theoretical and practical approaches to physics highlights the need to discuss and investigate lab learning environments, especially because they are notably complex and address many different goals, such as developing students' technical and practical skills, designing experiments, analyzing data, communicating results and developing students' epistemological beliefs (American Association of Physics Teachers 1998; Kozminski et al. 2014). Faced with this complexity, instructors need to choose specific learning goals, and these choices need to be communicated to students.

Furthermore, developing students' experimental skills requires substantial time and effort (Etkina et al. 2008), and these competencies are often challenging to assess (May 2023). To promote students' development over an extended period, clear and objective assessment is needed (Ketonen et al. 2023). Rubrics offer an option for both summative and formative assessment purposes (Etkina et al. 2006). Developing an objective rubric takes time. Decisions need to be made on which learning goals are to be formally assessed and how to describe different levels of achievement (Pols and Dekkers 2024). If developed properly, students should be able to use rubrics to guide their own improvements while trying to master experimental skills (Faletič and Planinšič 2020).

Ideally, learning goals and the assessment thereof are well aligned (Biggs and Tang 2011), but this can be difficult to realize because of the inherent complexity of lab learning environments. Challenges include the restrictions imposed by limited time and resources (May 2023), seemingly low student and instructor motivation, insufficient scientific and didactic competencies of instructors and a lack of continuity in teaching staff. These challenges may drive the use of highly instructed 'cookbook' activities (May 2023) that are not representative of scientific investigation (Koretsky et al. 2011) and thereby undermine students' epistemological beliefs (Walsh et al. 2022).

In this manuscript, we address questions regarding learning goals in lab courses (Sect. 3.1), their assessment (Sect. 3.2) and the interplay between these two aspects (Sect. 3.3) in order to create a collaborative overview of this research field. It will give an extended summary of research results and practical experiences to support the development of learning-goal-centred physics lab courses and to identify open questions in this field of research.

2 Method

The content for this manuscript is an overview of research results from the literature and practical experiences gathered before, during and after a panel discussion on the three topics (learning goals in lab courses, their assessment and the interplay between these two aspects) during the fourth World Conference on Physics Education in 2024 in Kraków, Poland. This panel discussion follows an earlier workshop that was held online during the Malta seminar in 2020, which addressed lab work in teacher education (Bearden et al. 2022). Both events were initiated by I.B. under the international umbrella of the GIREP Thematic Group Laboratory Based Teaching in Physics (LabTiP) (Bearden and Logman 2025).

To encourage members of the community to engage, a call went out for contributions to the discussion before the conference. This resulted in three proposals for questions to include in the discussion. A preliminary synthesis based on the three topics and these three questions were prepared and reviewed by all authors. This preliminary synthesis was then presented and discussed in Kraków among about 40 members of the community. The discussions were organized by topic, with topics introduced and chaired by M.A., S.F. and P.L. with many members of the community contributing their practical experiences and research findings to the discussions. All authors made notes, and two of the panel discussants (S.S. and K.D.) joined in writing this manuscript. The notes were combined, analyzed and relevant references added. This final synthesis combines practical experiences, theoretical frameworks and research results of the community.

3 Results

3.1 Open Questions on Learning Goals

Historically, typical goals for lab courses include reinforcing physics concepts and increasing students' interest and motivation. More recently, developing scientific experimental skills and problem-solving abilities, fostering an understanding of the nature of science and cultivating expert-like habits of mind and attitudes have been added as goals (American Association of Physics Teachers 1998; Trumper 2003; Hofstein and Lunetta 2004; Kozminski et al. 2014; Geschwind et al. 2024). This reflects a shift in the focus of physics lab education towards experimental processes (inquiry-based learning) and reasoning skills (e.g. critical thinking, sense-making) (May 2023). The results on lab learning goals can be used to explain how these can be achieved when lab work is designed with different foci (Etkina et al. 2006; Zwickl et al. 2014; Holmes et al. 2017; Wilcox and Lewandowski 2017; Walsh et al. 2022) and how to transform a lab course (Etkina et al. 2006; Zwickl et al. 2013; Logman and Kautz 2021).

There was general agreement on the need to rethink lab course learning goals in light of findings from the literature and their own experiences. Several members of the community shared their efforts to shift the focus from teaching specific experiments to fostering general experimental skills. Examples included contributions from The Netherlands (Bradbury and Pols 2020; Logman and Kautz 2021; Logman 2024; Pols and Dekkers 2024), Denmark, Germany (Alemani 2023; Alemani 2025) and the USA (Etkina et al. 2008; Zwickl et al. 2013; Holmes et al. 2017; Holmes and Wieman 2018; Robbins and Burkholder 2025). These activities emphasize the critical role of continued research in laboratory education.

Even if this is a crucial initial step (May 2023), transforming a lab course involves more than just setting new learning objectives; it also requires the development of a curriculum with activities aligned to these goals and the evaluation of students' learning outcomes. New instructional materials need to be created for students, such as laboratory manuals or rubrics for (self-)assessment (Pols and Dekkers 2024). In addition, new equipment may be needed, lab spaces may need to be refurbished to enable collaborative work and teaching assistants will need to be trained. Despite the inherent complexity of the transformation towards experimental processes and reasoning skills, numerous examples in literature demonstrate that it can be successfully accomplished (e.g. Zwickl et al. 2013; Holmes and Wieman 2018; Etkina et al. 2021; Logman and Kautz 2021; Alemani 2023; Logman 2024; Pols and Dekkers 2024; Robbins and Burkholder 2025). These redesigned courses not only improve student learning but also enhance instructor satisfaction and student engagement. Additional reports of best practices, research on their effect on student learning and a vibrant community of instructors and researchers are needed to improve the transfer of experiences and concepts between institutions and extract general design strategies despite the variety of site-specific requirements and implementations. Regardless of the approach, the multifaceted and interconnected nature of the task demands careful coordination throughout the implementation process. Transformed lab courses are more authentic and provide opportunities for students to undertake the processes physicists engage in during research. The Decoding the Disciplines methodology can also support students in developing a scientific mindset as they simulate how experts develop an answer to a research question (Pinnow 2016). Transformation often involves incorporating inquiry-based learning approaches, which can vary in their level of inquiry (e.g. structured, guided, open or authentic inquiry) (Buck et al. 2008). Many instructors start with lower levels of inquiry and gradually increase complexity by scaffolding the course (Sokołowska 2024).

Focusing on inquiry shifts the emphasis of the lab course and requires posing questions with answers unknown to students and even instructors. Such activities typically demand more time and may require multiple sessions, and student-led projects are increasingly being used to allow students to experience the full research process. In The Netherlands, and subsidized by a Dutch grant, educational material has been published open access to facilitate this transition, including a manual on how to do so, not just for physics but across the natural sciences (Bradbury et al. 2023; De Putter et al. 2025).

Rubrics that assess procedural and cognitive skills, as well as the use of authentic laboratory notebooks in which students record their experimental work, including thought processes, challenges and solutions, can be effective tools in such lab courses (Stanley and Lewandowski 2018).

Can learning goals be implemented bottom-up at the supervisor level?

Despite empirical evidence for the effectiveness of competence-based physics lab courses, many lab courses still follow historically developed structures pragmatically using the given infrastructure (May 2023). Even if staff are interested in modernizing the didactic concept, personal and financial resources might interfere with these ambitions.

At the Technische Universität Dresden in Germany, most of the experiments are supervised by advanced undergraduate and PhD students as regularly done in higher education institutes (Ryan 2014). A hands-on workshop, consisting of low-threshold activities, introduces these lab work supervisors to essential didactic concepts with direct application to the experiments and creates room for discussion and peer feedback. To compensate for large turnovers of supervisors, and to promote the exchange of experiences among experienced and new supervisors, the workshop is offered every semester.

An integral feature of the workshop is the development of learning goals for the individual experiments by their supervisors and an alignment of the different responsibilities with these goals, i.e. the design and assessment of an entrance test, the supervision of students during the experiment and the assessment of reports. Over the years, the supervisor's feedback on the workshop has been positive. Despite taking only 3.5 h, the participants in this workshop reflect on aspects of their work and apply didactic concepts to their experiments to promote learning and save time. However, participation is voluntary, and it is mostly the already motivated supervisors who attend the workshop. Still, it is a resource-saving solution that may be used to implement learning goals in physics lab courses in a bottom-up strategy.

A member of the community from Italy gave an example of a top-down approach to lab course reconstruction, where the entire department is involved with work coordinated by a task force (Zwickl et al. 2013). When jointly defining learning goals, experiments, etc., the task force is aware of keeping support from all people involved and research implemented changes.

Should lab courses focus on concepts, skills or both, and why?

Lab work that focuses on concepts is often confirmatory and closely tied to lecture topics (Smith et al. 2020). Skills-based labs can cover competencies from technical skills with minimal abstraction (e.g. soldering or using an oscilloscope) and/or higher-order cognitive processes, such as modelling (Zwickl et al. 2015). To practice the latter, students' certainly need to apply their physics content knowledge as well.

Experimental education extends much beyond the traditional aim of reinforcing physics concepts. In fact, research indicates that such lab work can lead to the development of non-expert-like attitudes (Zwickl et al. 2014, Wilcox and Lewandowski 2017) and even questionable research practices in attempts to confirm expected results (Smith et al. 2020). To prepare students for independent and longer-lasting lab work, it is essential to incorporate a broader range of experimental competencies into the learning goals of the lab, especially those that can be best and perhaps uniquely practiced in a lab setting.

In addition, experience in restructuring lab courses to foster experimental abilities shows that it is unrealistic to effectively pursue goals reinforcing physics concepts and developing experimental skills. However, the distinction between skills-based and purely conceptual learning goals may not be necessary. A member of the community from The Netherlands described a way of combining them successfully while thinking back and forth between the two. Although in his research projects, learning goals have not (yet) been measured, implementing thinking back and forth did not pose large challenges to instructors (Spaan et al. 2023). This provides a way to modify highly instructed 'cookbook' activities without needing to completely redesign them.

How can we ensure that students achieve the intended learning goals?

It is crucial to design activities within the individual experiments that address specific subgoals (objectives/outcomes) (American Association of Physics Teachers 1998). Moreover, the objectives of each activity should be explicitly communicated to students from the outset, along with a clear rationale for their significance both within the curriculum and in broader professional contexts (Hofstein and Lunetta 2004).

Reflection and repetition are essential if students want to improve their skills (Hofstein and Lunetta 2004; Smith and Holmes 2021). Thus, students should have multiple opportunities to practice each subgoal, rather than each subgoal being tied to a single activity.

As discussed by Etkina et al., (2008), there is a need for structured guidance to support students, with the degree of autonomy gradually increasing over time to prepare them for independent work in the lab. That said, cases of full open inquiry and immersing the students in an authentic research laboratory setting can also achieve good results (Van Willigen et al. 2020; Hoehn et al. 2021; Goesaert and Logman 2024).

Instructors should provide space and time for students' decision-making and provide opportunity for iteration and self-improvement (of equipment, measurement procedures, data quality, etc.) (Zwickl et al. 2015; Yasemin Kalender et al. 2021). Furthermore, it is essential for instructors to have a shared vision on the lab course's overarching learning goals as well as its specific objectives and outcomes. However, achieving this alignment can be difficult, especially when instructors are only involved for a short period and are not carefully introduced to the lab course's instructional approach by additional training.

3.2 Open Questions on Rubrics and Grading

Lab work should always be assessed in line with the goals of the course, because students are often pragmatic and will focus on learning what is to be assessed (Biggs and Tang 2011). Depending on the goals of the lab course, different assessment methods will be relevant. A traditional option is to grade student reports that describe the practical work and present the data collected. Another is for students to defend their findings in an oral examination. A third is to use rubrics to grade students' work during or after the lab sessions.

Rubrics consist of a table where lines represent elements to assess, sometimes called items, and columns represent the level of achievement reached by the student on a particular item. Each cell, therefore, contains a description to achieve a certain level on a particular aspect. Well-structured rubrics generally do not have overlapping items.

Rubrics can be used in two main ways: for assessing the students during their performance in the lab and afterwards. In both modalities, all or only selected rubric items can be used. If there are many lab sessions in a course, rubric items can be split between lab sessions and only the goals most prominent in each lab session are assessed. In project work, all rubric items can be used.

Rubrics can lead to improved performance, for example, by increasing transparency regarding expectations and can also be used by students for self-assessment (Panadero and Jonsson 2013). Research on the use of rubrics to assess lab work found that a class's abilities generally reach saturation at about 75% of full score, on average after the third lab session in which the ability has been assessed (Etkina et al. 2008; Buggé 2023). Research also found that using rubrics reduces the time spent providing feedback to students compared to commenting on students' reports. When students have access to rubrics, the quality of the reports also increases (Faletič and Planinšič 2020).

After students receive feedback, they can be given the opportunity to improve their reports. In case of multiple lab sessions, students usually get the opportunity to improve a fixed number of times, usually once or twice. In case of a single lab session or project work, students may be allowed to improve an unlimited number of times within a limited time period until they meet a threshold for the score on each item (Etkina et al. 2008; Faletič and Planinšič 2020; Buggé 2023). When using rubrics in multiple lab sessions, an improvement can generally only occur in the next lab session. In any case, it is important that students have an incentive to reflect on the received scores and an opportunity to improve; otherwise, improvement is unlikely to occur (Holmes 2014).

An example of research-based rubrics to assess students' scientific abilities is the Rutgers Scientific Abilities Rubrics (RSA rubrics) (Rutgers Physics and Astronomy Education Research group 2025). Various non-research-based rubrics are also available (e.g. Bradbury et al. 2023).

How can we resolve student difficulties with rubrics and related assessment issues in pre-service teachers?

How do we get students to do what we expect them to do based on the rubrics? Or, how do we get students to understand the rubrics as instructors intend correctly? Students learn to interpret the rubrics through training, similarly to how they learn expectations in every course. One approach is to discuss with each student how they plan to improve their score on a particular rubric item after each of the first few assessments. The question puts in focus the different goals of labs, some focusing on developing specific experimental skills and others focusing on general experimental skills. A connecting concept between the two can be developing habits of mind (American Association of Physics Teachers 1998). One such habit of mind is performing the experiment as best as one can in given circumstances. Experimental uncertainties were suggested as a connecting ability that could be used to test this. If a student minimizes the experimental uncertainties, correctly assesses them and can explain where they come from, this is an indicator of a well-performed experiment.

How can we efficiently assess students' work? How can we encourage students to reflect on given feedback?

We treat these two questions together because in the context of rubrics, they are closely related. The goals of assessment are likely to give feed-forward to students to improve their performance. Research indicates that reflection is crucial for improvement (Sadler and Good 2006). Efficiency would be measured by getting students' work as close as possible to the instructors' learning goals while using the least amount of instructors' time. The RSA rubrics are designed with feed-forward in mind, and their wording has been researched to maximize feed-forward while minimizing the number of words used (Rutgers Physics and Astronomy Education Research group 2025). Implementation of the RSA rubrics is generally done with at least one opportunity for students to improve, either in the same lab session or in a next lab session, and should include an explicit reflection from students. To this end, the descriptors in the rubrics provide criteria for a score of 'adequate' and elaborate on what might be the reasons for a score below 'adequate'.

Experiences of the members of the community were in line with research (Rutberg et al. 2023): it is productive to train instructors and students in the use of rubrics. For instructors, team-scoring with a more experienced instructor is a good strategy. Students, meanwhile, could prepare a list of proposed changes in response to their scores that are discussed with instructors to clarify expectations. It is also important for students to reflect on the process by which they get to the end result. A useful conceptualization coming from the community is that the final product is the result of considerations and decisions. Considerations and dilemmas may be considered the process, and the decisions may be considered the end result. So explicitly asking the students for their considerations is a way to probe the process. Students can be asked to explain their decision-making process in the reflection and also describe the reasoning that led to particular decisions/results. This highlights the importance of training sessions mentioned above, where the reflection of students and their decision-making process can be investigated (Yasemin Kalender et al. 2021).

3.3 Interplay between Learning Goals and Assessment

Ideally, learning goals, lab activities and assessment are constructively aligned within the lab course curriculum and its individual sessions (Biggs and Tang 2011) and should be coherent over different courses. But to implement this, instructors face various constraints such as limited time, resources and instructor availability. Aligning what students should learn with what students actually learn takes time (Zwickl et al. 2013; Logman and Kautz 2021; Alemani 2025): one needs to establish the desired learning goals, implement instructional approaches and test those to see whether students do improve. For instance, community members from Leiden, Potsdam and Copenhagen shared their experience in redesigning lab courses by realigning learning goals and assessment, which took, in each case, about 7 years.

Some educational systems are more grade-focused than others, and in such systems, students can be strategic about where they place effort and focus on achieving a satisfactory grade. This has been seen, for example, in quantum mechanics (Johansson et al. 2018), but also exists in lab work when embedded openness is negligible (Phillips et al. 2021). A focus on grades can be a barrier to students' engagement with inquiry tasks (DeFeo et al. 2021), and the experience of reproducing established results can effectively discourage academic risk taking (Hübner and Pfost 2024).

Given the difficulty of assessing students' improvement on specific learning goals, one may wonder whether skill-development (and even task-completion) in practical work should even be assessed or evaluated by instructors. Or can students be prompted to work on skills development and left to judge whether they have done so to their own satisfaction? One way to prompt students to work on skills development without summative assessment may be through game elements. Game elements introduce goals, rules and a feedback system, but can also provide students with autonomy and the flexibility to choose what to work on and when (Matallaoui et al. 2017). Game elements, such as Activity Bingo (Glessmer et al. 2023), do not need to contribute to a student's grade, and students can be trusted to be honest about the tasks they complete. The instructor's time to evaluate can thus be very low: one only needs to discuss items in detail when a student has a question.

Rewarding students for self-assessed competencies may help make transferable skills development relevant and realistic, avoiding either introducing 'false' Intended Learning Outcomes in early courses or leaving such development as unstated Hidden Curriculum (Snyder 1970), with perfect skills suddenly, and unrealistically, expected to have been mastered when students complete thesis and project work (Barrie 2007). In some respects, self-assessment within a game element format is simply formative feedback, but it can go beyond this and emphasize the assessment of students' own needs and standards, providing students with freedom to choose where they think they could improve and gain recognition for working on this area.

At the level of individual lab courses or sessions, a pass-fail boundary may be effective for lifting focus from task-completion and reproduction of already established values (like c, g and h). However, although the transferable skills may be known to instructors, their presence, relevance and importance may not be recognized by students, and the progressive development is hard to communicate in a way that is genuine.

4 Summary and Future Perspectives

The 90-minute panel discussion with about 40 participants covered several aspects of learning goals, their assessment and the interplay between the two, providing a wealth of research results from the literature and practical experiences related to the development of learning-goals-centred physics lab courses.

It is clear that instructors are rethinking laboratory learning goals towards experimental skills in light of research findings. At least two different but interesting ways of implementing learning goals exist: a bottom-up approach driven by those running the lab sessions, and a coordinated collaborative approach involving an entire department. Recommendations to promote student achievement of learning goals also exist. An interesting thing to consider is that the distinction between skills-based and conceptual-based labs may not be necessary.

Rubrics are often used to assess students' performance in revised lab courses. These are either adopted from the literature (e.g. Rutgers Physics and Astronomy Education Research group 2025) or developed by instructors to make them context-specific (e.g. Bradbury et al. 2023). Rubric items mostly use a general phrasing of a level of achievement, but these can be made more specific when a specific lab session calls for it. While not all items in a rubric need to be assessed in every lab session, repetition is needed to provide opportunities for students to improve, so most items should be assessed several times in a lab course. Aligning a full undergraduate physics lab curriculum requires considerable time, with several examples taking up to seven years. This time scale may be reduced with research into lab learning environments and vibrant networks within which knowledge and experiences can be shared. An interesting point to consider is whether students can assess themselves. This may increase the inclusivity of lab work by legitimizing working on skills that instructors expect students to have already mastered, but that not all students will have had opportunities for, due to systematic inequities in pre-university education. By letting students choose where they want to improve, this way of self-assessment can also promote self-regulated learning.

Several questions remain unanswered. For example, questions regarding the qualification of instructors, the assessment of upper-division labs, engagement of students and an expansion on the discussion on assessment. Among others, development of physics lab courses requires scientific analysis of the following questions: What learning occurs in relation to other areas (e.g. physics concepts) when lab work is explicitly designed to support learning in one specific area (e.g. skills)? What effects do different goals and assessment of lab work have on students' self-efficacy and motivation in and beyond the immediate context? How do the results differ between physics students, physics teacher students and non-physics students, and how should the didactic concept take these differences into account to create a favourable learning atmosphere?

It is clear that physics lab education in Europe struggles with the same questions as researchers and instructors in the USA and skills-based labs and using rubrics for assessing students in lab courses seem to be the way forward. At the moment, there is a lot of activity in physics lab education research, but it is fragmented and not always documented in terms of publications or conference contributions. The potential of national and international collaboration on lab courses is shown by the first results of a lab taxonomy survey by Geschwind et al. (2024), Bearden et al.'s (2022) position paper on laboratory work in teacher education and open access educational material in The Netherlands (De Putter et al. 2025). Such promising outcomes call for more national and international collaborations in the field, but collaborative efforts to address open issues (such as the qualification of instructors, assessment of upper-division labs, engaging students and alternative assessment methods) will need to be low-threshold and specific to instructors' concerns.

We do not claim that this is an exhaustive overview of research results and practical experiences concerning the development of learning-goals-centred physics lab courses. Rather, the panel session format was an efficient and effective method to create a collaborative outline of the current state of thinking and development efforts, among attendees at the fourth World Conference in Physics Education in 2024.

References

Alemani M (2023) The redesign of an introductory physics laboratory course. Il Nuovo CimO C. https://doi.org/10.1393/ncc/i2023-23181-y

Alemani M (2025) A Scaffolded curriculum to Foster experimental skills acquisition in an introductory physics lab course. In: Ješková Z, Hanč J (eds) Advancing physics education. Challenges in physics education. Springer, Cham. https://doi.org/10.1007/978-3-031-94207-5_14

American Association of Physics Teachers (1998) Goals of the introductory physics laboratory. Am J Phys 66(6):483–485. https://doi.org/10.1119/1.19042

Barrie SC (2007) A conceptual framework for the teaching and learning of generic graduate attributes. Stud High Educ 32(4):439–458. https://doi.org/10.1080/03075070701476100

Bearden I, Dvořák L, Planinšič G (2022) Work group 2 position paper: experiments and laboratory work in teacher education. J Phys Conf Ser 2297(1):012008. https://doi.org/10.1088/1742-6596/2297/1/012008

Bearden I, Logman PSWM (2025) Laboratory based teaching in physics (LabTiP). GIREP. https://www.girep.org/thematic-groups/laboratory-based-teaching-in-physics/. Last accessed 13 July 2025

Biggs J, Tang C (2011) Teaching for quality learning at university. McGraw-Hill education, UK

Bradbury FR, Pols FJ (2020) A pandemic-resilient open-inquiry physical science lab course which leverages the maker movement. El J for Res in Sci Math Ed 24(3):60–67. https://ejrsme.icrsme.com/article/view/20416

Bradbury FR, Logman PSWM, Mesman S, Strømme MH (2023) Supporting students' scientific open inquiries in labs (SssOIL) collection. Edusources. https://edusources.nl/mate. rials/045 66738-bd6f-49c6-a2d9-f64a3885394b/supporting-students-scientific-open-inquir, ies-in-labs-sssoil-collection. Last accessed13 July 2025

Buck LB, Bretz SL, Towns MH (2008) Characterizing the level of inquiry in the undergraduate laboratory. J Coll Sci Teach 38(1):52–58

Buggé D (2023) Improving scientific abilities through lab report revision in a high school investigative science learning environment classroom. Phys Rev Phys Educ Res 19(2):020166. https://doi.org/10.1103/PhysRevPhysEducRes.19.020166

De Putter L, Bradbury FR, Logman PSWM, Mesman S, Strømme MH, Jonker MJ (2025) Playbook open inquiry in labcourses. https://maken.wikiwijs.nl/200332/Playbook_Open_Inquiry_in_labcourses. Last accessed 13 July 2025

DeFeo DJ, Tran TC, Gerken S (2021) Mediating students' fixation with grades in an inquiry-based undergraduate biology course. Sci Educ 30:81–102. https://doi.org/10.1007/s11191-020-001 61-3

Etkina E, van Heuvelen A, White-Brahmia S, Brookes DT, Gentile M, Murthy S, Rosengrant D, Warren A (2006) Scientific abilities and their assessment. Phys Rev ST Phys Ed Res 2(2):020103. https://doi.org/10.1103/PhysRevSTPER.2.020103

Etkina E, Karelina A, Ruibal-Villasenor M (2008) How long does it take? A study of student acquisition of scientific abilities. Phys Rev ST Phys Ed Res 4(2):020108. https://doi.org/10.1103/PhysRevSTPER.4.020108

Etkina E, Brookes DT, Planinsic G (2021) The investigative science learning environment (ISLE) approach to learning physics. J Phys Conf Ser 1882(1):012001. https://doi.org/10.1088/1742-6596/1882/1/012001

Faletič S, Planinšič G (2020) How the introduction of self-assessment rubrics helped students and teachers in a project laboratory course. Phys Rev Phys Educ Res 16(2):020136. https://doi.org/10.1103/PhysRevPhysEducRes.16.020136

Geschwind G, Alemani M, Fox MFJ, Logman PSWM, Tufino E, Lewandowski HJ (2024) Development of a global landscape of undergraduate physics laboratory courses. Phys Rev Phys Educ Res 20(2):020117. https://doi.org/10.1103/PhysRevPhysEducRes.20.020117

Glessmer MS, Latuta L, Saltalamacchia F, Daae K (2023) Activity bingo nudging students to make the most out of fieldwork. Oceanography 36(2–3):225–228. https://doi.org/10.5670/oceanog.2023.217

Goesaert WM, Logman PSWM (2024) Easy method to establish the dispersion relation of capillary waves on water jets. Am J Phys 92(2):93–99. https://doi.org/10.1119/5.0144849

Hoehn JR, Fox MFJ, Werth A, Borish V, Lewandowski HJ (2021) Remote advanced lab course: a case study analysis of open-ended projects. Phys Rev Phys Educ Res 17(2):020111. https://doi.org/10.1103/PhysRevPhysEducRes.17.020111

Hofstein A and Lunetta VN (2004), The laboratory in science education: foundations for the twenty-first century. Sci Ed 88:28–54. https://doi.org/10.1002/sce.10106

Holmes NG (2014) Structured quantitative inquiry labs: developing critical thinking in the introductory physics laboratory (T). University of British Columbia. https://open.library.ubc.ca/collections/ubctheses/24/items/1.0165566

Holmes NG, Olsen J, Thomas JL, Wieman CE (2017) Value added or misattributed? A multi-institution study on the educational benefit of labs for reinforcing physics content. Phys Rev Phys Educ Res 13(1):010129. https://doi.org/10.1103/PhysRevPhysEducRes.13.010129

Holmes NG, Wieman CE (2018) Introductory physics labs: we can do better. Phys Today 71(1):38–45. https://doi.org/10.1063/PT.3.3816

Hübner V, Pfost M (2024) Academic risk taking and teaching quality in higher education. Learn Instr 90:101877. https://doi.org/10.1016/j.learninstruc.2024.101877

Johansson A, Andersson S, Salminen-Karlsson M, Elmgren M (2018) "Shut up and calculate" the available discursive positions in quantum physics courses. Cult Stud of Sci Educ 13:205–226. https://doi.org/10.1007/s11422-016-9742-8

Ketonen L, Lehtinen A, Koskinen P (2023) Assessment designs of instructional labs: a literature review and a design model. Phys Rev Phys Educ Res 19:020601. https://doi.org/10.1103/PhysRevPhysEducRes.19.020601

Koretsky M, Kelly C, Gummer E (2011) Student perceptions of learning in the laboratory: comparison of industrially situated virtual laboratories to capstone physical laboratories. J Eng Educ 100:540–573. https://doi.org/10.1002/j.2168-9830.2011.tb00026.x

Kozminski J, Lewandowski H, Beverly N, Lindaas S, Deardorff D, Reagan A, Dietz R, Tagg R, Eblen-Zayas M, Williams J, Hobbs R, Zwickl B (2014) AAPT recommendations for the undergraduate physics laboratory curriculum. https://www.aapt.org/Resources/upload/LabGuidlinesDocument_EBendorsed_nov10.pdf

Logman PSWM, Kautz J (2021) From Dublin descriptors to implementation in bachelor labs. J Phys Conf Ser 1929(1):012065. https://doi.org/10.1088/1742-6596/1929/1/012065

Logman PSWM (2024) Engaging theoretically primed students in second year lab courses. J Phys Conf Ser 2727(1):012021. https://doi.org/10.1088/1742-6596/2727/1/012021

Matallaoui A, Hanner N, Zarnekow R (2017) Introduction to gamification: foundation and underlying theories. In: Stieglitz S, Lattemann C, Robra-Bissantz S, Zarnekow R, Brockmann T (eds) Gamification. Progress in IS. Springer, Cham. https://doi.org/10.1007/978-3-319-45557-0_1

May JM (2023) Historical analysis of innovation and research in physics instructional laboratories: recurring themes and future directions. Phys Rev Phys Educ Res 19(2):020168. https://doi.org/10.1103/PhysRevPhysEducRes.19.020168

Panadero E, Jonsson A (2013) The use of scoring rubrics for formative assessment purposes revisited: a review. Educ Res Rev 9:129–144. https://doi.org/10.1016/j.edurev.2013.01.002

Pinnow E (2016) Decoding the disciplines: an approach to scientific thinking. Psychol Learn Teach 15:94–101. https://doi.org/10.1177/1475725716637484

Phillips AML, Sundstrom M, Wu DG, Holmes NG (2021) Not engaging with problems in the lab: students' navigation of conflicting data and models. Phys Rev Phys Educ Res 17(2):020112. https://doi.org/10.1103/PhysRevPhysEducRes.17.020112

Pols CFJ, Dekkers PJM (2024) Redesigning a first year physics lab course on the basis of the procedural and conceptual knowledge in science model. Phys Rev Phys Educ Res 20:010117. https://doi.org/10.1103/PhysRevPhysEducRes.20.010117

Robbins M, Burkholder EW (2025) Development and evaluation of decision-based Laboratories for non-Science Majors. Phys Teach 63(6):492–495. https://doi.org/10.1119/5.0179162

Rutberg J, Jammula D, Ahmed S (2023) Implementation of an investigative science learning environment based laboratory course taught by novice instructors. Phys Rev Phys Educ Res 19(2):020153. https://doi.org/10.1103/PhysRevPhysEducRes.19.020153

Rutgers physics and astronomy education research group (2025) Rubrics. https://sites.google.com/site/scientificabilities/rubrics. Last accessed 13 July 2025

Ryan BJ (2014) Graduate teaching assistants; critical colleagues or casual components in the undergraduate laboratory learning? An exploration of the role of the postgraduate teacher in the sciences. Eur J Sci Math Educ 2(2):98–105. https://doi.org/10.30935/scimath/9403

Sadler PM, Good E (2006) The impact of self-and peer-grading on student learning. Educ Assessment 11(1):1–31. https://doi.org/10.1207/s15326977ea1101_1

Smith EM, Stein M, Holmes NG (2020) How expectations of confirmation influence students' experimentation decisions in introductory labs. Phys Rev Phys Educ Res 16(1):010113. https://doi.org/10.1103/PhysRevPhysEducRes.16.010113

Smith EM, Holmes NG (2021) Best practice for instructional labs. Nature Phy 17:662–663. https://doi.org/10.1038/s41567-021-01256-6

Snyder BR (1970) The hidden curriculum. MIT Press, Cambridge, Mass

Sokołowska D (2024) Inquiry for meaningful learning. In: Faletič S, Pavlin J (eds) Teaching and learning physics effectively in challenging times. Challenges in physics education. Springer, Cham. https://doi.org/10.1007/978-3-031-72541-8_1

Spaan W, Oostdam R, Schuitema J, Pijls M (2023) Thinking-back-and-forth in practical work experienced by students: identifying evidence-informed characteristics of good practices in secondary education. Res Sci Techn Educ 43(1):294–311. https://doi.org/10.1080/02635143.2023.2268005

Stanley JT, Lewandowski HJ (2018) Recommendations for the use of notebooks in upper-division physics lab courses. Am J Phys 86(1):45–53. https://doi.org/10.1119/1.5001933

Trumper R (2003) The physics laboratory—a historical overview and future perspectives. Sci Ed 12:645–670. https://doi.org/10.1023/A:1025692409001

Van Willigen J, Loman C, Thibaudier P, Fokkema DBRA, Hijmans TW (2020) Uranium fission and plutonium production in the undergraduate lab. Am J Phys 88(3):200–206. https://doi.org/10.1119/10.0000206

Walsh C, Lewandowski HJ, Holmes NG (2022) Skills-focused lab instruction improves critical thinking skills and experimentation views for all students. Phys Rev Phys Educ Res 18(1):010128. https://doi.org/10.1103/PhysRevPhysEducRes.18.010128

Wilcox BR, Lewandowski HJ (2017) Developing skills versus reinforcing concepts in physics labs: insight from a survey of students' beliefs about experimental physics. Phys Rev Phys Edu Res 13(1):010108. https://doi.org/10.1103/PhysRevPhysEducRes.13.010108

Yasemin Kalender Z, Stump E, Hubenig K, Holmes NG (2021) Restructuring physics labs to cultivate sense of student agency. Phys Rev Phys Educ Res 17(2):020128. https://doi.org/10.1103/PhysRevPhysEducRes.17.020128

Zwickl BM, Finkelstein N, Lewandowski HJ (2013) The process of transforming an advanced lab course: goals, curriculum, and assessments. Am J Phys 81(1):63–70. doi:https://doi.org/10.1119/1.4768890

Zwickl BM, Hirokawa T, Finkelstein N, Lewandowski HJ (2014) Epistemology and expectations survey about experimental physics: development and initial results. Phys Rev ST Phys Ed Res 10(1):010120. https://doi.org/10.1103/PhysRevSTPER.10.010120

Zwickl BM, Hu D, Finkelstein N, Lewandowski HJ (2015) Model-based reasoning in the physics laboratory: framework and initial results. Phys Rev ST Phys Ed Res 11(2):020113. https://doi.org/10.1103/PhysRevSTPER.11.020113

Cultural Understanding of Quantum Physics Through a Historical and Pedagogical Reconstruction of Old Quantum Theory and Early Quantum Mechanics

Luisa Lovisetti and Marco Giliberti

Abstract In a society where technological applications risk overshadowing the curiosity-driven aspects of physics, understanding historical and epistemological aspects becomes crucial for a meaningful and cultural introduction of Quantum Physics (QP). In this sense, we believe that presenting only Quantum Mechanics (QM) when introducing QP (especially at high school) may prove insufficient, and we emphasize the importance of dealing also with the Old Quantum Theory, as it provides insights into the cultural relevance of QM, facilitates a deeper reflection on the Nature of Science, and aids also in understanding the physics-related QM problems. In this work, we will discuss these aspects.

Keywords Cultural understanding of physics · Old quantum theory · Quantum mechanics · History of physics · Physics education

1 Introduction: Historical Aspects for a Meaningful Introduction of Quantum Physics

As Dirk ter Haar wrote in 1966, in his book *The Old Quantum Theory*,

> Nowadays, there is a tendency among university students to learn physics exclusively from textbooks, without familiarizing themselves with the original literature and, as a result, without fully understanding how the discipline has grown and developed. (ter Haar 1966, back cover).

Indeed, students and teachers at all levels are often unfamiliar with the original works and the historical development of physical theories, including Quantum Mechanics (QM). And even most Physics Education Research (PER) groups consider

---

L. Lovisetti (✉) · M. Giliberti
Department of Physics "Aldo Pontremoli", University of Milan, Milan, Italy
e-mail: luisa.lovisetti@unimi.it

D. Sokołowska and E. McLoughlin (eds.), *Embracing Changes Together in Physics Education*, Challenges in Physics Education,
https://doi.org/10.1007/978-3-032-27385-7_12

that, for an effective introduction to Quantum Physics (QP) (at both school and university levels), starting with historical questions and aspects is not an appropriate choice since they suggest that an approach that directly addresses the physical aspects is preferable, being more effective (Michelini and Stefanel 2021). Actually, if the goal is to lead students to an operational understanding of QM, even at an elementary level, it is often deemed unnecessary to discuss historical aspects (Michelini and Stefanel 2021), especially those related to the Old Quantum Theory (OQT). This belief arises from the following ideas: a) covering both the OQT and QM would require too much time (Michelini and Stefanel 2021); b) the OQT does not contribute to a better understanding of QM and might even lead to misconceptions (Wittmann et al. 2002; Kalkanis et al. 2003; Fischler and Lichtfeldt 1992); c) the "pseudo-historical" approach commonly proposed by textbooks and primarily based on the OQT is not effective from a didactic standpoint (Michelini and Stefanel 2021). Indeed, there are many examples showing that the presentation of OQT topics in an uncritical and pseudo-historical manner can hinder students' later understanding and appropriation of genuinely quantum concepts (see, for example, Wittmann et al. (Wittmann et al. 2002)).

We advocate for a broader perspective, rooted in a long-standing Italian and European tradition that emphasizes the formative aspects of education. Therefore, we believe that certain carefully selected historical aspects, appropriately presented, are fundamental to a genuine understanding of QM. Studying history and observing how quantum theory evolved from early models and attempts allows us to understand why QM is the theory we know today, enabling a deeper grasp of its formal structure (Lovisetti et al. 2023). The approach we propose, of which we have already proven the effectiveness (Lovisetti 2024; Giliberti and Lovisetti 2025), is not a historical approach per se, but one that puts history in the service of physics, using historical elements to foster a deeper understanding of physics content, to highlight the historical and epistemological reasons underlying a profound conceptual shift—such as that brought about by QM (Giliberti and Lovisetti 2024; Levrini et al. 2014).

Our framework is culturally close to the so-called Discipline-Culture Framework. However, it differs from it in its explicit lack of reference to the structure of scientific theories divided into nucleus, body, and periphery, as proposed by Galili (Galili 2022). While we consider the aforementioned epistemological reflections to be of great interest, we believe that an approach that can genuinely speak to a wide audience and be implemented in classroom practice must be as free as possible from explicit epistemological references—particularly in order to be accepted by physics teachers. We believe that such a reconstruction should be done not only for QM but for all topics, as it not only provides cultural significance to QM (and, more generally, to physics) in itself (Weissman et al. 2019; Hadzidaki et al. 2000), but it also facilitates a deeper reflection on the Nature of Science (NOS) (Stadermann and Goedhart 2020; Stadermann and Goedhart 2021).

In fact, the NOS refers to the understanding of the fundamental characteristics, principles, and processes that define and shape scientific knowledge and its development. Understanding the NOS is crucial for both scientists and the general public, as it promotes a more informed engagement with scientific information, fostering critical thinking and a deeper appreciation for the dynamic and evolving nature of scientific knowledge (Lovisetti and Giliberti 2025b). In this regard, we perceive the need and urgency to reconstruct the crucial role of physics as a cultural promoter of scientific knowledge in society (Giliberti et al. 2025). This need is clearly expressed also in the GIREP thematic group "Cultural Understanding of Physics." Notably, a cultural understanding of QM demands even more considerable and immediate effort. This urgency arises from a society increasingly reliant on scientific knowledge and quantum technologies, where the application of physical knowledge risks overshadowing the curiosity-driven aspects of physics (Nita et al. 2023; Vermaas 2017; Fox et al. 2020).

Therefore, we believe that, for a cultural understanding of QP, discussing only QM may not be sufficient, and it should be complemented by a suitable presentation of the OQT to provide a gradual, reasoned, and culturally meaningful introduction to the quantum aspects. This does not mean that it is not possible to meaningfully address issues related to the NOS in education even without the use of history (see for instance Faletič et al. (Faletič et al. 2024)), but rather that the learning of the NOS in a broader cultural landscape is facilitated by an appropriate use of history in teaching.

2 Theoretical Background: The Role of the OQT in QP Education

Traditionally, in high-school education and textbooks, QP is often reduced to a presentation primarily focused on the OQT (Stadermann et al. 2019; Stadermann et al. 2021; Krijtenburg-Lewerissa et al. 2017). Early research on students' learning has highlighted that this approach, as traditionally presented, is not effective for a deep and meaningful understanding of the concepts. As early as 2002, for example, several articles published in Volume 70, Issue 3, of the *American Journal of Physics* expressed severe criticism regarding the misconceptions generated by such teaching methods.

Moreover, recent technological advancements have increasingly favored a "pragmatic" presentation of QM, focusing on aspects more directly related to technology. As a result, the study of the effects of a thorough and thoughtful teaching of the OQT on learning has largely been neglected or relegated to a secondary role. In support of this, we can cite the entire Volume 21, Issue 1, of *Physical Review* from 2025, in which the adjective "cultural" appears in only a single article, which is not at all related to the OQT; and also the Special Collection "Celebrating the International Year of Quantum Science and Technology" of the *American Journal of Physics* of 2025 (Volume 93, Issues 1 and 2), where the importance of the OQT is discussed only in our work (Giliberti and Lovisetti 2025) and in that of Kragh (Kragh 2025).

We therefore believe that further analysis regarding the effects of the OQT on learning should be done. Furthermore, we contend that many of the negative criticisms about learning are partially distorted and incorrectly contextualized, as insufficient attention might have been given to a clear and meaningful physical presentation of the fundamental aspects of the OQT. It is indeed quite predictable that a poorly structured and shallow approach to physics can lead to disappointing results, but we do not believe that a comprehensive analysis of the effectiveness of teaching the OQT, when presented properly, has been conducted. Additionally, even where studies conducted on one or more models have been accurate, they cannot be considered sufficient to fully understand the problem: only an adequate number of approaches and attempts to solve the OQT can provide an effective evaluation and draw concrete conclusions regarding its actual usefulness in physics education. Therefore, in our view, in many cases, the judgment that teaching the OQT is counterproductive has been made without a sufficiently broad and in-depth evaluation; in this work, we thus propose to examine in more detail the importance of a correct reformulation of the OQT for students' learning.

From previous research, we have already found that it is possible to create a well-done and comprehensible reconstruction of the OQT for high-school students and that a correct reconstruction of the OQT aids in understanding the physics related to QM problems (Giliberti and Lovisetti 2025; Lovisetti and Giliberti 2025a).

3 OQT or Not OQT, that Is the (Research) Question

We can now, therefore, deepen our research with two more punctual questions:

1. In what aspects, can the student-addressed reconstruction of the OQT be particularly effective?
2. In which sense, discussing the OQT could help in developing a deeper awareness of the NOS?

4 Conceptual Frameworks, Methods, and Approaches

Our framework is based on the so-called "Model of Educational Reconstruction." This model, developed by Duit et al. (Duit et al. 2012), distinguishes between the evolution of scientific knowledge and the learning process. While scientific knowledge emerges through abstraction and reduction in institutional contexts, education requires a partial reversal of this process to make complex ideas more accessible and meaningful.

This is particularly relevant in the teaching of QP, where the historical development plays a crucial role in conceptual understanding. A key aspect of this approach is *elementarization*, which involves a) identifying fundamental scientific principles for instruction, b) simplifying complex ideas while preserving their scientific integrity, and c) structuring learning into sequenced instructional elements, guiding students from intuitive notions to formal understanding. This model integrates empirical research on students' preconceptions, interests, and attitudes to create enriched learning experiences.

The model consists of three phases:

1. analyzing educational significance, identifying key concepts and their broader implications,
2. investigating students' and teachers' perspectives, considering cognitive and affective dimensions, and,
3. designing and evaluating learning environments, developing instructional materials, and assessing their impact.

The application we have made of this model to QP underscores the importance of the OQT as a bridge to QM. The OQT, despite its flaws and limitations, provides essential insights into the transition from classical to QP. Concepts such as energy quantization in the black-body problem or the Bohr model of the atom, as well as the discussions concerning the structure and characteristic of QM formulas, the intrinsic probabilistic nature of quantum theory or the early wave-particle duality help students develop an intuitive grasp of a set of quantum principles as an intermediate step before confronting the full formalism of QM with its axioms (for details, see Giliberti and Lovisetti (Giliberti and Lovisetti 2025) and Lovisetti and Giliberti (Lovisetti and Giliberti 2025a)). A thorough consideration of the OQT in teaching can help students connect the abstract postulates of QM with their physical meaning (Giliberti and Lovisetti 2025). By discussing also the OQT, we provide a historically and conceptually grounded path that enhances understanding, making QM more coherent and intellectually accessible. Furthermore, the interdisciplinary significance of QM extends beyond physics, influencing fields such as information science, philosophy, and technology. Integrating the OQT into QM education strengthens these connections, offering students not just technical knowledge, but also a deeper appreciation for the evolution of scientific thought.

4.1 Methods

Since 2022, we have developed an instructional strategy to teach selected OQT content within a broader cultural framework for QM. This approach emphasizes not only physical concepts but also integrates historical and cultural perspectives, enriching students' understanding of the theory's development and its impact on science and society. Moving away from traditional methods (proposed at school and in textbooks) that present quantum theory in isolation, our strategy aims to offer a deeper, more contextualized educational experience. The instructional strategy is grounded in two key principles: iterative refinement and theoretical grounding. The methodology follows a cyclical process of continuous design, testing, and revision, as in the Design-Based Research. Each cycle incorporates feedback and empirical data to refine teaching materials and adjust methods, ensuring content is aligned with theoretical frameworks and responsive to students' learning experiences. This iterative approach allows for the systematic improvement of materials and a rigorous evaluation of their impact on students' conceptual understanding and skill development.

Data collection throughout the process included questionnaires, written responses to the questions related to the analyzed texts (provided by groups of 3–4 students each), written problems given to students (carried out individually), as well as *Kahoot!* tests and Google Form questionnaires, classroom recordings, and student interviews. A qualitative analysis was conducted on the interviews and group responses related to the analyzed texts, focusing on the modes of text comprehension, the awareness of the NOS, and the concept of observable. This provided insights into teaching and learning interactions, helping us refine the approach. The methodology's cyclical nature ensured that materials were continuously refined based on real classroom experiences. Evaluations, including pre- and post-tests and classroom observations, allowed us to assess the impact on students' grasp of QP concepts like quantization (in its different interpretations) and the intrinsic probabilistic nature of QM. The results showed the effectiveness of the strategy in enhancing conceptual understanding, critical thinking, and problem-solving skills in QP (Lovisetti 2024). In this study, we answer another question, that is, whether discussing the OQT can help develop a deeper awareness of the NOS.

4.2 *Our Approach*

To test the approach proposed, several trials were designed, implemented, and tested in the last three years, involving high-school students and teachers (participating mostly as learners or, for a few of them, as both learners and teachers of students), and university master's-degree students. Indeed, starting from our research questions and from the analysis of the literature, to address these issues, between 2022 and 2025, we have organized two editions of the educational laboratory "Old (but Gold) Quantum Theory," the course "Introduction to Quantum Mechanics," and the teachers' training course "Principles and Equations of Physics III: Quantum Mechanics," with 320 high-school students and 241 teachers overall (on a voluntary basis). The activities were also tested at the university level, in four trials with 69 students of master's degrees in physics or mathematics, during the courses of "Preparation of Didactic Experiences 2" (Lovisetti 2024; Giliberti and Lovisetti 2025). In general, the approach gives rise to a path divided into two parts, "OQT" and "Principles of QM," each lasting no more than 15 hours. Moreover, although most of the students' work is carried out during class time, we could estimate homework to account for approximately an additional 15 hours.

In our approach, we adopted an inquiry-based learning method to engage students in a process of active exploration and critical thinking. Our inquiry-based approach was evident in the structure of the lessons, which involved a variety of methods that encouraged exploration and active engagement. We also organized experimental activities where students could test some of the theoretical ideas and conceptual aspects and compare their results with predictions. In addition to experimental activities, we used commented readings of excerpts from original papers. Instead of simply presenting these texts to the students, we asked them to engage actively by working in group assignments. In these assignments, we guided students to read the original scientific texts critically and then collaborate to extract the key ideas, reflecting on how the concepts evolved and how they were influenced by the experimental data available at the time. By encouraging active reading and comprehension, we created an environment where students could directly engage with the historical context of scientific discoveries and gain a deeper understanding of the challenges faced by scientists in their quest to explain the quantum world.

In every step of this process, we stressed the importance of explicit reasoning. We consistently asked students to explain their thinking, justify their conclusions, and provide evidence from experiments or from the readings to support their ideas. The inquiry-based approach was particularly effective in encouraging students to take ownership of their learning. Rather than passively absorbing information, they became active participants in the learning process, constantly asking questions, making predictions, and testing their ideas against experiments and data (for instance, single-slit diffraction and double-slit interference of light, atomic and molecular spectra, the charge-to-mass ratio, and the analysis of the Franck-Hertz experiment). This hands-on, investigative approach allowed them to experience firsthand how quantum theory emerged from a series of experimental discoveries and how new data can sometimes force a reevaluation of existing models. Here, we used the expression "hands-on" in a broad sense, meaning that students actively worked on data coming from historical experiments (for example, those of Geiger and Marsden related to Rutherford's atomic model), as well as on the reading and interpretation of selected excerpts from important historical papers.

The interdisciplinary lens of our approach further enriched students' understanding. We did not just focus on the physical principles themselves; we also discussed the cultural and historical context in which these ideas were developed. Ultimately, our aim was not just to teach QM concepts but to help students develop critical thinking skills that would allow them to understand the evolution of scientific knowledge and scientific reasoning within a broad context.

5 OQT for a Better Understanding of the Physics Related to QM Problems

In order to understand how the OQT helps in better understanding why QM developed in the way we know it today, with its specific characteristics, we focus here on two aspects taken as examples (but others could also be considered, as in Giliberti and Lovisetti (Giliberti and Lovisetti 2025)).

5.1 Classical Formulas

The formulas used to express physical quantities employed by Planck for explaining the black-body spectrum, those used by Einstein for explaining the photoelectric effect, those used by Bohr for his atomic model, those for interpreting the Franck-Hertz experiment, and so forth, are all conventional mechanical formulas formally identical to their classical counterparts. To them, are "simply" added appropriate, albeit ad hoc, quantization hypotheses, and these formulas give correct results from a physical perspective (for details, see Giliberti and Lovisetti (Giliberti and Lovisetti 2025) and the related supplementary material).

5.2 Why Certain Physical Quantities Are Not "Observables," Based on Experimental Facts

In QM, there are certain quantities that can be measured but are not called "observables." Observables are represented by operators, and their values correspond to the eigenvalues of these operators. For example, position, momentum, energy, etc., are observables. Electric charge, on the contrary, is not. This is because, when we perform a measurement on a specific kind of particle (i.e., an electron), we obtain only a single result. An analogous consideration can be made for the mass.

If a linear operator corresponding to electric charge were introduced (which can certainly be done, although it is not typical in elementary QM), we would therefore need to describe the system in the Hilbert space, that is, the eigenspace corresponding to the single eigenvalue that gives the particle's charge, thus effectively making the operator unnecessary (or useless). The same can be said concerning mass. The fact that measurements taken on particles of a specific type yield a unique mass and charge comes from classical physics: certain quantities, such as charge-to-mass ratio, charge (as measured in Millikan's experiment), and mass, consistently provide a single, fixed value. This value ultimately determines the type of beam or particle we are examining.

In this sense, by examining experimental data, we can determine whether the quantity we are measuring is a QM observable or not. Certain quantities are considered by defining the system we are dealing with, so that, for example, if we measured two different values for the charge, we would think of two distinct particles (not as the same particle in two different states). The same goes for the mass.

6 OQT for a Deeper Awareness of the NOS

The expression "NOS" refers to the fundamental characteristics and principles that underpin scientific inquiry and the development of scientific knowledge. It encompasses the ideas about how science operates, the ways scientific knowledge is generated, validated, and modified over time, and the influence of cultural, historical, and social contexts on scientific discoveries.

As already mentioned, we believe it crucial to integrate physics education into a broader cultural framework, emphasizing the NOS. Although past decades of physics education research underscored the importance of integrating the NOS into curricula (Akgun and Kaya 2020; Cofré et al. 2019; Erduran et al. 2021; Irzik and Nola 2014; Knorr-Cetina 1999; Lederman et al. 2002), a notable absence of explicit guidance for incorporating physics history and philosophy into courses still remains. In fact, although the NOS has long been regarded as a fundamental component of scientific education, empirical studies consistently report a worldwide limited representation of the NOS in physics textbooks (Abd-El-Khalick et al. 2017).

From our perspective, discussing the OQT helps students develop a deeper awareness of the NOS. The numerous discussions held with the students, in an inquiry-based approach, served to clarify a series of observations and considerations regarding the role of the OQT for a better understanding of the NOS (Lovisetti 2024; Giliberti and Lovisetti 2025). In the following, we will report some of the comments provided by students during group activities, interviews, and questionnaires, denoting them generally with "S." In evaluating the responses, we must certainly take into account that the students who participated in the trials did so on a voluntary basis, and were, therefore, generally interested and motivated. The excerpts provided have been selected from among the best in terms of completeness and clarity. However, this should not lead us to believe that only a minority of students have internalized the NOS. In fact, even if expressed in a less clear and detailed manner, many students conveyed similar opinions and answers.

First of all, the OQT illustrates how scientific knowledge evolves incrementally. This shows students that scientific theories are often refined and built upon by successive contributions and ideas from multiple scientists: this fact was grasped by 79% of the high-school students who joined our trials. For example, to the question "Did the OQT help you understand the evolution of physics and scientific theories? And how?", one of the participants answered:

> S. "One of the key concepts I have grasped is that science is a dynamic and ever-evolving process. By studying the OQT, I have understood that science is not a static body of knowledge, but rather an ongoing process of discovery and revision. For example, when Planck

> introduced the concept of energy quantization for the oscillators to explain black-body radiation, he had to abandon the previous notion of continuous energy and accept that it was quantized into discrete quanta. This illustrates that science relies on accepting new experimental evidence and the ability to amend existing theories accordingly."

Another student commented:

> S. "Studying the OQT, I have come to realize that science is not just a collection of facts, but also a process of constructing models that can be subjected to experimental verification. For instance, Bohr's atomic model successfully predicted the spectral lines of hydrogen, but it was later expanded and modified to account for new discoveries and anomalies. This demonstrates that science relies on the formulation of hypotheses and models that can be empirically tested, and the ability to adapt and modify these models based on new evidence is essential for scientific progress."

And a further student observed:

> S. "It is interesting how the OQT was not just created by one person but evolved over time. Planck started with his work on black-body spectrum, introducing the idea that the energy of the oscillators within a black-body cavity is quantized. Then, Einstein did not simply take Planck's idea and apply it to a new context; he gave it a completely new epistemological significance and deep physical meaning. By explaining the photoelectric effect, Einstein showed that light itself consists of discrete quanta, fundamentally reshaping the way we understood light and energy. This shows that scientific knowledge is not just the application of an existing idea to different situations, but also the transformation and evolution of concepts as new challenges and contexts emerge. Each new scientific development builds upon and refines earlier concepts, changing their meaning and broadening their scope over time."

From our perspective, we are struck by the precision of the responses, especially considering that they were given by high-school students, as well as the reasoning provided, which does not appear to be mere rote repetition of previously discussed points. The "linguistic" dimension of these responses seems to us to show how the learning of the NOS is rooted in a broad cultural landscape. This is particularly notable since these aspects related to the NOS were not explicitly addressed and discussed by the tutors during the course (we recall that the approach used was inquiry-based).

Overall, most students' responses reflected an understanding of key NOS concepts such as the evolving nature of scientific knowledge, the role of models and hypotheses, the importance of empirical testing, and the collaborative nature of scientific discovery. These insights indicate that the students have internalized the idea that science is a dynamic, iterative process that evolves in response to new evidence, and that scientific knowledge is always subjected to refinement and modification.

Another aspect related to the NOS concerns the development of scientific theories. Indeed, the OQT faced several anomalies and limitations, such as its inability to explain the fine structure of spectral lines or the anomalous Zeeman effect. These limitations eventually led to the development of the more comprehensive framework of QM. This shows students that scientific theories are always provisional and subjected to revision or replacement in light of new evidence. 72% of the students grasped the provisional aspects of the OQT models. For example, let us consider the following question given in a questionnaire: "In what way do the flaws of Bohr's model help us understand how scientific process proceeds?". One of the students answered that:

> S. "Understanding the limitations and anomalies inherent in Bohr's atomic model was an eye-opening experience, highlighting the importance of acknowledging the provisional nature of scientific theories and embracing uncertainty in the pursuit of knowledge. Bohr's model, while revolutionary in its time and successful in explaining certain phenomena like the discrete spectral lines of hydrogen, faced significant challenges that could not be fully resolved within its framework. One is the inability to account for the fine structure of spectral lines in more complex atoms or ions, which deviated from its predictions. As I delved deeper into the historical context, I realized that scientific progress often involves grappling with anomalies and contradictions that push the boundaries of existing theories."

The student's recognition of the anomalies faced by Bohr's model, such as the unexplained fine structure of spectral lines, demonstrates an important element of the NOS: anomalies and contradictions are not failures but opportunities for further scientific exploration. The process of identifying and addressing such discrepancies is what drives scientific progress. As the student mentions, Bohr's model did not fully explain all phenomena, yet it was seen as a starting point for deeper inquiry, which ultimately led to the development of QM. This points to the idea that scientific inquiry is not a smooth journey but often involves grappling with unresolved issues that stimulate further questioning and refinement.

Last but not least, the OQT shows how scientific knowledge and ideas are deeply influenced by and intertwined with the social, economic, political, technological, and cultural contexts of their time (Lovisetti 2024). 73% of the students recognized these deep interconnections. For example, at the end of the section on the black-body spectrum, the following written question was assigned "Please highlight an aspect of today's lesson that particularly caught your attention." One of the students provided this answer:

> S. "The study of black-body radiation and the development of Planck's law were not just theoretical breakthroughs, but also tied to practical needs in society, especially in the context of industrial and technological advancement. At the time, the world was undergoing rapid industrialization, and there was a growing demand for more efficient lighting. The issue of how objects absorb and emit radiation became important for improving incandescent light bulbs and other lighting technologies. The quest to understand the spectrum of black-body radiation was directly linked to improving practical devices, showing how physics was not just an abstract science but also had real-world applications shaped by societal needs."

Taking into consideration a similar question posed after the laboratory activity with spectral lamps, another student commented that:

> S. "The discovery of spectral lines and their interpretation in terms of atomic structure were not just steps in the development of quantum theory; they were also influenced by technological advances in spectroscopy. In the late-nineteenth and early-twentieth centuries, spectroscopic techniques were becoming increasingly important in both chemistry and astronomy. Scientists needed precise ways to analyze elements and their compositions, which led to the development of more sophisticated experimental tools. The work of Balmer, Rydberg, and Bohr in explaining spectral lines was directly connected to these advancements. This made me realize that the progress of physics is not only shaped by abstract theoretical reasoning, but also by the availability of better instruments and the practical demands of other scientific fields (for example, in this case, chemistry)."

These kinds of responses underscore a critical element of the NOS: scientific knowledge is not developed in a vacuum but is deeply influenced by the socio-economic and technological context of its time. In these cases, the study of black-body radiation and the development of Planck's law, or the discovery and interpretations of spectral lines, were not purely abstract theoretical pursuits but were also directly driven by practical needs arising from industrialization and technological progress.

This highlights how societal needs and advancements can guide the direction of scientific research, with theoretical breakthroughs often emerging from attempts to solve real-world problems. The responses from the students show that they understood the idea that scientific knowledge is not just a collection of abstract truths, but a body of knowledge that is shaped by the social, economic, and technological contexts in which it is developed. This insight into the interconnection between science and society is a crucial aspect of the NOS, highlighting the contextual nature of scientific progress. The students' reflections indicate that they grasped the fact that scientific theories and innovations are often driven by practical needs, and that theoretical and technological advancements are often interdependent, reflecting the dynamic relationship between science, technology, and society. This recognition helps students understand that science is a human endeavor, influenced by the values, priorities, and circumstances of the time in which it is conducted.

A further observation (that we believe useful to report) was the following, provided by one of the high-school teachers in the written test given at the end of the lesson on the photoelectric effect:

> T. "Exploring the history of the photoelectric effect has shown me how scientific progress is deeply influenced by experimental discoveries and technological advancements. At the turn of the twentieth century, the development of electric lighting and early vacuum tubes required a deeper understanding of how materials interacted with light. Scientists studying these effects were motivated not only by curiosity but also by the demand for more efficient technologies. Hertz was the first to observe that ultraviolet light could enhance electrical sparks, but no one fully understood why. Hallwachs and Righi then demonstrated that light could cause certain metals to lose their charge, hinting at a deeper connection between light and electricity. Lenard's experiments provided crucial quantitative data through three experimental laws, but it was Einstein's interpretation that finally provided a clear explanation: Light is composed of quantized packets of energy (that we now call photons), and a photoelectron can only be ejected if a single quantum of light has enough energy to overcome the material's work function. This insight revolutionized our understanding of light and matter, laying the foundation for QM. What I find fascinating is how this discovery, made 120 years ago in attempts to understand fundamental physics, later had enormous technological implications, from the development of photocells to modern solar panels. It really highlights how scientific knowledge and technological progress evolve together, often in unexpected ways."

This last observation provides an important reflection on how scientific discoveries from the past have had a lasting impact on modern technologies, such as solar panels and photodiodes. A key aspect of science is that its practical applications, often perceived as modern and innovative, are deeply rooted in theories and discoveries formulated decades, or even centuries, ago. The connection between the photoelectric effect and solar energy, for example, highlights how understanding the behavior of light and the interaction between light and electrons, which initially seemed like purely theoretical matters, has evolved into a concrete technology that now powers a significant portion of our energy needs. By recognizing this link, participants come to understand that scientific discoveries do not exist in a temporal or theoretical vacuum, but are the result of an evolutionary process that connects past observations with present innovative solutions. The solar panel technology, based on the photoelectric effect, is a clear example of how a seemingly distant discovery, like the photoelectric effect explored by Hertz, Hallwachs, Lenard, and Einstein, turned out to be crucial for technological advancements in the field of renewable energy. In this way, historical scientific discoveries are directly connected to modern social and economic needs, such as sustainable energy and reducing environmental impact (thus including reflections on "ethical" issues or matters of current relevance that are of immense importance to society as a whole). Moreover, this type of reflection helps students understand how science and technology are interconnected and how theories can evolve into applications that transform society. Students recognize that modern technologies are not only the result of recent discoveries but also the outcome of a long process of knowledge building that has developed over time, making scientific innovations tangible and essential for the progress of our daily lives. In this way, they realize that many of the practical advancements in science have stemmed from curiosity-driven questions, thus highlighting the importance of fundamental research.

Based on the examples and observations just presented, we therefore believe that providing a presentation of the OQT in which the physics contents are commented on and enriched with historical perspectives (Kragh 1992) is valuable for fostering a deeper understanding of the NOS (Lovisetti and Giliberti 2025a). In this sense, taking into account all of the above-mentioned aspects, we believe that for a NOS-based approach, textbook-like simplifications are not appropriate, since they do not allow for a culturally logical presentation of the conceptual organization of QP based on an interpretative and not purely descriptive attitude.

7 Conclusions

In this study, we have shown how an introduction to QP that includes the OQT may encourage a deeper understanding of both QM and the NOS.

The approach we propose is not simply a chronological overview of scientific developments but rather a means to foster critical awareness and a more organized comprehension of quantum concepts. By analyzing the implemented teaching strategies and the results obtained, we have indicated that discussing the OQT helps students understand how QM evolved into its current form, with its distinct characteristics.

Additionally, our findings suggest that engaging with the OQT can lead to a richer understanding of the NOS. These insights imply that teaching QM through an approach that integrates the OQT, along with historical and epistemological elements, may contribute to a deeper and more nuanced appreciation of both the subject matter and the broader scientific process.

References

Abd-El-Khalick F et al (2017) A longitudinal analysis of the extent and manner of representations of nature of science in U.S. high school biology and physics textbooks. J Res Sci Teach 54(1):82–120. https://doi.org/10.1002/tea.21339

Akgun S, Kaya E (2020) How do university students perceive the nature of science?. Sci & Educ 29:299–330. https://doi.org/10.1007/s11191-020-00105-x

Cofré H et al (2019) A critical review of students' and teachers' understanding of nature of science. Sci & Educ 28:205–248. https://doi.org/10.1007/s11191-019-00051-3

Duit R et al (2012) The model of educational reconstruction—a framework for improving teaching and learning science. In: Jorde D, Dillon J (eds) Science education research and practice in Europe. Sense Publishers, Rotterdam, pp 13–37. https://doi.org/10.1007/978-94-6091-900-8_2

Erduran S et al (2021) Perceptions of nature of science emerging in group discussions: a comparative account of pre-service teachers from Turkey and England. Int J Sci Math Educ 19:1375–1396. https://doi.org/10.1007/s10763-020-10110-9

Faletič S et al (2024) In: Fazio C, Logman P (eds) Symposium on teaching and learning quantum physics. Physics Education Today. Springer, Cham, pp 55–72. https://doi.org/10.1007/978-3-031-48667-8_4

Fischler H, Lichtfeldt M (1992) Modern physics and students' conceptions. Sci Educ 14(2):181–190. https://doi.org/10.1080/0950069920140206

Fox MFJ, Zwickl BM, Lewandowski HJ (2020) Preparing for the quantum revolution: what is the role of higher education? Phys Rev Phys Educ Res 16:020131. https://doi.org/10.1103/PhysRevPhysEducRes.16.020131

Galili I (2022) Scientific knowledge as a culture. Springer, Cham, The Pleasure of Understanding. https://doi.org/10.1007/978-3-030-80201-1

Giliberti M, Lovisetti L (2024) Old quantum theory and early quantum mechanics. Springer, Cham, A Historical Perspective Commented for the Inquiring Reader. https://doi.org/10.1007/978-3-031-57934-9

Giliberti M, Lovisetti L (2025) Teaching quantum mechanics within a nature-of-science framework. Am J Phys 93(2):172–179. https://doi.org/10.1119/5.0210636

Giliberti M et al (2025) Cultural understanding of physics: instruments and methods. In: Ješková Z, Hanč J (eds) Advancing physics education. Innovations from early learning to higher education. Springer, Cham, pp 177–197. https://doi.org/10.1007/978-3-031-94207-5_11

GIREP thematic group "Cultural Understanding of Physics" https://www.girep.org/thematic-groups/cultural-understanding-of-physics-instruments-and-methods/

Hadzidaki P, Kalkanis G, Stavrou D (2000) Quantum mechanics: a systemic component of the modern physics paradigm. Phys Educ 35(6):386–392. https://doi.org/10.1088/0031-9120/35/6/302

Irzik G, Nola R (2014) New directions for nature of science research. In: Matthews M (ed) International handbook of research in history, philosophy and science teaching. Springer, Cham, pp 999–1021. https://doi.org/10.1007/978-94-007-7654-8_30

Kalkanis G, Hadzidaki P, Stavrou D (2003) An instructional model for a radical conceptual change towards quantum mechanics concepts. Sci Educ 87:257–280. https://doi.org/10.1002/sce.10033

Knorr-Cetina K (1999) Epistemic cultures: how the sciences make knowledge. Harvard University Press, Cambridge

Kragh H (1992) A sense of history: history of science and the teaching of introductory quantum theory. Sci Educ 1:349–363. https://doi.org/10.1007/BF00430962

Kragh H (2025) Quantum mechanics: a historical-linguistic perspective. Am J Phys 93(1):7–13. https://doi.org/10.1119/5.0202443

Krijtenburg-Lewerissa K et al (2017) Insights into teaching quantum mechanics in secondary and lower undergraduate education. Phys Rev Phys Educ Res 13:010109. https://doi.org/10.1103/PhysRevPhysEducRes.13.010109

Lederman NG et al (2002) Views of nature of science questionnaire: towards valid and meaningful assessment of learners' conceptions of the nature of science. J Res Sci Teach 39(6):497–521. https://doi.org/10.1002/tea.10034

Levrini et al (2014) Meeting the discipline-culture framework of physics knowledge: a teaching experience in Italian secondary school. Sci & Educ 23:1701–1731. https://doi.org/10.1007/s11191-014-9692-z

Lovisetti L (2024) Can old-quantum-theoretical description of physics be rendered coherent? A pilot experimentation for high school. Il Nuovo Cimento C 47(5):277. https://doi.org/10.1393/ncc/i2024-24277-6

Lovisetti L, Organtini G, Giliberti M (2023) Inducing the construction of formal axioms of quantum mechanics and fostering their comprehension by high school students: the effectiveness of a conceptual approach. Il Nuovo Cimento C 46(6):200. https://doi.org/10.1393/ncc/i2023-23200-1

Lovisetti L, Giliberti M (2025a) Can old-quantum-theoretical description of physical reality be considered worth teaching? In: Testa I, Michelini M, Esposito S (eds) Connecting physics education research and practice. Springer, Cham, pp 155–172. https://doi.org/10.1007/978-3-031-86609-8_10

Lovisetti L, Giliberti M (2025b) Thomson's alleged "plum pudding": a model that could never stand (yet persists in the minds and imagination of many physicists). Il Nuovo Cimento C 48(6):290. https://doi.org/10.1393/ncc/i2025-25290-y

Michelini M, Stefanel A (2021) In: Jarosievitz B, Sükösd C (eds) Approaches on T/L quantum physics from PER literature. Teaching-Learning Contemporary Physics. Springer, Cham, pp 3–17. https://doi.org/10.1007/978-3-030-78720-2_1

Nita L et al (2023) The challenge and opportunities of quantum literacy for future education and transdisciplinary problem-solving. Res Sci Technol Educ 41(2):564–580. https://doi.org/10.1080/02635143.2021.1920905

Stadermann HKE, Goedhart MJ (2021) Why and how teachers use nature of science in teaching quantum physics: research on the use of an ecological teaching intervention in upper secondary schools. Phys Rev Phys Educ Res 17:020132. https://doi.org/10.1103/PhysRevPhysEducRes.17.020132

Stadermann HKE, Goedhart MJ (2020) Secondary school students' views of nature of science in quantum physics. Int J Sci Educ 42(6):997–1016. https://doi.org/10.1080/09500693.2020.1745926

Stadermann HKE, van den Berg E, Goedhart MJ (2019) Analysis of secondary school quantum physics curricula of 15 different countries: different perspectives on a challenging topic. Phys Rev Phys Educ Res 15:010130. https://doi.org/10.1103/PhysRevPhysEducRes.15.010130

Stadermann HKE, van den Berg E, Goedhart MJ (2021) How high schools teach quantum physics—a cross-national analysis of curricula in secondary education. J Phys Conf Ser 1929:012045. https://doi.org/10.1088/1742-6596/1929/1/012045

ter Haar D (1966) The old quantum theory. Pergamon Press Inc., New York

Vermaas PE (2017) The societal impact of the emerging quantum technologies: a renewed urgency to make quantum theory understandable. Ethics Inf Technol 19:241–246. https://doi.org/10.1007/s10676-017-9429-1

Wittmann MC, Steinberg RN, Redish EF (2002) Investigating student understanding of quantum physics: spontaneous models of conductivity. Am J Phys 70(3):218–226. https://doi.org/10.1119/1.1447542

Weissman EY, Merzel A, Katz A, Galili I (2019) Teaching quantum mechanics in high-school discipline-culture approach. J Phys Conf Ser 1287:012003. https://doi.org/10.1088/1742-6596/1287/1/012003

Reviewing 55 Years of Literature on Physics Education Using Natural Language Processing

Martina Caramaschi, Tor Ole B. Odden, and Olivia Levrini

Abstract Literature reviews play a crucial role in informing the Physics Education Research (PER) community about the evolution of teaching practices and educational interests over time. Large-scale data analysis techniques offer novel possibilities for conducting comprehensive literature reviews. In this work, we applied an unsupervised Natural Language Processing (NLP) method, called Latent Dirichlet Allocation (LDA), to a large dataset of articles from *The Physics Teacher* journal. Our goal was to identify and track thematic trends spanning nearly six decades (1963–2020). We found that the journal's content has balanced discipline-specific topics, pedagogy and laboratory methods and learning-theory perspectives, while placing increasing emphasis on students learning over time. We discuss the strengths of this approach and its limitations, including the need for contextual expert knowledge to interpret topics. Finally, we briefly present a follow-up comparative analysis to examine the influence of differing socio-educational contexts on physics education.

1 Introduction

Over the past several decades, there has been a growing interest in understanding the evolution of teaching practices and research trends in physics education. Physics Education Research (PER) has been concerned with studying how educational approaches in Physics have shifted in response to pedagogical paradigms, technological advancements and societal needs. By systematically analysing the literature in this field, researchers can gain insight into its historical development and identify patterns that inform its future trajectory. A notable example of this growing attention is the literature review by Girwidz et al. (2019), which analysed multimedia applications in physics teaching across 34 teacher-oriented journals in multiple European

M. Caramaschi (✉) · O. Levrini
Department of Physics and Astronomy "A. Righi", University of Bologna, Bologna, Italy
e-mail: martina.caramaschi@studio.unibo.it

T. O. B. Odden
Center for Computing in Science Education, University of Oslo, Oslo, Norway

D. Sokołowska and E. McLoughlin (eds.), *Embracing Changes Together in Physics Education*, Challenges in Physics Education,
https://doi.org/10.1007/978-3-032-27385-7_13

languages. This study highlights the value of reviewing broad corpora of educational publications to detect recurring themes, preferred strategies and transnational trends. However, traditional review methods face significant limitations, including the inability to efficiently analyse large volumes of literature and the challenges of identifying and quantifying long-term trends. Recent advancements in Machine Learning (ML) and Natural Language Processing (NLP) have introduced powerful tools that can support qualitative analyses by addressing these limitations. ML and NLP techniques enable researchers to process extensive datasets, identify thematic patterns and visualise trends over time.

The use of ML and NLP methods for textual data analysis in the context of science education has grown in recent years. For example, these methods have been applied by Wulff and colleagues to analyse preservice teachers' written reflections by filtering specific elements of the teachers' reasoning and clustering into themes. In this case, the quantitative and data-driven techniques supported and enabled the identification of significant differences in different teachers' texts, supporting detailed formative assessment (Wulff et al. 2023).

More specifically, in the context of literature analyses in physics education, ML and NLP have already demonstrated significant utility in thematic analyses of conference proceedings and journal articles. For instance, a specific unsupervised machine learning technique called Latent Dirichlet Allocation (LDA) was applied to the Physics Education Research Conference (PERC) Proceedings from 2001 to 2018 to uncover ten distinct themes within the field (Odden et al. 2020). This analysis revealed the evolution of PER topics over time, from early qualitative, theory-building studies of student understanding to a focus on problem solving in the late 2000s, and, more recently, to sociocultural perspectives emphasising communities of practice, student identities and institutional change. These automated methods are valuable for enabling quantitative, replicable insights into the trajectory of PER, which would be challenging to achieve through manual reviews alone.

Similarly, LDA has been used to analyse over a century of articles published in the journal *Science Education*, categorising research into 21 distinct topics within three thematic groups: science content, teaching-focused topics and student-focused topics (Odden et al. 2021). In this case, the analysis not only traced the rise and fall of these topics over time but also shed light on the field's historical development, the influence of national policies and intellectual cross-pollination across topics.

By quantitatively identifying long-term trends and uncovering novel patterns or clusters within a massive corpus of educational literature, these approaches highlight the transformative potential of ML and NLP for mapping research landscapes that might otherwise remain inaccessible.

1.1 Aims and Research Question

Our work aimed to analyse a wide corpus of more than 7000 papers published on *The Physics Teacher*, a peer-reviewed academic journal that has served as a pivotal platform for physics education since its inception in 1963. The journal is recognised as a significant reference point in the field, particularly within the North American context, and publishes articles on innovative teaching strategies, laboratory techniques, demonstrations, curriculum developments and insights into the teaching and learning of physics at the high school and introductory college levels.

We decided to analyse more than 55 years of published articles, from its inception in 1963–2020. The analysis of this long-standing publication history provides a unique opportunity to explore what topics have been of interest during its history and their evolution. Moreover, by incorporating NLP and ML methods into the qualitative analyses conducted by PER experts, we aimed to uncover and explore the key topics that have defined the journal's contributions to the field, quantify their trends and visualise their progression.

Thus, our research question was: 'What topics are present in the history of *The Physics Teacher* journal papers published from 1963 to 2020, and to what extent are they prevalent each year?'

2 Methodology

The methodology consisted of developing a thematic analysis of the articles published in *The Physics Teacher*. The approach that we adopted in this qualitative analysis integrates expert knowledge with quantitative methods, leveraging NLP techniques and ML methods. In particular, we used the Latent Dirichlet Allocation (LDA) algorithm to infer topics that characterise the content of the articles. The methodological process involved several phases: a pre-processing stage to transform the articles from PDF format into a corpus of words, the application of LDA to discover topics as distributions over words and documents and an interpretation phase where topic results were contextualised considering the distribution of topics across words and documents, contextual information and relevant literature on physics teaching.

These analyses were implemented using Python-based Jupyter Notebooks, ensuring transparency and reproducibility. In the following, the methodology is described in detail.

2.1 Background

Several studies on the use of ML and NLP methods for research purposes in Physics Education have been carried out in the last decade, especially for the analysis of textual data. Odden et al. (2024) provided an overview of works in which supervised and unsupervised NLP methods were applied both for inductive and deductive analyses in PER.

In this field, the role of the researcher in dealing with textual data analysis is crucial. As Caramaschi et al. (2025) show, specific conditions need to be considered when dealing with textual data for research in the context of science education. These are the specificity of the language, whose meaning is well known to a researcher in the field. Secondly, the awareness of the context in which the text was collected or to which it refers, which allows researchers to know which parts of the text are more relevant than others in relation to a given research question.

For this reason, our work explicitly includes both the model and the researcher as fundamental and active roles. The choice of the researcher's approach towards the use of ML and NLP methods in this study is guided by Grimmer and Stewart's principle (Grimmer and Stewart 2013). This principle states that quantitative methods should augment human expertise, rather than replace it.

In this work, both the physics education expert and the computational data-driven model are recognised as peer actors that can offer necessary and unique contribution: on one side, the expert welcomes the results provided by the model, which they themselves helped to set up based on meaningful choices, considering elements such as the research question, context and data. On the other, the model offers the possibility to explore a vast dataset and identify patterns based on statistical rules, providing an unprecedented quantitative contribution to the researcher.

2.2 *Introduction to LDA*

Latent Dirichlet Allocation (LDA) is a widely used NLP technique designed to uncover hidden thematic structures within large collections of texts (Blei et al. 2003; Odden et al. 2021). It belongs to the family of *topic modelling*, which is often an unsupervised method. This means that LDA models are not trained on pre-labelled data; instead, they identify patterns in the text based solely on statistical relationships.

LDA is based on the Bag-of-Words (BoW) model, which represents each document as a list of unique words and their frequencies, ignoring the order of the words. Each document is transformed into a high-dimensional vector, where every dimension corresponds to a unique word across the entire corpus. The value in each dimension indicates how often that word appears in the document. For example, a document discussing physics education might be represented with high frequencies for words like 'students', 'teaching' and 'learning'.

LDA assumes that each document in the corpus is a mixture of latent topics, and that each topic is represented as a statistical distribution of words (Blei et al. 2003). The fundamental idea is that semantically related words tend to appear together. For instance, a topic about laboratory experiments might include high-probability words like 'equipment', 'measurements' and 'data', while a topic on pedagogy might include words like 'learning', 'students' and 'instruction'.

By identifying clusters of frequently co-occurring words, LDA can infer these topics without prior knowledge of their existence. Thus, this method provides two main outputs:

1. A set of topics, where each topic is defined as a list of words more likely to appear together than others. Every word may belong to multiple topics, but the probability of a word being strongly associated with unrelated topics is minimal, helping to separate coherent themes.
2. A representation of each document as a mixture of these topics, allowing researchers to determine the prominence of each topic in a given document or across the entire corpus.

These topics are designed to be distinct and ideally orthogonal to each other in semantic space, meaning that the model aims to capture non-overlapping themes. Each topic is characterised by its specific distribution of words, while each article is represented as a combination of these topics with varying weights, reflecting the mixture of themes present in its text. This means that some articles may be strongly associated with a single topic, showing, for example, a 90% weight for that topic, which indicates that the article predominantly focuses on one theme. Conversely, other articles may exhibit a more balanced mixture of topics, such as weights of 30%, 40%, and 20%, reflecting a more diverse thematic content within the same text.

2.3 Analysis Approach

As for the specific phases that characterised the analysis of data and its interpretation, we will present them in a linear sequence, although some phases have been carried out iteratively together with others. Some phases have been refined and completed in the light of choices or results found in subsequent phases. For example, the choice of hyperparameters relating to vocabulary, which determines which portion of words is considered or not, is made after training some initial models. This is because it is only in the light of preliminary results that the researcher is able to decide whether the parametric setting should be changed or need for further adjustments and subsequent iterations.

Dataset and Data Collection. We selected *The Physics Teacher* (TPT) as the dataset for our analysis. TPT is a peer-reviewed journal dedicated to teaching methods, pedagogical strategies and physics content explanations, primarily targeting high school and introductory college-level physics educators. Its extensive focus on physics education makes it a valuable resource for identifying thematic trends in the field. The dataset comprises all articles published in TPT from its inception in 1963 through 2020, comprising over 15,000 PDF documents. This data corpus was then filtered to remove advertisements and journal business like announcements and editorials, resulting in a corpus of 7287 PDF documents. The data collection process was completed in early 2020, meaning that the dataset includes partial results for that year.

Preprocessing. This part included several steps:

- PDF-to-Text Conversion: All PDFs were converted into text documents using Adobe Acrobat. The extracted text was organised into a structured table, including metadata such as article titles, publication years, authors and DOIs.
- Creation of the Words and Bigrams Corpus: Each document was transformed into a corpus containing individual words and meaningful bigrams (e.g. 'physics education', 'teaching strategies') to preserve important multi-word expressions relevant to the analysis.
- Text Cleaning: Non-essential elements, such as metadata, headers, footers and non-text characters, were removed to focus solely on the article content.
- Stopword Removal: Common English stopwords (e.g. 'the', 'and' 'or') were removed, as they do not provide meaningful semantic information for topic modelling.
- Tokenization and Lemmatization: Each article was split into tokens (individual words), and each token was lemmatized to its root form. For example, variations such as 'teaching', 'teacher' and 'taught' were all standardised to 'teach'.

Following these steps, the text of each article was reduced to a 'bag-of-words' representation, which has the advantage of reducing the complexity of the documents. However, it is assumed that this representation is sufficient to capture the semantic meaning of the documents, allowing the LDA model to identify meaningful patterns in the input corpus of words.

LDA Model Development. For model training, we varied the number of topics and tuned the hyperparameters to ensure that the topics generated were both coherent and interpretable. One important aspect of hyperparameter tuning involved filtering out rare words. We removed words that appeared in fewer than 15 documents, as these were deemed to contribute minimally to the overall meaning of the documents. This decision was based on the assumption that very rare words are unlikely to represent the central themes of the articles. Additionally, this step helped eliminate potential errors such as typos, which could occur in very infrequent words that do not contribute substantially to the content. Still, more widespread words across the documents, such as 'student' or 'physics', were excluded because they are too general to highlight specific themes or provide meaningful differentiation between topics. These common words risk diluting the distinctiveness of the topics, as they do not emphasise anything specific to the areas of interest in physics education. To address this, we set the 'no above' hyperparameter to 0.40, meaning that words appearing in more than 40% of the documents were excluded. This threshold ensured that only words with a meaningful contribution to the topics were retained, avoiding overly generic terms that could obscure the identification of distinct themes.

To determine the optimal number of topics for the LDA model, we performed an *elbow analysis*. This approach involved calculating coherence scores for 10 randomly initialised models at each topic count, ranging from 6 to 24 topics. The coherence score, a measure of topic quality and interpretability, helped identify a range where the model's performance stabilised. Based on this analysis, we determined that 15 topics provided the best balance between coherence and granularity.

We also complemented this quantitative evaluation with expert judgement. By inspecting the models with varying numbers of topics, we found that the 15-topic model generated meaningful and interpretable themes that were most relevant to Physics education. This combination of statistical methods and domain-specific expertise ensured that the chosen model aligned with the goals of our analysis.

After fixing the number of topics and tuning the hyperparameters, we proceeded to select the 'best' model. The use of the term 'best' is necessary because LDA is a probabilistic model that is initialised with a random seed, meaning that each model run could yield slightly different results. To identify the most consistent and reliable model, we performed multiple runs of the LDA model with different random seeds.

Specifically, we trained 100 models with identical parameters, varying only the random seed. To determine the best model, we computed the centroids of the topics for each of the 100 runs. The centroid represents the average distribution of words across topics. We then calculated the average distance between the topics of each model and their respective centroids. The models with the smallest average distance to the centroids were selected as the 'best' models. This approach ensures that the chosen models (which would then be evaluated qualitatively to decide which could become the best candidate) are the most stable and consistent, minimising the variability introduced by the random initialisation process.

Topic Interpretation and Validation. Once the LDA model generated a candidate set of topics, we assigned preliminary labels by examining each topic's top-weighted words. However, since the algorithm outputs only word distributions without explicit meanings, it is essential for researchers to integrate contextual knowledge and examine concrete examples to interpret these results accurately.

To this end, we reviewed additional information from the dataset, including the titles, abstracts, publication years and authors of the top 10 most representative articles associated with each topic. For instance, articles with a very high prevalence of Topic 1 in their text (e.g. over 85%) were read and analysed to interpret the detailed meaning of that topic. This deeper contextual analysis helped us understand how the identified words coalesce into coherent and meaningful themes.

This iterative process merged researcher expertise with the model's output. The combination of expert knowledge in physics education and the insights from the LDA model was used to achieve two critical goals:

1. Identifying the Best Model: Based on our evaluation, the model initialised with seed number equal to 304 was selected as the most reliable and interpretable.
2. Interpreting Topics: Expert knowledge of physics education informed the final categorisation of each topic, ensuring that the labels accurately reflected the themes relevant to the field.

This collaborative approach ensured that the final topics were both data-driven and aligned with the nuanced understanding of the context and informed by the research question.

3 Results

The LDA analysis identified 15 topics, as previously mentioned. Two of them were removed because they were related to journal business (e.g. awards, announcements) and did not provide insights relevant to the research question. Table 1 presents the remaining 13 topics, showing the labels we assigned, their most representative words (in decreasing order of weight) and a summary of the main contents identified in the top 10 most representative articles for each topic, which were instrumental in characterising them.

The analysis enabled us to quantify the average prevalence of each topic in the journal year-by-year (Fig. 1). For instance, Topic 1 about particle physics shows a consistent prevalence when the journal was founded, which decreased until 1975. In the following years, topics 10 and 11 about pedagogy and laboratorial activities in Physics grew their prevalence, having their maximum prevalence between 1980 and 2000, before stabilising (Fig. 1).

The year-by-year prevalence analysis revealed distinct temporal patterns in the topics. By observing the 13 topics, their meaning and trends, and considering the reference literature (Meltzer and Otero 2015), it seems reasonable to group them in three categories:

1. *Content-Focused Topics*: These include classical physics themes such as particle physics, electromagnetism and optics. Their prevalence has been steady or slightly decreasing over time.
2. *Pedagogy, Laboratory and Data Analysis-Focused Topics*: Emerging prominently around 1980, these topics, such as pedagogy-focused material and physics laboratories, gained prevalence until stabilising after 2000.
3. *Learning-Theory-Related Topics*: These began trending in the late 1980s and have become the most prevalent topic in recent years. They emphasise the interest towards active learning, student autonomy and self-assessment, highlighting the growing focus on students' learning processes.

The grouping of topics into three macro-categories is further evident in Fig. 2, which shows the prevalence of each topic per year in a normalised area graph. In this graph, the sum of all topics prevalence is equal to 1 for each year. Based on this figure, we hypothesise that the journal's focus was primarily oriented towards content-focused physics topics up until 1980. Between the 1980s and 2000, there was a significant rise in the prevalence of topics dedicated to pedagogy. Finally, from 1990 onward, there was another shift towards topic 13, which is about the use of research-based materials or articles that draw from learning theories. This topic grew significantly, becoming the most prevalent topic.

Beyond this general overview, some topics merit closer attention for the insights they offer into the evolution of physics education. Topic 10, labelled *Pedagogical Material*, is particularly notable. It reflects a broad pedagogical effort to make physics more comprehensible and relatable to learners. Interestingly, mechanics appears to be the primary ground on which this pedagogical innovation takes place, perhaps due to

Table 1 Assigned topic names, top-weighted words and summaries of representative article contents

#	Name	Most representative words of the topic	Contents of most representative articles
1	Particle physics	Energy particle, electron, atom, state	Particle physics concepts, weak interactions, high-energy phenomena, particle decays and strategies for teaching these advanced topics
2	Electromagnetism, current and circuits, physics of solid state: Teaching and experiments	Circuit, current voltage, resistance battery	Use of AC current and wave generators, specific teaching apparatus (e.g. FET electrometer, transistors, oscilloscope) and experiments on semiconductors and basic circuit laws (ohm's, Kirchhoff's)
3	Thermology and impacts on earth system	Temperature, energy, heat, water, radiation	Energy and environmental physics: Refrigeration, radiation, temperature, nuclear and solar energy, climate-related topics (greenhouse effect, carbon dioxide) and microwave radiation
4	Astronomy and planets: Study and activities	Earth, star, sun, planet, astronomy	Celestial geometry, sky effects, simulations of moon phases, orbital eccentricity, eclipses, astronomical inquiry, Kepler's laws, milky way, asteroids and orbital dynamics
5	Sound and light waves and optics: Teaching and experiments	Frequency, wave, sound, signal, amplitude	Acoustic beat analysis, combination tones, music Fourier analysis, resonance, sound properties, acoustic spectra, sound measurements and demonstrations using ripple tanks, smartphones or digital tools
6	Colour and light: Studying and experiments	Light, colour, laser, wavelength, spectrum	Fluorescence, spectra, colours, prisms and light, polarisation, Rayleigh scattering, displaying spectra, apps for colour study (addition and subtraction), rainbow
7	Electromagnetic, magnetism, induction and magnetic field	Magnet, charge field, magnetic field, coil	Biot-savart law, magnetic forces, magnets, solenoids, teaching faraday's and Lenz laws, experiments for studying earth's magnetic field, symmetry in electromagnetism, GEOMAG
8	Thermodynamics, temperature, entropy	Water, air, pressure, volume, liquid	Marotte's bottle, entropy, barometer, siphons, Bernoulli's principle, Torricelli's law, experiments for teaching pressure, viscous resistance
9	Optics: Teaching and experiments	Image, lens, object, mirror, camera	Building different kinds of lenses, mirrors, studying images formed by lenses, reflection, pinhole glasses and eye, optic camera, holography, optical illusions

(continued)

Table 1 (continued)

#	Name	Most representative words of the topic	Contents of most representative articles
10	Pedagogical material	Mass, eq, velocity, equation, motion	Intuitive or simple approaches to physics laws such as elastic collisions, energy of a body, momentum, rotation, oscillator motion, motion with friction, harmonic motion, spring-mass motion and Newton's laws
11	Mechanics and momentum: Activities and demonstrations	Ball, motion, demonstration, weight, speed	Gyroscope, independence of motions, rotational motion demonstrations, activities on angular momentum, measures of centripetal forces, spinning and tippy top
12	Data, tools and software for physics laboratory	Data, measurement, graph, programme, calculate	Experiments and data relationships, calculators and graphing calculators, software for laboratory physics, video tools for motion analysis, software for plotting, animation and visualisations, data manipulation, regressions, video clips for monitoring motion and computer-based plotting techniques
13	Learning theories for physics	Learn, question, group activity, teacher	Problem solving, peer learning, physics learning, effects of student wellness on learning, best practices for concept inventories/for surveys on attitudes and beliefs of students in physics, introductory physics courses material, interactive activities, active learning, students' self-diagnoses of mistakes, self-evaluation, flipped classrooms, students' autonomy for learning improvement, real-time feedback tool for teachers, emphasising skills than results

its foundational role in physics and its intuitive connection to everyday phenomena. The increasing prominence of this topic, particularly from the 1980s to early 2000s, may also reflect a broader political and educational agenda: During this period, there was a strong institutional push to improve scientific literacy and to present physics in a way that was not only rigorous, but also engaging and accessible (Meltzer and Otero 2015). These efforts aligned with the need to cultivate scientific vocations in response to societal and economic pressures in the late Cold War era.

Topic 11 (*Mechanics and momentum: activities and demonstrations*) and Topic 12 (*Data, tools and software for Physics laboratory*) further highlight the evolution of educational practices in physics. Topic 11 reveals the importance of hands-on learning and the growing reliance on physical demonstrations and experiments in classrooms. It reached its peak between the mid-1980s and early 2000s, a period that corresponds to an increased emphasis on active learning and inquiry-based laboratory

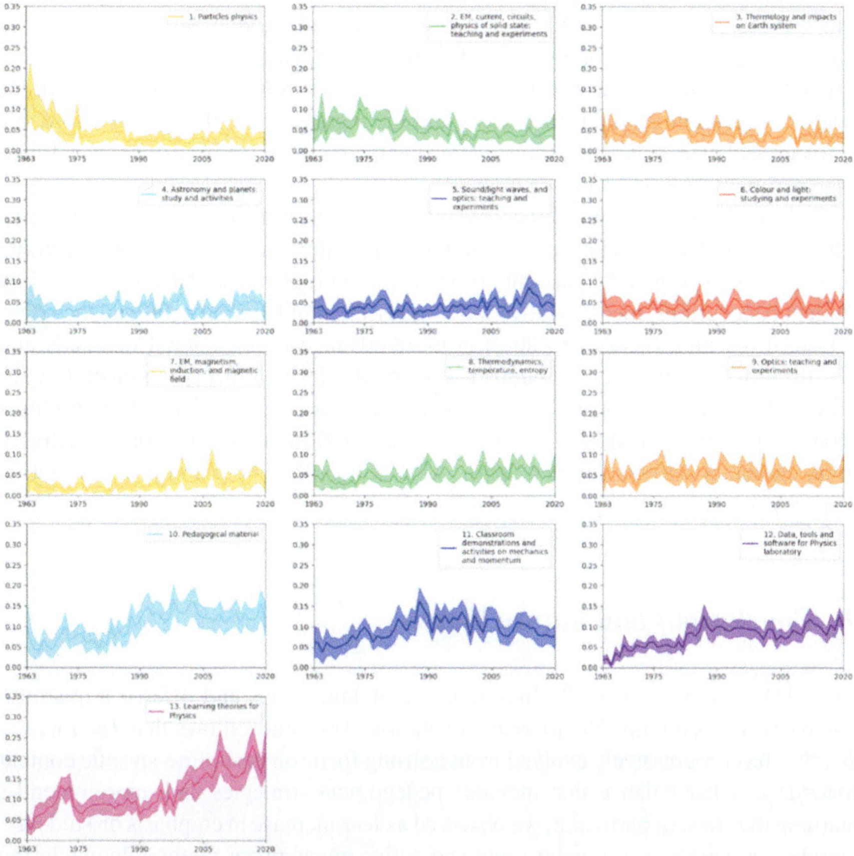

Fig. 1 Average prevalence in the articles of each topic year by year

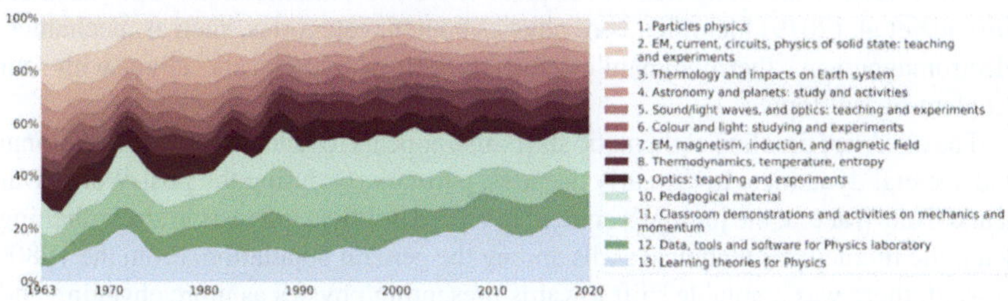

Fig. 2 Stacked area plot of all topics, grouped according to their thematic clusters: physics content topics (red, top), pedagogical/laboratorial topics (green, middle), learning theories/PER topic (blue, bottom)

experiences. In contrast, Topic 12 shows a broader and more gradual diffusion over time. It focuses on the integration of digital tools, video analysis and computer simulations, with a clear rise from the early 2000s onwards. This pattern aligns well with technological advancements and the increasing availability of affordable software and data acquisition tools, which enabled teachers to implement new forms of experimentation and analysis in class.

Finally, Topic 13 (*Learning Theories for Physics*) captures one of the most significant shifts in the journal's history: the rise of research-based teaching and learning strategies. This topic takes insights from research on physical education, including student conceptions, metacognitive tools, assessment of learning gains and the application of learning theories to develop more advanced instructional materials and practices. Its steady and continuous rise since the 1990s reflects the institutionalisation of Physics Education Research (PER) as a recognised field. It highlights a broader epistemological turn in physics education, from a focus on content delivery to an emphasis on how students learn, what beliefs they hold and how instruction can be adapted to support meaningful learning outcomes.

4 Conclusions and Future Works

The LDA model successfully highlighted thematic trends and offered a quantitative basis for examining the journal's evolution. This study shows that *The Physics Teacher* has progressively evolved from a strong focus on discipline-specific content towards a richer balance that includes pedagogical strategies and, more recently, learning theories. In particular, we observed a clear increase in emphasis on students' learning processes, assessment tools and active engagement methodologies in the most recent decades.

These results are consistent with prior literature classifications. For instance, Girwidz et al. (2019) identified core physics content categories, such as mechanics, electromagnetism, thermodynamics and optics, which closely align with our 'content-based' topics.

The observed timeline of thematic shifts also appears to reflect broader educational and societal dynamics. In the early decades, particularly during the post-World War II and Cold War era, the journal's content mirrored political interests in strengthening scientific literacy and technical skills among the general population. From the 1980s onward, there was a notable shift towards presenting physics as more engaging and accessible, likely driven by the need to boost enrolment in science education and respond to changing educational paradigms. More recently, the rise of research-based strategies, concept inventories and metacognitive tools reflects a focus on how students learn, not just what they learn, an evolution aligned with advances in cognitive science and educational research.

The application of LDA proved valuable for large-scale literature review, revealing thematic trends that align with shifts in educational priorities within the physics education community. However, interpreting the context and significance of certain topics required expert input, emphasising the complementary role of domain knowledge in such analyses. Nonetheless, this work confirms the potential of NLP tools in PER for inductive, scalable and reproducible literature analysis.

A continuation of this work has involved extending the methodology to another journal (*Physics Education*), performing comparative investigations (Caramaschi and Odden 2025). The analysis has enabled the exploration of cross-cultural similarities and differences in educational approaches, highlighting how the evolution of physics education themes may reflect and be shaped by distinct educational priorities and political contexts.

References

Blei DM, Ng AY, Jordan MI (2003) Latent Dirichlet allocation. J Mach Learn Res 3:993–1022. http://www.jmlr.org/papers/volume3/blei03a/blei03a.pdf

Caramaschi M, Odden TOB (2025) Analyzing the history of physics education in the USA and Europe through natural language processing. Phys Rev Phys Educ Res 21:020153. https://doi.org/10.1103/wfvw-hkyy

Caramaschi M, Zanellati A, Levrini O (2025) Comparing textual data analysis methods in science education research. In: Cakmakci G, Fatih Tasar M (eds) Connecting science education with cultural heritage. https//doi.org/https://doi.org/10.1007/978-3-031-90944-3

Girwidz R, Thoms LJ, Pol H, López V, Michelini M, Stefanel A, Greczyło T, Müller A, Gregorcic B, Hömöstrei M (2019) Physics teaching and learning with multimedia applications: a review of teacher-oriented literature in 34 local language journals from 2006 to 2015. Int J Sci Educ 41(9):1181–1206. https://doi.org/10.1080/09500693.2019.1597313

Grimmer J, Stewart BM (2013) Text as data: the promise and pitfalls of automatic content analysis methods for political texts. Polit Anal 21(3):267–297. https//doi.org/https://doi.org/10.1093/pan/mps028

Meltzer DE, Otero VK (2015) A brief history of physics education in the United States. Am J Phys 83:447–458. https//doi.org/https://doi.org/10.1119/1.4902397

Odden TOB, Marin A, Caballero MD (2020) Thematic analysis of 18 years of physics education research conference proceedings using natural language processing. Phys Rev Phys Educ Res 16(010142):1–25 https//doi.org/https://doi.org/10.1103/PhysRevPhysEducRes.16.010142

Odden TOB, Marin A, Rudolph JL (2021) How has science education changed over the last 100 years? An analysis using natural language processing. Sci Ed 105:653–680. https//doi.org/https://doi.org/10.1002/sce.21623

Odden TOB, Tyseng H, Mjaaland JT, Kreutzer MF, Malthe-Sørenssen A (2024) Using text embeddings for deductive qualitative research at scale in physics education. Phys Rev Phys Educ Res 20:02015. https//doi.org/https://doi.org/10.1103/PhysRevPhysEducRes.20.020151

Wulff P, Westphal A, Mientus L, Nowak A, Borowski A (2023) Enhancing writing analytics in science education research with machine learning and natural language processing—formative assessment of science and non-science preservice teachers' written reflections. Front Educ 7:1061461. https//doi.org/https://doi.org/10.3389/feduc.2022.1061461

Addressing Climate Change Through Physics Education

From Alternative Conceptions to Meaningful Understanding through Physics-Based Climate Change Education

Thomas Schubatzky and Sarah Wildbichler

Abstract Climate change represents a critical global issue requiring comprehensive educational responses that foster scientific understanding and informed action. Yet, students frequently hold alternative conceptions that hinder their comprehension of key climate phenomena. This chapter demonstrates how physics-based climate change education can address such challenges by grounding instruction in core physical principles. Drawing on a systematic literature review and two empirical studies, we first synthesize prevalent student conceptions regarding the greenhouse effect, the carbon cycle, and other foundational topics. We then present the development and validation of a concept inventory (CCCI-422) used with over 700 students and preservice teachers in Austria and Germany to assess conceptual understanding. Finally, we describe and evaluate two instructional interventions: (1) a set of research-informed tutorials for upper secondary students developed in the Erasmus+ project ENGAGING, and (2) a design-based learning environment on the greenhouse effect for lower secondary students. Both interventions explicitly address common misconceptions and foster substantial learning gains. The findings highlight the potential of physics education to support meaningful conceptual development and lay the groundwork for informed engagement with climate change.

Keywords Climate change education · Conceptions · Conceptual understanding

1 Introduction

Climate change poses one of the most significant global challenges of our time. Yet, despite widespread recognition of its urgency, numerous alternative conceptions (Wildbichler et al. 2025a) and fragmented conceptions (Schubatzky et al.

T. Schubatzky (✉) · S. Wildbichler
Department of Subject-Specific Education and Department of Experimental Physics, University of Innsbruck, Innsbruck, Austria
e-mail: thomas.schubatzky@uibk.ac.at

D. Sokołowska and E. McLoughlin (eds.), *Embracing Changes Together in Physics Education*, Challenges in Physics Education,
https://doi.org/10.1007/978-3-032-27385-7_14

2024a) persist among students. Addressing these alternative conceptions effectively is critical for fostering meaningful understanding and empowering students to engage confidently and competently in climate-related decision-making and actions. Physics education, with its emphasis on fundamental scientific principles, quantitative reasoning, and systems thinking, offers valuable tools and frameworks to deepen students' understanding of climate change phenomena. This chapter explores how physics-based climate change education can bridge students' alternative conceptions and scientific understanding, providing learners with a solid foundation for both scientific literacy and informed climate action.

2 Climate Change and Physics Education

Climate Change Education (CCE) refers to educational efforts aimed at equipping learners with the knowledge, skills, attitudes, and agency to understand and respond to the complex challenge of climate change (Chang 2022). It encompasses more than just conveying scientific facts; rather, it involves fostering a multidimensional literacy that includes understanding the physical mechanisms of the climate system, socioeconomic and ethical dimensions of climate action, and the capacity to critically evaluate information, make informed decisions, and participate in mitigation and adaptation efforts (UNESCO 2017; Monroe et al. 2019; White et al. 2024).

From a science education perspective, CCE must strike a balance between conceptual understanding and action-oriented learning. Conceptual knowledge about key climate processes—such as the greenhouse effect, the carbon cycle, and climate system feedback (Schubatzky et al. 2024b)—serves as a foundation for informed reasoning and decision-making. However, this knowledge must be embedded in meaningful contexts and put to use in the evaluation of, e.g., climate change mitigation measures in private and public spheres (Kranz et al. 2022). Effective CCE also acknowledges the uncertainty and complexity inherent in climate science and aims to support students to tackle these challenges, including understanding the role of scientific modeling, evidence, and uncertainty.

Moreover, CCE has a strong citizenship dimension. Oftentimes, it is framed as a form of Education for Sustainable Development (ESD) (UNESCO 2017), with goals that go beyond cognitive achievement and include fostering agency, hope, and engagement (Chang 2022; White et al. 2024). Therefore, CCE is inherently interdisciplinary, bridging natural sciences, social sciences, and humanities. In this contribution, we adopt a physics education perspective on CCE, focusing on how students come to understand the scientific underpinnings of climate change as a crucial, though not sufficient, component of climate literacy.

Physics plays a fundamental role in enabling a scientific understanding of climate change. Many of the core processes underlying climate change—such as radiative energy transfer, the interaction of electromagnetic radiation with matter, thermodynamics, and feedback mechanisms—are rooted in physics. A physics-based approach to climate change education therefore allows learners to understand climate change not merely as a social or political issue, but as a phenomenon grounded in well-established scientific principles.

Within CCE, physics contributes especially to explaining the greenhouse effect, the role of trace gases in Earth's' energy balance, and the mechanisms that govern Earth's climate system. For instance, understanding how greenhouse gases absorb and re-emit infrared radiation, why their relatively low concentration can still lead to significant warming, or how albedo and cloud feedback influence climate sensitivity, all require physical models and quantitative reasoning (Kranz et al. 2022; Tasquier et al. 2016; Schubatzky et al. 2024b).

Furthermore, physics education can support critical engagement with climate models and data, helping students understand how physical laws underpin simulations and projections (Tasquier et al. 2016; Tasquier and Pongiglione 2017). This involves not only conceptual knowledge but also overarching competencies, such as interpreting graphs, evaluating model assumptions, and understanding the limitations of scientific predictions.

Despite its importance, physics is often perceived as abstract or detached from everyday concerns. Integrating climate-related contexts into physics instruction thus offers a powerful opportunity to increase relevance and interest, while supporting meaningful learning. In this sense, climate change can serve both as an application area for physics concepts and as a context for promoting scientific literacy as well as climate agency (White et al. 2024).

This chapter is therefore guided by the overarching question:

(1) *How can physics-based climate change education support students in developing a scientifically grounded understanding of key climate concepts?*

To address this, we explore three sub-questions:

(1a) What alternative conceptions and learning difficulties do students hold regarding the physical basis of climate change?

(1b) How can these conceptions be systematically assessed using a validated instrument?

(1c) How can instructional interventions be designed and evaluated to effectively address these conceptions in lower and upper secondary education?

To answer these questions, we combine insights from a systematic review, a validation study of a concept inventory, and two developmental research projects. However, before describing these studies, we provide a description of key content areas.

3 Key Content Areas for Physics-Based Climate Change Education

Based on a review of the literature and interviews with climate scientists (Schubatzky et al. 2023; Schubatzky et al. 2024b), six content areas emerge as central for developing a scientific understanding of climate change in science education:

(a) **The Greenhouse Effect**: This is the cornerstone of scientific explanations of climate change. Students must understand the difference between incoming solar radiation and outgoing terrestrial radiation, the role of greenhouse gases in absorbing infrared radiation, and how the natural and anthropogenic greenhouse effects differ. Alternative conceptions in this area are especially widespread, as described in the next section.
(b) **The Earth's Atmosphere**: Understanding the composition and structure of the atmosphere is essential. Students should grasp the relative proportions of gases, the vertical layering of the atmosphere, and the specific roles that different atmospheric components play in climate processes, such as greenhouse gas distribution, ozone formation, and aerosols.
(c) **The Carbon Cycle**: As a dynamic system involving exchanges between the atmosphere, biosphere, oceans, and geosphere, the carbon cycle is essential for explaining CO_2 accumulation. Physics thereby contributes through mass conservation principles and system modeling.
(d) **The Climate System and Feedback Mechanisms**: Feedback mechanisms such as the ice-albedo feedback or cloud feedback are essential for explaining the non-linear and potentially accelerating effects of climate change. These topics require understanding of radiative balance and thermodynamics and are often underemphasized in science curricula as well as studies (Wildbichler et al. 2025a).
(e) **Weather and Climate Distinctions**: Differentiating between short-term variability (weather) and long-term trends (climate) is crucial. Students frequently conflate the two, undermining their ability to interpret data and understand the evidence for climate change. This topic bridges physical timescales, measurement, and statistical reasoning.
(f) **Nature of Climate Science:** Understanding how scientific knowledge about climate change is generated and validated is fundamental. Students should become familiar with scientific methodologies specific to climate science, such as data collection, climate modeling, interpretation of climate proxies, uncertainty and probability in predictions, and the peer-review process. Grasping the nature of climate science thereby helps students appreciate the robustness and complexity of climate science.

Together, these areas form a coherent framework for physics-based CCE from our perspective. These five content areas not only represent the scientific underpinnings of climate change, but also serve as relevant areas for investigating students' conceptual understanding. Research shows that each of these areas is associated with specific alternative conceptions and learning difficulties. In the following section, we provide an overview of students' ideas about these scientific underpinnings of climate change, based on a systematic review of the literature (Wildbichler et al. 2025a). This review sheds light on where students struggle, what alternative conceptions are prevalent, and what this implies for the design of effective instruction.

4 Students' Ideas about the Scientific Underpinnings of Climate Change

To synthesize research findings about students' ideas about the scientific underpinnings of climate change and to provide an overview, we conducted a systematic literature review (Wildbichler et al. 2025a) according to PRISMA guidelines (Page et al. 2021) and previous reviews with a similar thematic focus (e.g., Bhattacharya et al. 2021). Details of the review process can be found in (Wildbichler et al. 2025a). In the final analysis, 44 studies published between 2013 and 2021 were included. Analysis and synthesis were conducted according to qualitative content analysis (Kuckartz 2016). Analyzed studies reported general student ideas about Global Warming or Climate Change, sources of and reasons for students' ideas, as well as student ideas about the scientific concepts of the greenhouse effect, the carbon cycle, the Earth's atmosphere, weather and climate, the climate system, the nature of climate science, and climate change mitigation. Moreover, several reported student ideas imply a general confusion of environmental issues like pollution or biodiversity loss with climate change (Wildbichler et al. 2025a). In the following, we provide a short overview of findings about student ideas concerning the described content areas.

Global warming, climate change. Throughout the analyzed studies, student ideas about Global Warming (GW) or Climate Change (CC) are frequently reported, referring to a general understanding without discussing specific mechanisms or phenomena (Wildbichler et al. 2025a). Regarding causes of global warming or climate change, most students attribute CC/GW to anthropogenic factors such as industry, transportation, and energy consumption (e.g., Chang and Pascua 2016), while some also mention natural causes like solar activity or volcanoes (e.g., Chang and Pascua 2016; Dawson 2015). Several studies highlight that students are aware of CC/GW and its importance (e.g., Holthuis et al. 2014). Consequences acknowledged by students include rising temperatures, impacts on human health, species extinction, and ecosystem changes like glacier melting and sea level rise (Wildbichler et al. 2025a).

(a) Greenhouse Effect.

The analyzed literature describes students' conceptions regarding the mechanisms of the greenhouse effect, the underlying physical principles and scientific concepts, such as electromagnetic radiation, greenhouse gases, the natural and enhanced greenhouse effect, the impacts of the greenhouse effect, as well as a general confusion of the greenhouse effect with other phenomena. In the analyzed literature, different student conceptions regarding the origin of the energy or radiation responsible for the greenhouse effect are reported. These conceptions include the scientifically accurate description that the energy or radiation is emitted by the Earth (e.g., Dawson 2015). However, other conceptions are also described, according to which the energy or radiation is reflected by the Earth or directly originates from the Sun (e.g., Jarrett and Takacs 2020). Students' ideas about the overall process of the greenhouse effect can be described either as an increased heat input, a decreased heat output, or as "heat production" (Wildbichler et al. 2025a). The increased heat input is attributed either to the destruction of the ozone layer or the ozone hole, leading to more radiation reaching the Earth's surface, or to an intensified energy release from the Sun (Wildbichler et al. 2025a). The decreased heat output is explained by the trapping of heat, radiation, gases, or fossil fuels by a distinct layer in the atmosphere, by a kind of blanket or insulation of greenhouse gases, or by the (potentially altered) atmosphere in general (Wildbichler et al. 2025a). The idea of "heat production" describes the warming of the Earth and/or its atmosphere due to hot objects such as factories or car exhaust emissions (e.g., (Chang and Pascua 2016)).

(b) Carbon Cycle.

Throughout analyzed studies, students' ideas about the carbon cycle vary significantly. Some recognize its necessity for life, while others perceive it as harmful to nature (Sivrikaya 2020). Alternative conceptions about carbon compounds are common, with many students identifying only CO_2 and failing to recognize carbon's presence in various forms (e.g., Düsing et al. 2019). There is also a tendency to overemphasize atmospheric carbon, with some believing it exists almost exclusively in the atmosphere (e.g., Jarrett & Takacs). Students frequently misunderstand CO_2's properties, sometimes distinguishing between "natural" and "unnatural" CO_2, or assuming that CO_2 from fossil fuels cannot be absorbed by plants (Niebert and Gropengießer 2014). Additionally, difficulties in grasping the conservation of matter and differentiating between matter and energy (e.g., Jarrett and Takacs 2020) further highlight conceptual gaps in students' understanding of the carbon cycle.

(c) Earth's Atmosphere.

Students' ideas about the atmosphere include an overestimation of the proportion of greenhouse gases (e.g., Jarrett and Takacs 2020), with some believing that a "healthy" atmosphere contains no CO_2 (Niebert and Gropengießer 2014) or that CO_2 forms an upper atmospheric layer (Niebert and Gropengiesser 2013). There is also widespread confusion regarding the ozone layer, with students often misattributing its depletion to greenhouse gases, CO_2, or pollution in general (e.g., Dawson 2015). Additionally, some misunderstand its protective function, associating it with shielding against heat or solar radiation rather than specifically blocking UV radiation (e.g., Chang and Pascua 2016). Changes to the atmosphere are frequently described in broad and imprecise terms, with students perceiving it as being harmed (Chang and Pascua 2016), dirtied (Düsing et al. 2019; Mutlu and Nacaroglu 2019), or even "killed" (Hestness et al. 2016) due to human activity. Furthermore, some believe that atmospheric oxygen levels are decreasing (Kilinc et al. 2013), reflecting broader alternative conceptions about atmospheric composition and climate processes.

(d) Weather and Climate.

Analyzed studies also highlight students' alternative conceptions about weather and climate, particularly regarding timescales. Bodzin et al. (2014) found that, while most eighth graders recognized climate changes more slowly than weather, many underestimated the required time period. Dawson (Dawson 2015) observed that confusion about weather and climate was more prevalent among younger students. Students connect changes in weather patterns, extreme weather events, changes relating to seasons, and changes in precipitation with climate change (e.g., (Mutlu and Nacaroglu 2019)).

(e) Climate System.

Students' ideas about the climate system reveal alternative conceptions about its causes, consequences, and underlying mechanisms. Many believe that climate change primarily affects distant regions, such as the poles, while some students assume temperature changes are felt uniformly across the globe (e.g., Hestness et al. 2016). Regarding the complexity of the climate system, students acknowledge its intricate nature and the presence of unknown factors (Schauss and Sprenger 2021). However, there is little understanding of feedback effects, with most students unfamiliar with positive feedback loops (e.g., Chang and Pascua 2016). Similarly, knowledge of the albedo effect is limited, and it is sometimes confused with heat capacity (Bodzin et al. 2014). Students also struggle with the idea that small amounts of greenhouse gases significantly influence the climate (Niebert and Gropengiesser 2013).

(f) Nature of Climate Science.

Few of the analyzed studies discuss students' ideas about the nature of climate science. For example, Tasquier et al. (Tasquier et al. 2016) found that students lacked knowledge and language regarding scientific modeling in the context of Climate Change (CC). Many could not define what a model is, with only a few providing examples like the atomic or DNA model. Schauss and Sprenger (2021) observed that students had a basic understanding of climate models but struggled with the complexities of climate systems and uncertainties. Some students believed that climate projections were influenced by scientists' opinions rather than objective data.

Based on these findings regarding students' ideas about the scientific underpinnings for climate change, researchers suggest that learners face multiple challenges when developing an understanding of climate change. This includes cognitive difficulties, such as transferring knowledge between domains. Another factor is context-related, referring to where and how students learn about climate change, and is most relevant to daily life experiences, media, public discourse, and formal education. Additionally, some challenges are topic-related, as climate change concepts are often abstract and complex, making them difficult to grasp without direct experience (Wildbichler et al. 2025a). Thus, to successfully foster a scientific understanding of climate change and the underlying concepts, science teaching needs to be mindful of these difficulties and alternative ideas.

5 Evaluating Student Understanding of the Scientific Underpinnings of Climate Change

To evaluate students' conceptual understanding of the scientific underpinnings of climate change, the Climate Change Concept Inventory with 42 items (CCCI-422) was developed and validated. This instrument aims to identify both scientifically accurate conceptions and alternative conceptions across five key content areas as described above. Nature of climate science was not part of the test instrument as it would have shifted the focus of the test; in addition, it would have been difficult to assess students' understanding of the nature of climate science using single-select items.

The development process followed established guidelines for test construction (AERA 2014) and involved multiple steps. First, open-ended questions were created for each content area and used in over 50 interviews with students and university learners to collect authentic conceptions and alternative conceptions. These responses informed the design of items, including both scientifically correct answers (attractors) and plausible distractors based on students' common ideas.

Several pilot studies were conducted to iteratively refine the test. Early versions were tested with high school students, preservice teachers, and university students. Item statistics, distractor analysis, and Rasch modeling were applied to select a robust item set. The final instrument includes 35 multiple-choice items, each linked to one of the five content areas. Items cover both conceptual understanding and factual knowledge, particularly in the area of atmospheric composition.

A major validation study with over 780 students and 48 preservice physics teachers from Austria and Germany confirmed the quality and usefulness of the instrument. The CCCI-422 showed acceptable reliability values (e.g., person reliability of 0.72 overall, 0.65 for students, and 0.81 for preservice teachers) and good model fit under the Rasch framework. Importantly, Differential Item Functioning (DIF) analysis revealed minimal bias across groups, with only one item (about greenhouse gas proportion) showing substantial DIF in favor of the student group.

The study also tested for instructional sensitivity and made an argument for external validity. For example, students who received instruction on climate change performed significantly better than those who had not. Likewise, preservice teachers outperformed both student groups, and climate science experts performed best, with large effect sizes observed between groups (Cohen's $d > 1.3$ in all pair-wise comparisons). Additionally, a pre-post study in school and university settings revealed significant learning gains after targeted instruction, demonstrating the instrument's ability to detect conceptual growth.

The results support using the CCCI-422 either as a one-dimensional measure of overall climate science understanding or as a five-dimensional tool for more fine-grained insights into specific content domains. A dimensional analysis showed both models are viable, with the five-dimensional model providing detailed feedback suitable for instructional design and intervention studies.

In sum, the CCCI-422 is a validated, theory-based assessment tool that enables researchers and educators to diagnose conceptual understanding, track learning progress, and design effective climate change instruction grounded in students' thinking.

Using this instrument, we investigated how well German A-level students understand the scientific underpinnings of climate change (Schubatzky et al. 2024b). Data were collected from 501 students across five urban secondary schools in North Rhine-Westphalia. The results revealed a relatively low level of conceptual understanding: on average, students answered only 39% of the items correctly.

The study uncovered a range of widespread alternative conceptions. For instance, many students significantly overestimated the proportion of CO_2 in the atmosphere, often believing it to be one of the main components alongside oxygen. A large number also assumed that greenhouse gases form a distinct layer in the upper atmosphere, rather than being well-mixed. Another common confusion was between the ozone hole and the greenhouse effect, or assuming a reflection-based mechanism of the greenhouse effect.

Although students were often familiar with basic climate-related terminology, their understanding of the underlying physical mechanisms was limited. The lowest performance was observed in the domains of Earth's atmosphere and the greenhouse effect, while scores were somewhat higher in the area of weather versus climate, likely due to its closer connection to everyday experience.

We concluded that these conceptual difficulties are not only a matter of individual learning, but also point to structural gaps in the curriculum. They emphasize the need for a stronger integration of physical science foundations into climate change education. The study thus offers valuable insights for improving curricula, teaching materials, and instructional strategies in science education contexts focused on climate literacy.

Both the systematic literature review (Wildbichler et al. 2025a) and the study by Schubatzky et al. (2024b) reveal that even older students struggle with the scientific concepts underlying climate change. Hence, the development of effective instructional approaches facilitating a better understanding is necessary. Ideally, evaluations and redesign based on students' understanding should already be included in the developmental process. Two examples of developmental research aiming to address this need will be presented in the following section.

6 Supporting Conceptual Understanding about the Scientific Underpinnings of Climate Change

Results from our studies about students' conceptions show that a big proportion of lower as well as upper secondary students does not possess a scientifically adequate understanding of the scientific underpinnings of climate change. Hence, climate phenomena and underlying science concepts should be addressed in science education. Taking into account the specific needs of students in lower and upper secondary education, we developed learning materials in two developmental research projects. The first project addresses the key content areas for physics-based climate education as described in Sect. 3 by developing tutorials for upper secondary education. The second project aims at fostering lower secondary students' understanding of the greenhouse effect using instructional explanation to potentially reach all students in basic education.

6.1 Fostering Conceptual Understanding in Upper Secondary Education with Tutorials in Climate Change

Although the urgency of climate change has led to increased attention on climate change education (Wildbichler et al. 2025a; Monroe et al. 2019; Kranz et al. 2022; Bhattacharya et al. 2021), research reveals significant gaps in supporting students' conceptual understanding of the underlying scientific mechanisms. Numerous studies point out that students hold fragmented and often incorrect ideas about climate change phenomena (Schubatzky et al. 2024a). However, only a limited number of instructional approaches or materials explicitly target these challenges—particularly in a structured and evidence-based manner.

One major gap in the existing landscape is the lack of dedicated, research-informed instructional materials that systematically address conceptual difficulties. Many initiatives in CCE focus on raising awareness or promoting action (Bhattacharya et al. 2021), but fewer focus on how students learn scientific ideas about climate change and where they struggle from a learning process perspective (Niebert and Gropengießer 2014; Reinfried and Tempelmann 2014). This is especially evident in lower secondary education, where age-appropriate, scientifically grounded materials are scarce. Additionally, there is often no clear progression in how core concepts are introduced, revisited, and built upon.

Finally, there is a need for empirically tested instructional interventions that take students' prior ideas seriously and offer specific learning trajectories anchored in the scientific content.

In response to these gaps, our work focuses on developing research-based instructional materials that align with students' conceptual difficulties, particularly in lower and upper secondary science education. These efforts aim to close the "last mile" between research on student conceptions and classroom instruction, ultimately enabling learners to develop a coherent and scientifically grounded understanding of climate change.

To address knowledge gaps and alternative conceptions of students in upper secondary education, we developed Tutorials in Climate Change in the Erasmus+ Project ENGAGING, a cooperation of the Universities of Dresden, Bochum, Leuven, and Innsbruck. For a detailed description of the structure of these learning materials and their effectiveness in fostering student understanding, see (Wildbichler et al. 2025b).

The aim of Tutorials in Climate Change was to support students' learning processes concerning science concepts which are particularly difficult for them. Hence, designing tutorials (McDermott and Shaffer 2002; Heron et al. 2004) is a promising approach that gives students an opportunity to practice solving qualitative problems, explain their reasoning, and thereby effectively deepen their conceptual understanding.

Building on the findings of the systematic literature review (Wildbichler et al. 2025a) and the study investigating A-level students' conceptions of the scientific underpinnings of climate change (Schubatzky et al. 2024b), we identified ten conceptual ideas that were especially difficult for students to grasp. These ten ideas are connected to the most difficult items from the CCCI-422 test instrument for each of the five concept areas underlying a scientific understanding of climate change as described by Schubatzky et al. (Schubatzky et al. 2023). The percentage of correct responses was particularly low for these items. The ten ideas addressed by Tutorials in Climate Change are, with the concept area mentioned in brackets, (1) distinguishing warm periods and ice ages (weather and climate), (2) greenhouse gases are trace gases (Earth's atmosphere), (3) the role of solar radiation for the greenhouse effect (greenhouse effect), (4) the mechanisms underlying the greenhouse effect (greenhouse effect), (5) comparing natural and anthropogenic greenhouse effect (greenhouse effect), (6) cloud feedback as an example of feedback effects (climate system), (7) the carbon cycle as a (simplified) cycle (carbon cycle), (8) the role of oceans for the carbon cycle (carbon cycle), (9) water vapor's role as a greenhouse gas (Earth's atmosphere), and (10) differences and connections between weather and climate (weather and climate).

By addressing these ideas that pose conceptual difficulty to students, our tutorials aim to support and enrich teaching about climate change in science classrooms. They can be implemented individually or combined and take less than 45 minutes to implement per tutorial. All ten tutorials and a teacher guide to support implementation are available in English, German, and Dutch on the project website www.engagingforclimate.eu.

The ten Tutorials in Climate Change were tested with N = 69 preservice teachers from Austria, Belgium, and Germany in a pre-post format using the test instrument CCCI-422 (Schubatzky et al. 2023). After participating in the implementation of the tutorials, this group positively developed their conceptual understanding of climate change with an average normalized gain of 0.45 (for further elaboration, see Wildbichler et al. 2025b).

6.2 Addressing the Greenhouse Effect in Lower Secondary Education Using Instructional Explanation

Based on the outlined problem, we also developed a learning environment on the greenhouse effect for lower secondary education. As a methodological framework, we applied Design-Based Research (e.g., McKenney and Reeves 2013). Within this framework, we pursue two main goals: the development of a learning environment and the investigation of learning processes based on this learning environment (Sandoval and Bell 2004). To increasingly optimize the alignment between the learning environment and the learners, five design cycles were conducted. Initially, theoretically and empirically grounded design conjectures were formulated, from which design criteria were derived (Obczovsky et al. 2025). These design criteria define measures and strategies for implementing the design conjectures effectively.

The design conjectures led to several fundamental principles for implementing the learning environment:

(a) Since the greenhouse effect is a new and complex phenomenon for secondary school students, self-explaining is difficult, and prior knowledge is likely limited; instructional explanation (Kulgemeyer 2018) serves as the core structure of the learning arrangement. In addition to cognitive activation and constructive support, diagnosing understanding and learning tasks are essential components.
(b) The content of the instructional explanation is defined by the scientific core ideas of the greenhouse effect (Shepardson et al. 2017).
(c) As the learning environment combines verbal explanations with visual representations, its implementation follows multimedia principles aimed at reducing cognitive load, based on Mayer andMoreno (2003).

To ensure a transparent and systematic reflection of the design conjectures and criteria, they were categorized as either "strong" or "weak." Design conjectures based on meta-analyses are considered "strong," while those derived from non-generalizable individual studies are labeled as "weak." However, within this categorization, design conjectures can also be situated along a spectrum between "strong" and "weak." The analysis focuses on weak design assumptions to gather arguments for refining, revising, or potentially discarding those (Wildbichler et al. in preparation).

Each design cycle consisted of a development phase, in which the learning environment or parts of it were created or revised based on the (revised) design conjectures and criteria, an implementation phase and an evaluation phase, in which findings from the implementation were analyzed and the learning environment was evaluated. In the implementation and evaluation phases, qualitative and quantitative investigations were used in a mixed methods design (Wildbichler et al. in preparation). An overview of the methods used in each design cycle is provided in Fig. 1.

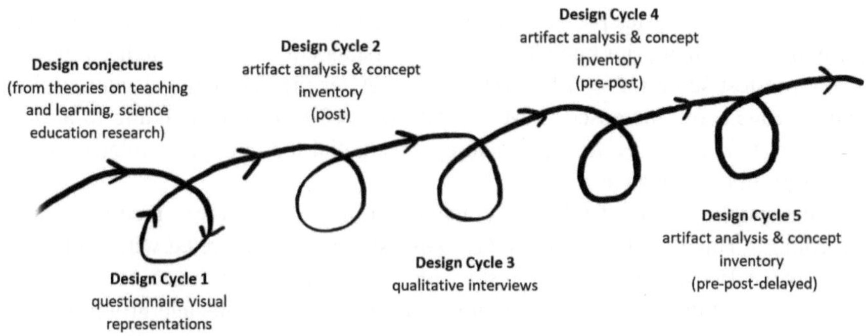

Fig. 1 Methods used in each design cycle (own graphic)

In the final version, the learning environment consisted of three parts addressing Earth's energy balance, the role of greenhouse gases, and the natural and the anthropogenic greenhouse effect. Each part was introduced with an explanation using visual representations and an audio explanation, intending to use both channels so as not to overload learners' working memory (Mayer and Moreno 2003). Students then had to solve several learning tasks to practice and deepen their understanding in accordance with requirements of good science explanations (Kulgemeyer 2018). Moreover, in each part of the learning environment, students had to either conduct an experiment and analyze data or participate in an energy theater exercise (Daane et al. 2014) using embodied cognition (Anderson 2003) to facilitate the development of conceptual understanding.

A questionnaire on visual representations of phenomena underlying climate change based on insights from physics education research (Haagen-Schützenhöfer and Hopf 2020; Zloklikovits and Hopf 2021; Budimaier and Hopf 2024) and several qualitative analyses of student artifacts and interviews, according to qualitative content analysis [14], were used to refine and revise design conjectures. Based on this, the actual embodiment in the learning environment was (re-)designed, leading to the final version (Wildbichler et al. in preparation).

Additionally, quantitative analyses using parts of the test instrument CCCI-422 (Schubatzky et al. 2023) were conducted to evaluate the development of students' understanding in design cycles 2, 4, and 5 (Wildbichler et al. in preparation). In design cycle 5, the learning environment adhered to the previously described structure. Therefore, to gain insights into students' conceptual understanding of the greenhouse effect after working with the learning environment in its final structure, the following analysis draws on data collected during design cycle 5. In this design cycle, the learning environment was tested with N = XX Austrian eighth-grade students from two "Mittelschulen" (lower track schools in Austria) and two "Gymnasien" (higher track schools in Austria). For the analysis, items from the CCCI-422 (Schubatzky et al. 2023) addressing the greenhouse effect are considered, as this was the concept area addressed by the learning environment. The maximum achievable score was eight points. The test covered a broad range of difficulty in the questions that are difficult even for physics pre-service teachers. Since the students participating were in eighth grade, their prior knowledge was understandably limited, which is reflected in low pre-test score ($M = 1.86$, $SD = 1.36$). After the instructional intervention, their average score increased to M = 3.54 ($SD = 1.82$), corresponding to a large effect size (*Cohen's d = 1.05*), indicating substantial learning gains relative to the variability in performance.

The normalized gain of $\langle g \rangle = 0.27$, while falling into the "low to medium" range according to Hake's framework (Hake 1998), should be interpreted with caution in this context. Because the test targets a wide conceptual range, much of which lies beyond the eighth grade, the maximum achievable gain for this age group is naturally constrained. Thus, the normalized gain may underestimate the actual learning progress.

These results indicate that our learning environment on the greenhouse effect fosters student learning and improves their conceptual understanding. Notably, after participating in the implementation of the learning environment, the eighth-grade students participating in our study, on average, performed better than A-level students in the study by Schubatzky et al. (Schubatzky et al. 2024b). The average score of N = 501 A-level students was M = 1.98, SD = 1.21 [7], while the average score of N = 80 eighth-grade students participating in our study was M = 3.54, SD = 1.82. A statistical comparison between the two groups using Welch's t-test revealed a large and significant effect, $t(\sim 90) = -7.41$, $p < 0.001$, with a Cohen's d of 1.19, indicating that the difference in performance is not only statistically significant but also with a large effect. Hence, we conclude that developmental research provides the opportunity to design learning environments that effectively foster students' understanding of complex scientific concepts like the greenhouse effect.

7 Conclusion and Outlook

In this contribution, we have emphasized the importance of fostering students' conceptual understanding of the scientific underpinnings of climate change as a central element of science and physics education. Drawing on empirical findings from a systematic review and our own studies involving German and Austrian students, we identified key areas where learners struggle—ranging from the greenhouse effect and the role of greenhouse gases to the dynamics of the carbon cycle and feedback mechanisms within the climate system. The development and validation of the CCCI-422 concept inventory has provided valuable insights into these challenges and highlighted substantial gaps in students' understanding, even at the upper secondary level. However, we acknowledge that our study was conducted in German-speaking countries (Austria and Germany), and findings may not fully generalize to other educational systems or cultural contexts.

To address these challenges, we developed targeted instructional materials—including the Tutorials in Climate Change for upper secondary education and a learning environment on the greenhouse effect for lower secondary education. These interventions were grounded in research on students' conceptions and designed to support teachers. First evaluations show that such materials can support conceptual change and improve understanding, especially when common alternative conceptions are explicitly addressed and learners are provided with opportunities to reflect, reason, and apply their knowledge.

However, conceptual understanding is only one pillar of Climate Change Education (Chang 2022). As climate change represents a complex socio-scientific issue with far-reaching ethical, political, and personal dimensions, education must also address broader affective, behavioral, and action-oriented goals. This includes fostering agency, defined as students' perceived and actual capacity to engage meaningfully with climate change—whether through personal behavior, civic participation, or professional ambitions (White et al. 2024).

Educational interventions that focus solely on the development of knowledge—without connecting this knowledge to climate action—risk producing what has been described as "well-informed inaction." (Jensen 2002). In contrast, when conceptual understanding is embedded in meaningful contexts, and when learners are supported in making connections between scientific knowledge, societal discourse, and their own actions, they are more likely to develop a sense of ownership, efficacy, and engagement.

Looking ahead, we see three major challenges and opportunities for future work:

(a) Bridging Levels: Conceptual understanding must be supported across educational stages—from lower to upper secondary and into teacher education. This requires coherent learning progressions and carefully designed transitions between age groups and disciplines.
(b) Integrating Knowledge and Action: There is a need for instructional designs that interweave scientific understanding with discussions of uncertainty, responsibility, and action. This includes both individual and collective dimensions of agency and should empower students rather than overwhelm them (Kranz et al. 2022).
(c) Connecting Research and Practice: Translating insights from learning research into classroom-ready materials remains a key task. Collaborations between researchers, teachers, and curriculum developers are essential to ensure that new materials are feasible, adaptable, and grounded in the realities of science teaching.

In summary, supporting learners in developing a robust, coherent understanding of climate change is essential—but not sufficient—for climate literacy. We must also equip them to critically engage with societal debates and participate in shaping a sustainable future. To do so, physics education must evolve into an integrative, empowering, and forward-looking endeavor that addresses the complexity of climate change—and the potential of students to meet it.

References

AERA (2014) Standards for educational and psychological testing. American Educational Research Association, Washington, DC

Anderson ML (2003) Embodied cognition: a field guide. Artif Intell 149(1):91–130. https://doi.org/10.1016/S0004-3702(03)00054-7

Bhattacharya D, Carroll Steward K, Forbes CT (2021) Empirical research on K-16 climate education: a systematic review of the literature. J Geosci Educ 69(3):223–247. https://doi.org/10.1080/10899995.2020.1838848

Bodzin AM, Anastasio D, Sahagian D, Peffer T, Dempsey C, Steelman R (2014) Investigating climate change understandings of urban middle-level students. J Geosci Educ 62(3):417–430. https://doi.org/10.5408/13-042.1

Budimaier F, Hopf M (2024) Evaluation of a new teaching-learning sequence on the particulate nature of matter using crystal structures. Phys Rev Phys Educ Res 20(2). https://doi.org/10.1103/PhysRevPhysEducRes.20.020104

Chang C-H (2022) Climate change education: knowing, doing and being. Taylor & Francis

Chang C-H, Pascua L (2016) Singapore students' misconceptions of climate change. Int Res Geogr Environ Educ 25(1):84–96. https://doi.org/10.1080/10382046.2015.1106206

Daane AR, Wells L, Scherr RE (2014) Energy theater. Phys Teach 52(5):291–294. https://doi.org/10.1119/1.4872412

Dawson V (2015) Western Australian high school students' understandings about the Socioscientific issue of climate change. Int J Sci Educ 37(7):1024–1043. https://doi.org/10.1080/09500693.2015.1015181

Düsing K, Asshoff R, Hammann M (2019) Tracing matter in the carbon cycle: zooming in on high school students' understanding of carbon compounds and their transformations. Int J Sci Educ 41(17):2484–2507. https://doi.org/10.1080/09500693.2019.1686665

Haagen-Schützenhöfer C, Hopf M (2020) Design-based research as a model for systematic curriculum development: the example of a curriculum for introductory optics. Phys. Rev. Phys. Educ. Res. 16(2):20152. https://doi.org/10.1103/PhysRevPhysEducRes.16.020152

Hake RR (1998) Interactive-engagement versus traditional methods: a six-thousand-student survey of mechanics test data for introductory physics courses. Am J Phys 66(1):64–74. https://doi.org/10.1119/1.18809

Heron, P., Shaffer, P. S., and McDermott, L. C. 2004. Research as a guide to improving student learning: an example from introductory physics: invention and impact, proceedings of a course, curriculum, and laboratory improvement conference

Hestness E, McGinnis JR, Breslyn W (2016) Examining the relationship between middle school students' sociocultural participation and their ideas about climate change. Environ Educ Res 6(1):1–13. https://doi.org/10.1080/13504622.2016.1266303

Holthuis N, Lotan R, Saltzman J, Mastrandrea M, Wild A (2014) Supporting and understanding students' epistemological discourse about climate change. J Geosci Educ 62(3):374–387

Jarrett L, Takacs G (2020) Secondary students' ideas about scientific concepts underlying climate change. Environ Educ Res 26(3):400–420

Jensen BB (2002) Knowledge, action and pro-environmental behaviour. Environ Educ Res 8(3):325–334. https://doi.org/10.1080/13504620220145474

Kilinc A, Yeşiltaş NK, Kartal T, Demiral Ü, Eroğlu B (2013) School students' conceptions about biodiversity loss: definitions, reasons. Results and Solutions Res Sci Educ 43(6):2277–2307. https://doi.org/10.1007/s11165-013-9355-0

Kranz J, Schwichow M, Breitenmoser P, Niebert K (2022) The (un)political perspective on climate change in education—a systematic review. Sustainability 14(7):4194. https://doi.org/10.3390/su14074194

Kuckartz U (2016) Qualitative Inhaltsanalyse: Methoden, praxis, Computerunterstützung. 3, überarbeitete Auflage edn. Grundlagentexte Methoden. Beltz Juventa, Weinheim, Basel

Kulgemeyer C (2018) Towards a framework for effective instructional explanations in science teaching. Stud Sci Educ 54(2):109–139. https://doi.org/10.1080/03057267.2018.1598054

Mayer RE, Moreno R (2003) Nine ways to reduce cognitive load in multimedia learning. Educ Psychol 38(1):43–52. https://doi.org/10.1207/S15326985EP3801_6

McDermott LC, Shaffer PS (2002) Tutorials in introductory physics. University of Washington

McKenney S, Reeves T (2013) Conducting educational design research. Routledge

Monroe MC, Plate RR, Oxarart A, Bowers A, Chaves WA (2019) Identifying effective climate change education strategies: a systematic review of the research. Environ Educ Res 11(3):1–22. https://doi.org/10.1080/13504622.2017.1360842

Mutlu F, Nacaroglu O (2019) Examination of perceptions of gifted students about climate change and global warming. J Balt Sci Educ 18(5):780–792

Niebert K, Gropengiesser H (2013) Understanding and communicating climate change in metaphors. Environ Educ Res 19(3):282–302. https://doi.org/10.1080/13504622.2012.690855

Niebert K, Gropengießer H (2014) Understanding the greenhouse effect by embodiment—Analysing and using students' and scientists' conceptual resources. Int J Sci Educ 36(2):277–303. https://doi.org/10.1080/09500693.2013.763298

Obczovsky M, Bernsteiner A, Haagen-Schützenhöfer C, Schubatzky T (2025) Systematizing decisions in design-based research: from theory to design. Sci Educ 109(2):523–536. https://doi.org/10.1002/sce.21915

Page MJ, McKenzie JE, Bossuyt PM, Boutron I, Hoffmann TC, Mulrow CD, Shamseer L, Tetzlaff JM, Akl EA, Brennan SE, Chou R, Glanville J, Grimshaw JM, Hróbjartsson A, Lalu MM, Li T, Loder EW, Mayo-Wilson E, McDonald S, McGuinness LA, Stewart LA, Thomas J, Tricco AC, Welch VA, Whiting P, Moher D (2021) The PRISMA 2020 statement: an updated guideline for reporting systematic reviews. Syst Rev 10(1):89. https://doi.org/10.1186/s13643-021-01626-4

Reinfried S, Tempelmann S (2014) The impact of secondary school students' preconceptions on the evolution of their mental models of the greenhouse effect and global warming. Int J Sci Educ 36(2):304–333. https://doi.org/10.1080/09500693.2013.773598

Sandoval WA, Bell P (2004) Design-based research methods for studying learning in context: introduction. Educ Psychol 39(4):199–201. https://doi.org/10.1207/s15326985ep3904_1

Schauss, M., Sprenger, S. 2021. Conceptions of Uncertainties in the Context of Climate Change. International Research in Geographical and Environmental Education 30(4):332–347 (16 Seiten). https://doi.org/10.1080/10382046.2020.1852782

Schubatzky T, Haagen-Schützenhöfer C, Wackermann R, Wöhlke C, Wildbichler S (2024a) Navigating the complexities of student understanding: exploring the coherency of students' conceptions about the greenhouse effect. Sci Educ 108(4):1134–1161. https://doi.org/10.1002/sce.21867

Schubatzky T, Wackermann R, Haagen-Schützenhöfer C, Wöhlke C (2024b) How well do German A-level students understand the scientific underpinnings of climate change? Sustainability. https://doi.org/10.3390/su16177264

Schubatzky T, Wackermann R, Wöhlke C, Haagen-Schützenhöfer C, Jedamski M, Lindemann H, Cardinal K (2023) Entwicklung des Concept-Inventory CCCI-422 zu den naturwissenschaftlichen Grundlagen des Klimawandels. ZfDN 29(1). https://doi.org/10.1007/s40573-023-00159-8

Shepardson, D. P., Roychoudhury, A., Hirsch, A. 2017. Using conceptual and physical models to develop students' mental models of the greenhouse effect.In: D. P. Shepardson, A. Roychoudhury, A. Hirsch, (eds) teaching and learning about climate change: a framework for educators. Routledge, New York, pp 85–105

Sivrikaya SO (2020) Carbon cycle learning of chemistry students: modeling method. JEL 9(4):80. https://doi.org/10.5539/jel.v9n4p80

Tasquier G, Levrini O, Dillon J (2016) Exploring students' epistemological knowledge of models and modelling in science: results from a teaching/learning experience on climate change. Int J Sci Educ 38(4):539–563. https://doi.org/10.1080/09500693.2016.1148828

Tasquier G, Pongiglione F (2017) The influence of causal knowledge on the willingness to change attitude towards climate change: results from an empirical study. Int J Sci Educ 39(13):1846–1868. https://doi.org/10.1080/09500693.2017.1355078

UNESCO (2017) Education for sustainable development goals: learning objectives. UNESCO, Paris

White PJ, Ardoin NM, Eames C, Monroe MC (2024) Agency in the Anthropocene: education for planetary health. Lancet Planet Heath 8(2):e117–e123. https://doi.org/10.1016/S2542-5196(23)00271-1

Wildbichler S, Haagen-Schützenhöfer C, Schubatzky T (2025) Teaching the greenhouse effect in lower secondary education: a design-based research project. Physics Education. 61(1):015017. https://doi.org/10.1088/1361-6552/ae1c8b

Wildbichler S, Schubatzky T, Haagen-Schützenhöfer C (2025a) Students' ideas about the scientific underpinnings of climate change: a systematic review of the literature // students' ideas about the scientific underpinnings of climate change: a systematic review of the literature. Stud Sci Educ 61(1):1–53. https://doi.org/10.1080/03057267.2024.2395206

Wildbichler S, Wackermann R, DeCock M, Ivanjek L, Micoloi M, Pospiech G, Schubatzky T (2025b) Tutorials in climate change. Phys Teach 63(6):414–417. https://doi.org/10.1119/5.0224098

Zloklikovits S, Hopf M (2021) Evaluating key ideas for teaching electromagnetic radiation. J Phys Conf Ser 1929(1):12063. https://doi.org/10.1088/1742-6596/1929/1/012063

Addressing Dimensions of Climate Literacy in Physics Education: Examples from Four Studies

Sarah Wildbichler, Matthias Fasching, Magdalena Micoloi, and Thomas Schubatzky

Abstract In a world shaped by anthropogenic climate change, climate literacy becomes an important goal of science education. In this article, four studies from physics education research are presented, each focusing on one or two dimension(s) of climate literacy. The first study presents a Design-Based research project addressing lower secondary students' conceptual understanding of the greenhouse effect. The second study provides insights into the development and evaluation of "Tutorials in Climate Change," with special emphasis on the interconnectedness of conceptual understanding and willingness to teach about climate change. In the third study, the development of the test instrument "CTCC" (Critical Thinking in Climate Change) testing critical thinking skills in the context of climate change is described and analyzed. Finally, the fourth study explores in-service teachers' perceptions of teaching climate action as a socio-scientific issue. Combined insights from the four studies presented in this contribution can inform physics education practice and physics education research on several levels; for example, how climate literacy can be addressed, or which research designs are useful in targeting different levels of climate change education.

Keywords Climate literacy · Climate education · Conceptual understanding · Willingness to teach · Critical thinking · Climate action · Socio-scientific issues

---

S. Wildbichler (✉) · T. Schubatzky
Physics Education Research Group, University of Innsbruck, Innsbruck, Austria
e-mail: sarah.wildbichler@uibk.ac.at

T. Schubatzky
e-mail: thomas.schubatzky@uibk.ac.at

M. Fasching
Austrian Educational Competence Centre Physics, University of Vienna, Vienna, Austria
e-mail: matthias.fasching@univie.ac.at

M. Micoloi
Physics Education Research, TU Dresden, Dresden, Germany
e-mail: magdalena.micoloi@tu-dresden.de

D. Sokołowska and E. McLoughlin (eds.), *Embracing Changes Together in Physics Education*, Challenges in Physics Education,
https://doi.org/10.1007/978-3-032-27385-7_15

1 Introduction

Amid the current period of anthropogenic climate change, with overwhelming scientific evidence of its reality and seriousness (Eyring et al. 2023) and an increasing manifestation of its consequences (Pörtner et al. 2022), climate literacy becomes a central part of science literacy (USGCRP 2009). According to the USGCRP, a climate literate person (USGCRP 2009)

(a) "*understands the essential principles of Earth's climate system,*
(b) *knows how to assess scientifically credible information about climate,*
(c) *communicates about climate and climate change in a meaningful way, and*
(d) *is able to make informed and responsible decisions with regard to actions that may affect climate.*"

To foster climate literacy among students, science education needs to address each of these dimensions. The aim of this contribution is to showcase the connection of individual studies from physics education research that focus on different dimensions of climate literacy. This contribution synthesizes the findings of four studies presented at the WCPE 2024 symposium in Krakow.

Study 1 (Sect. 2) focuses on dimension (a) by describing a Design-Based Research project that evolves around the development of a learning environment on the greenhouse effect, one of the essential principles of Earth's climate system, for lower secondary education.

Study 2 (Sect. 3) addresses dimensions (a) and (c). It is centered on the development of materials fostering pre-service teachers' understanding of the essential principles of Earth's climate system as defined by Schubatzky et al. (2024), the implementation of these materials, and an analysis of the development of pre-service teachers' conceptual understanding and willingness to teach. In the analysis, the relationship of pre-service teachers' attitudes and their willingness to teach about climate change is also examined.

Study 3 (Sect. 4) targets dimensions (b) and (d) by focusing on critical thinking in the context of climate change. It describes and analyzes the development of a test instrument assessing critical thinking skills through items contextualized within scientific concepts underlying climate change, the test instrument "CTCC" (Critical Thinking in Climate Change).

Study 4 (Sect. 5) is centered around dimension (d) by examining teachers' willingness and intention to teach climate action and support students in making informed and responsible decisions in this regard. The analysis is anchored in the Theory of Planned Behavior and makes use of its constructs to gain insights into the relationship between teachers' experience in teaching the topic and their willingness and intention.

Each of these studies provides insights into the broader framework of climate literacy as defined by the USGCRP (2009). In Sects. 2, 3, 4 and 5, these individual studies are described in detail, including selected findings and a reflection on their contextualization within the USGCRP framework on climate literacy.

Finally, in Sect. 6, we discuss the consequences of these studies and summarize recommendations for promoting climate literacy.

2 Study 1: Fostering Conceptual Understanding of the Greenhouse Effect, an Essential Principle of Earth's Climate System

The focus of the first study presented in the symposium is fostering learners' understanding of the essential principles of Earth's climate system, the first dimension of climate literacy as defined by the USGCRP (2009). One of these essential principles of Earth's climate system is the greenhouse effect.

However, empirical studies on students' conceptions report numerous and widespread alternative ideas regarding the scientific principles of climate change and, in particular, the greenhouse effect (Wildbichler et al. 2024). Chang (2022) argues that alternative ideas must be addressed and developed into appropriate conceptions within the framework of school education to counteract a looming cycle of ignorance and inaction. To reach all students, the goal of fostering climate literacy should already be pursued in compulsory education.

The aim of this research project is therefore to develop a learning environment on the greenhouse effect for lower secondary education (compulsory education). Authors studying students' conceptions of the greenhouse effect attribute alternative ideas, among other factors, to misleading visual representations in media, textbooks, and other educational materials (Wildbichler et al. 2024). Therefore, a special focus is placed on effective visual representations of the processes underlying the greenhouse effect.

The outlined problem leads to the following research questions:

- What can an effective learning environment on the greenhouse effect, including visual representations, look like for lower secondary education?
- How do learners interact with the learning environment?
- How does learners' conceptual understanding develop during their interaction with the learning environment?

2.1 Research Design

Within the methodological framework of Design-Based Research, two goals are pursued: the development of a learning environment and the investigation of teaching and learning based on this learning environment (Sandoval and Bell 2004).

To optimize the alignment between the learning arrangement and learners' needs, the research design includes several sub-studies. Initially, theoretically and empirically grounded design conjectures were formulated, from which design criteria were

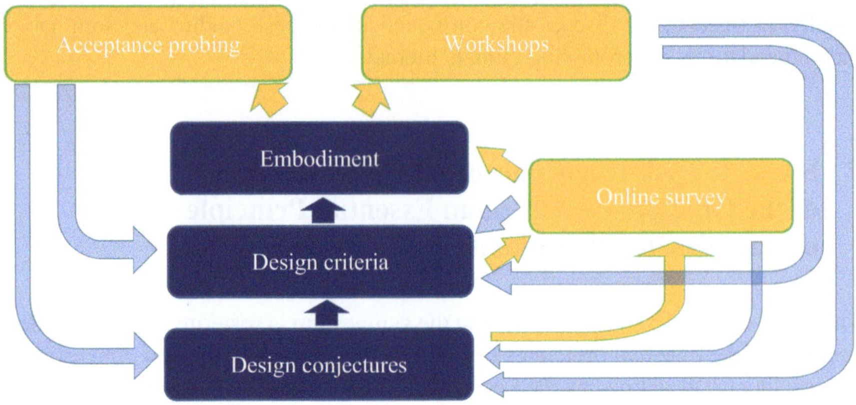

Fig. 1 Research design of the study

derived (Obczovsky et al. 2024). To ensure a transparent and systematic reflection on the design conjectures and criteria, design conjectures were categorized into "strong" and "weak" conjectures. For instance, a design conjecture based on meta-analyses is considered "strong," whereas one grounded in non-generalizable single studies is classified as "weak."

To strengthen "weak" design conjectures and validate design decisions concerning visual representations, a preliminary survey was conducted using an online questionnaire. This survey involved eighth-grade students ($N = 231$, convenience sample) and adults ($N = 100$, convenience sample), who were asked about the visual representation of several phenomena, such as radiation absorption. For this, multiple single-selected items were developed, allowing participants to choose one particular representation per question.

In the next step, an initial version of the learning environment was developed and subsequently tested in two workshops with ten students each ($N = 20$) and five acceptance-probing interviews (Wiesner and Wodzinski 1996) involving two students each ($N = 10$), all conducted with eighth-grade students. The aim of the workshops and acceptance probing was to gain insights into how students interact with the learning environment, identify obstacles, and uncover potential for improvement, with a focus on "weak" design conjectures. The entire research design is illustrated in Fig. 1.

2.2 Results

The design conjectures led to several fundamental principles for implementing the learning environment: (a) Since the greenhouse effect is a new phenomenon for lower secondary students, it is too complex for self-explanations, and students' prior knowledge is likely low, instructional explanation (Kulgemeyer 2018) serves as the foundational structure for the learning environment. In addition to cognitive activation and constructive support for students, diagnosis of understanding and learning tasks are essential components. (b) The content of the instructional explanation is defined by the scientific core concepts of the greenhouse effect (Shepardson et al. 2017). (c) As verbal explanations are linked with visual representations in the learning environment, the implementation adheres to multimedia principles to reduce cognitive load, as outlined by Mayer and Moreno (2003).

The data from the preliminary survey were analyzed using chi-square tests. For all items, one representation was selected significantly more frequently than the others, with one exception among the students. Selected results are shown in Table 1. The most frequently chosen representations were used for the implementation of the visual representations in the learning environment.

In the workshops, a multiple-choice test comprising six items from the "CCCI-422" test instrument (Schubatzky et al. 2023) was administered post-instruction. On average, students scored 4.8 out of a possible 6 points ($M = 4.8$; $SD = 1.0$). For the question describing the overall greenhouse effect, 15 out of 20 students selected the adequate response from five possible answers. Additionally, worksheets completed by students during the workshops were analyzed according to qualitative content analysis (Kuckartz and Rädiker 2022). The analysis revealed that the participating students adopted the representation methods used in the learning environment (sometimes with slight modifications) for their own visual representations. In students' written explanations, many scientifically accurate conceptions were identified. However, reflection-based explanations that omit processes such as radiation conversion also appeared repeatedly.

Table 1 Selected results from the online survey

Question	Most common answer	Results students		Results adults	
		p	Residual	p	Residual
Representation of infrared radiation		<0.01	4.33	<0.01	8.6
Representation of invisible radiation		0.01	3.84	<0.01	5.2
Absorption of radiation/ light		<0.01	10.75	<0.01	11
Representation of molecules		<0.01	14.09	<0.01	10.21

2.3 Discussion and Conclusion

The evaluation of the workshops indicates that the participating students accepted the provided visual representations, as they used these representations in their own depictions. Overcoming reflection-based ideas was identified as a particularly challenging aspect.

To gain more detailed insights into students' interactions with the learning environment, triangulate the results of the workshops and the preliminary survey and identify further obstacles, challenges, and potential improvements, the acceptance-probing interviews will be analyzed next according to qualitative content analysis (Kuckartz and Rädiker 2022). Based on the findings from the workshops and acceptance-probing interviews, the design conjectures and criteria will be revised, and the learning environment will be adjusted accordingly. For instance, increased focus will be placed on addressing and mitigating reflection-based ideas. According to the USGCRP (2009), a climate literate person, among other factors, "understands the essential principles of Earth's climate system." However, studies on students' conceptions indicate a lack of knowledge and understanding (Wildbichler et al. 2024). In the presented study, one possibility of designing an effective learning environment to foster climate literacy among students in lower secondary education is explored, aiming to potentially reach all students. We believe that developmental research, in the case of our study Design-Based Research, is a valuable tool for science education research to contribute to the achievement of societal goals like climate literacy by combining the application of previous research outcomes with the development, exploration, and improvement of a specific learning environment.

3 Study 2: Development and Evaluation of "Tutorials in Climate Change"

The focus of the second study presented in the symposium is on pre-service teachers' understanding of the essential principles of Earth's climate system and on developing pre-service teachers' skills for meaningful communication about climate and climate change, the first and second dimensions of climate literacy as defined by the USGCRP (2009). To address the essential principles of Earth's climate system, five concept areas that were described as crucial for developing an understanding of climate change are considered (Schubatzky et al. 2023): (1) Earth's atmosphere, (2) the difference between weather and climate, (3) climate as a system, (4) the carbon cycle, and (5) the greenhouse effect.

The study focuses on pre-service physics teachers, since when it comes to fostering conceptual understanding of the scientific principles underlying climate change among students, high-quality teaching is pivotal. Research has shown that in order to provide high-quality teaching, teachers need to have corresponding knowledge and skills (Förtsch et al. 2016). Thereby, teachers' conceptual understanding of climate change can be seen as one of the most fundamental basics, since profound content knowledge is considered as a prerequisite for the development of pedagogical content knowledge (Kulgemeyer and Riese 2018), which in turn is necessary to provide effective instructional explanations in classes (Kulgemeyer and Riese 2018). Furthermore, research has shown that teachers' knowledge of typical student learning difficulties also predicts student learning (Sadler et al. 2013).

Hence, the argument made in this study is that when it comes to teaching the scientific principles of climate change, teachers, first and foremost, need to develop their own conceptual understanding of climate change in addition to learning about typical learning difficulties, as previous research has shown that teachers often feel ill-prepared (Li et al. 2021). In turn, the hypothesis being pursued in this study is that a deepened understanding of climate change enhances teachers' willingness to teach about climate change, since, for example, Liu et al. (2015) and Plutzer et al. (2016) demonstrated that, in fact, a lack of understanding hinders teaching about climate change among science teachers.

To address these needs, ten so-called "Tutorials in Climate Change" were developed (McDermott and Shaffer 2002), which can be used in teacher education programs, but also for secondary school teaching.

3.1 Rationales for the Development of Tutorials in Climate Change

Several rationales guided the development of the tutorials. First, the tutorials address the five concept areas that were described as crucial for developing an understanding of climate change (Schubatzky et al. 2023). Second, the tutorials follow a modified anchored instruction approach, since research has emphasized the importance of situating instruction in contexts that are meaningful (Lovett and Shah 2007). Anchored instruction furthermore gives teachers and students the opportunity to work from a shared experience perspective. Third, concept cartoons were incorporated to support future teachers in overcoming their potential misconceptions and in turn also to learn about typical learning difficulties from school students (Kabapınar 2005). Finally, the tutorials were designed in such a way that they can also be used for teaching in upper secondary classes, so teachers are provided with similar learning experiences like their future school students. The Tutorials are described in more detail and are provided in Wildbichler et al. (2025).

3.2 Evaluation Design

The Tutorials in Climate Change were tested with $N = 71$ future physics teachers from Austria, Belgium, and Germany in a pre-post format. They were implemented in different seminars and also different formats (successively in two sessions or one tutorial each week over several weeks), resulting in about 3.5 h of intervention time. For the evaluation, the "CCCI-422" test instrument, a previously developed and validated climate change concept inventory (Schubatzky et al. 2023), was used, together with previously developed scales (Heuckmann et al. 2020) which were adapted for teaching about climate change.

3.3 Results

Results thereby show that the future teachers positively developed their conceptual understanding of climate change ($t(70) = 11.16, p < 0.001$) with a large effect size ($d = 1.32$) and independent of the course they participated in ($F(39,3) = 1.4$, $p = 0.26$). Furthermore, the pre-service teachers' willingness to teach about climate change significantly increased ($t(70) = 2.95, p < 0.001$) with a low-to-medium effect size ($d = 0.35$).

Due to the fact that the teachers' conceptual understanding increased with a large effect size and their willingness only increased with a low-to-medium effect size, we performed further analysis: the development of willingness to teach about a topic (when working with tutorials) might be dependent on additional variables of the pre-service teachers. Especially, previous research demonstrated that attitudes toward a topic predict whether teachers include the respective topic into their teaching (Heuckmann et al. 2020). Hence, a cross-lagged-panel model was calculated to investigate whether attitudes toward teaching climate change predict the development of willingness to teach about the topic. The model is shown in Fig. 2.

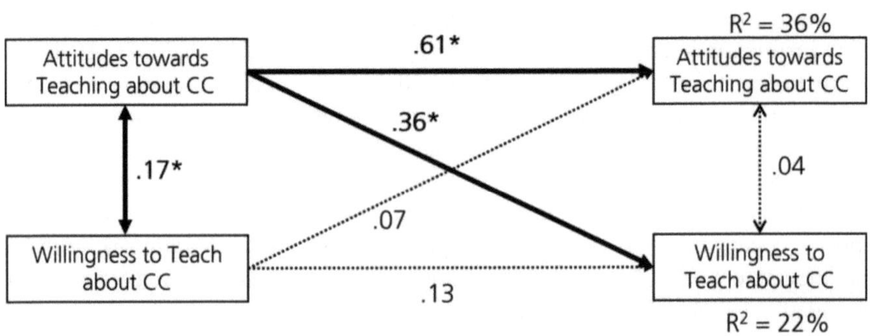

Fig. 2 Cross-lagged-panel model examining the relationships between attitudes toward teaching about climate change (CC) and willingness to teach about CC over time. Solid lines represent significant paths ($p < 0.05$), while dotted lines represent non-significant paths. Autoregressive paths indicate stability of attitudes (0.61*). Cross-lagged paths show that initial attitudes significantly predict later willingness (0.36*), whereas initial willingness does not significantly predict later attitudes (0.07). Synchronous correlations at Time 1 (0.17*) and Time 2 (0.04) are also included. The model explains 36% of the variance in attitudes and 22% in willingness at Time 2

The analysis shown in Fig. 2 shows that attitudes toward teaching about climate change predict the development of the pre-service teachers' willingness to teach about climate change, with about 22% of variance explained.

3.4 *Discussion and Conclusion*

The results demonstrate that, when working with "Tutorials in Climate Change," physics pre-service teachers positively developed their conceptual understanding of climate change. We interpret this as supporting evidence for the effectiveness of the tutorials. Furthermore, the results show that pre-service teachers' willingness to teach about climate change also developed positively. In accordance with previous research (Liu et al. 2015; Plutzer et al. 2016), we conclude that a lack of knowledge might in fact hinder teaching about climate change. However, the results of the cross-lagged-panel analysis demonstrated that attitudes toward teaching the topic also play a major role when it comes to pre-service teachers' willingness. In fact, the results demonstrate that a positive attitude toward teaching about climate change before engaging with "Tutorials in Climate Change" positively influenced the development of willingness to teach about the topic. For teaching practice, we therefore conclude that a course on teaching about climate change should first address why it is necessary to teach about climate change (and therefore their attitudes) before engaging in learning activities that address conceptual understanding.

There are, however, several limitations that are associated with this study. For example, the sample size of 71 pre-service teachers, while sufficient for the analyses conducted, is comparably small. Furthermore, the implementation of the "Tutorials in Climate Change" varied across the universities involved in the study. Although this variability could be viewed as a limitation due to the potential inconsistencies it introduces in the intervention, it can also be seen as a strength. It highlights that the positive outcomes are resilient and effective across diverse instructional contexts and teaching approaches.

Finally, we conclude that "Tutorials in Climate Change" support teachers in developing their conceptual understanding about the scientific underpinnings of climate change as well as their willingness to teach about climate change.

4 Study 3: Assessing Critical Thinking Skills for Climate Literacy—Evaluating Credibility, Focusing on Arguments, and Reviewing Decision-Making Skills

The focus of the third study presented in the symposium is on the development of the test instrument "CTCC" (Critical Thinking in Climate Change) to assess critical thinking skills in the context of climate change, dealing with the second and fourth dimensions of climate literacy as defined by the USGCRP (2009). Addressing climate change requires both solid content knowledge—as discussed in study 1 and study 2—and the development of critical thinking skills. It is vital to comprehend the fundamental principles and complex aspects of climate change, such as the greenhouse effect, the carbon cycle, and the Earth's atmosphere (Schubatzky et al. 2023, 2024). Content knowledge provides the necessary foundation for engaging with the topic at a meaningful level. However, climate change is not just a scientific challenge, but also a subject of public debate; it is often shaped by non-scientific sources, such as social media (Treen et al. 2020). By building critical thinking skills, conscious consumption and use of information should be fostered (Chang et al. 2022).

Critical thinking skills, such as (1) verbal reasoning, (2) argument analysis, (3) hypothesis testing, (4) likelihood and uncertainty analysis, and (5) problem solving and decision-making (Halpern 1998), enable individuals to navigate the highly complex and multifaceted issue of climate change, question their own assumptions, and justify their opinions with evidence. This capacity for critical self-reflection and rational debate is essential for fostering responsible, self-determined citizens who can contribute constructively to climate action (Fraude et al. 2021). Basically, content knowledge equips individuals with the "what" of climate change, while critical thinking provides the "how" to engage with the issue responsibly and effectively.

To understand the extent to which these skills are developed, a test instrument is needed to evaluate them. For assessing content knowledge, a test instrument already exists to identify students' difficulties in the context of climate change. The so-called "CCCI-422" test instrument (Schubatzky et al. 2023) focuses on five concept areas related to climate change, which are (1) Earth's atmosphere, (2) the difference between weather and climate, (3) climate as a system, (4) the carbon cycle, and (5) the greenhouse effect. Additionally, there are established tests for evaluating general critical thinking skills, such as those developed by Ennis, Facione, and Halpern (for an overview see Tiruneh et al. 2017). However, there is currently no instrument specifically designed to measure critical thinking skills within the context of climate change.

To address this gap, a new test instrument, the "CTCC" test (Critical Thinking in Climate Change), has been developed. This test is designed to address, among other things, the following research question: How can critical thinking be measured in the context of climate change among high school and university students?

4.1 Research Design

To make critical thinking in the context of climate change assessable, the test is grounded in Halpern's critical thinking framework, which includes the five core skills (Halpern 1998), and it integrates the conceptual areas covered in the "CCCI-422" (Schubatzky et al. 2023). The "CTCC" test evaluates critical thinking by having students to, for example, evaluate conclusions, assess probabilities, and test hypotheses, each within the respective contexts of climate.

To begin with the test development process, 35 questions were designed to combine critical thinking skills within the climate change context. These were evaluated through semi-structured interviews with high school students from Saxony and pre-service physics teachers at TU Dresden ($N = 16$) to generate realistic distractors for the items of the single-choice test. The interviews were transcribed, categorized, and analyzed using qualitative content analysis according to Kuckartz and Rädiker (2022). Based on this input, the first version of the "CTCC" test was developed, consisting of 41 items. Due to the large number of items, the test was divided into two booklets, each containing 26 items, with 11 anchor items. The first round of data collection involved students from four universities as well as 10th- and 11th-grade high school students, resulting in a total sample size of $N = 240$ (convenience sample). On average, participants completed their booklet in 40 min. The data were analyzed using Rasch data analysis (Rasch 1960), which enabled evaluation of the items and their alignment with the measurement model. Following this analysis, the test was revised and reduced to 32 items, reflecting a clear optimization of the content. In the second iteration, the revised test was again divided into two booklets, each containing 22 items, with 12 items serving as anchors. Participants took approximately 30 min to complete the test. A subsequent Rasch data analysis revealed further insights, prompting additional refinements.

4.2 *Results*

The results from both test iterations were promising, with high person and item reliability metrics. Table 2 shows an overview of different metrics for the first and second iteration. It is important to emphasize that for the second round, stricter cut-off values for Infit and Outfit values were applied (0.8–1.2 (Bond and Fox 2015)) compared to 0.5–1.5 in the first round, which was a reasonable fit (Boone et al. 2014). Only four items exceeded the stricter limits, leading to their removal for the next version of the test.

In addition, the item-person map for the second iteration shows a good coverage of the population. There are minor gaps in items, particularly for participants with lower ability levels. Compared to the first version, these gaps have already been reduced

Table 2 Overview of the different metrics for the first and second rounds of administering the "CTCC" test

Metrics	First round	Second round
Person reliability	0.85	0.75
Item reliability	0.97	0.96
Lowest infit value	0.64	0.77
Highest infit value	1.39	1.32
Lowest outfit value	0.47	0.62
Highest outfit value	1.68	1.48

and the duplication of items with similar difficulty has decreased. The exclusion of four additional items in the next round is expected to optimize the test. Further reductions in test length are needed to enhance usability while maintaining adequate psychometric properties.

4.3 Discussion and Conclusion

Despite the promising results, a limitation of the current study is the wide-ranging sample used to achieve sufficient sample size. Surveying a mixed group, both high school and university students at different academic levels, may have led to unpredictability that makes it difficult to interpret the results. Focusing on a more homogeneous group, such as high school graduates or first-year University students, could provide clearer insights and facilitate a further improvement of the test.

However, these initial analyses and refinements demonstrate that the test is an effective instrument for measuring critical thinking skills in the context of climate change. Even with stricter evaluation criteria (Bond and Fox 2015), the instrument has proven its applicability to the intended populations of high school and university students. The development of the test satisfactorily addresses the first research question.

5 Study 4: Investigating Teachers' Intention to Focus on Informed and Responsible Decisions for "Taking Climate Action" in Physics Classrooms

The focus of the fourth study presented in the symposium is on physics teachers' perspectives about teaching students to make informed and responsible decisions about actions that may affect climate, which refers to the last dimension of climate literacy as defined by the USGCRP (2009). Decision-making processes on mitigation and adaptation measures can be summarized as "taking climate action," for which students need to develop respective knowledge, abilities, and skills (USGCRP 2009; UN General Assembly 2015). Thereby, it is important to consider scientific and other dimensions (e.g., social or economic) in an integrative manner to prepare students for the complexity, ill-structuredness, and interdisciplinarity of such "Socio-Scientific Issues" (SSI) (Sadler 2011).

However, teachers are confronted with numerous challenges in teaching climate action as SSI. For example, they are overwhelmed by the complexity and interdisciplinarity of the topic or feel unsure due to a lack of instructional skills, the politically charged nature of the topic, as well as the learning habits of students (Chen and Xiao 2021; Borgerding et al. 2024). In view of these challenges, Fasching et al. (2025) investigated Austrian physics teachers' intention to teach climate action as SSI, which is a mandatory topic in Austrian physics education due to the national curriculum (Austrian Federal Ministry of Education, Science and Research 2023). Also, they examined teachers' willingness to teach the topic if there was free choice, and the relation between intention, willingness, and cognitive constructs derived from the Theory of Planned Behavior (TPB) (Fasching et al. 2025). The study determined, for example, that the relation between the intention to teach the topic on the one side, and the extent of how easy teaching the topic is perceived on the other side, depends on the type of school the teachers are working in (Fasching et al. 2025).

According to the TPB, physics teachers' cognitive prerequisites of their intention and willingness to teach the topic could also depend on whether they have already taught climate action as SSI or not (Ajzen 2011). This research issue has not been investigated yet. Thus, the data set from Fasching et al. (2025) is used to answer the research question "To what extent are the TPB constructs and their relation to teaching intention and willingness different for teachers with and without experience in teaching climate action as SSI?".

5.1 Research Design

The cross-sectional online survey took place between October 2023 and February 2024 (Fasching et al. 2025). In-service physics teachers in Austria were asked to voluntarily participate in answering a questionnaire consisting of seven TPB constructs regarding teaching climate action as SSI: the intention to teach the topic with the knowledge that it is a mandatory topic in the national curriculum (INT), the willingness to teach it if there was free choice (WIL), the perceived ease (ATT-E) and necessity (ATT-N) regarding teaching the topic, the subjective norms or perceived expectations from the professional environment (SN), as well as the perceived self-efficacy (PBC-SE) and autonomy (PBC-A) to teach the topic (Fasching et al. 2025). All constructs were assessed by 7-point rating scales. In addition, teachers also had to self-report whether they had taught the topic before or not (Fasching et al. 2025). Overall, 206 in-service physics teachers participated in the survey. Information on the psychometric properties can be found in (Fasching et al. 2025). Similar to the original study, the data is now analyzed by applying a Rasch rating scale model and comparing the teachers with Topic Teaching Experience (TTE) to those without TTE. This is done by conducting t-tests, multigroup linear regressions, and chi-square difference analysis. Therefore, the TAM, lavaan, and lm.beta packages are used in R-Studio (Rosseel 2012; Behrendt 2023; R Core Team 2023; Robitzsch et al. 2022).

5.2 Results

Table 3 lists all the results of the Welch tests. These results indicate that teachers with TTE show a higher willingness to teach the topic if they had a free choice, as

Table 3 Mean, standard deviation and Welch's t-test results of TPB constructs depending on topic teaching experience (TTE), see construct abbreviations in the methods section

TPB scales	Overall (*N* = 206)		With TTE (*n* = 132)		Without TTE (*n* = 74)		*t*	*df*	*p*	Cohen's *d*
	M	*SD*	*M*	*SD*	*M*	*SD*				
INT	2.99	1.69	3.15	1.64	2.71	1.75	1.75	143.31	0.083	0.26
WIL	2.21	1.36	2.46	1.28	1.77	1.39	3.51	141.16	<0.001	0.52
ATT-E	1.22	0.68	1.35	0.64	0.57	0.69	3.84	142.91	<0.001	0.56
ATT-N	2.80	1.23	2.97	1.15	2.49	1.32	2.61	134.59	0.010	0.39
SN	1.19	0.94	1.29	0.87	1.00	1.04	2.02	129.56	0.046	0.30
PBC-SE	1.94	1.28	2.25	1.11	1.37	1.37	4.73	127.51	<0.001	0.71
PBC-A	1.44	0.64	1.54	0.59	1.28	0.68	2.73	134.87	0.007	0.40

well as a higher perception of ease, necessity, self-efficacy, and autonomy regarding teaching the topic. In addition, they perceive significantly stronger expectations of their environment to teach the topic than teachers without TTE. In contrast, no difference regarding the teaching intention is indicated. The greatest difference between the two subgroups of teachers can be noticed for the perceived self-efficacy, and the smallest for the perceived expectations from the environment.

Chi-square difference analysis indicates that intention and willingness to teach climate action as SSI are not significantly better explained through grouping teachers regarding their TTE in the multiple linear regression analysis ($\Delta\chi^2 = 17.62$, $p = 0.128$).

5.3 Discussion and Conclusion

While the extents of the cognitive constructs differ largely between teachers with and without TTE, the connections to teaching intention and willingness seem not to be related to the past teaching experiences of teachers. Regarding the extents of the TPB constructs, only the teaching intention does not differ significantly. This may result from the fact that teaching climate action as SSI was introduced as a mandatory topic in the Austrian physics curriculum, which affects both groups equally. The connection between TTE and self-efficacy is comprehensible, as teachers with higher self-efficacy may be more likely to implement the topic, or in turn, develop self-efficacy from teaching the topic. Thereby, it is not possible to determine the more probable direction of this effect. Regarding the predictions of intention and willingness to teach climate action as SSI through TPB constructs, no difference due to TTE must be assumed.

In addition to the limitations raised by Fasching et al. (2025), three completive aspects should be highlighted: First, the limited sample size, resulting in only 74 teachers without TTE, could lead to actual effects remaining insignificant and undetected. Second, participants' TTE was assessed based on self-reports, which enables the risk of incorrect information. Third, effects of teachers' TTE on the relations between TPB constructs could be mediated by other factors such as teaching beliefs.

The aim of making informed and responsible decisions regarding climate action is challenging for both physics students and teachers. For supporting the topic's implementation in physics classrooms, it is important to investigate teachers' perceptions of teaching the topic and design further teacher education courses on this basis. For example, the knowledge that teachers with TTE perceive teaching climate action as easier than those without TTE can be used to bring participants from both subgroups into exchange about teaching methods. Therefore, future research may also focus on other variables than TTE or apply person-centered approaches (Woo et al. 2018).

6 Addressing Dimensions of Climate Literacy in Physics Education: Discussion and Concluding Remarks

The four studies presented in this contribution illustrate possibilities of addressing the dimensions of climate literacy as defined by the USGCRP (2009) in physics education research. Each study thereby focuses on one or two dimensions of climate literacy, addressing either students, teachers, or pre-service teachers. To foster climate literacy among all students and enable them to navigate a world shaped by anthropogenic climate change, we argue that education research needs to include multiple levels, i.e., the dimensions of climate literacy, but also various target groups, i.e., learners and teachers. Promoting climate literacy requires teachers who (1) have a solid conceptual understanding of the scientific principles underlying climate change (study 2) and are able to select suitable learning materials, for example, concerning effective graphic representations (study 1), (2) demonstrate critical thinking skills to assess available information (study 3), and (3) are willing to teach both the scientific underpinnings of climate change and climate action as SSI (study 4).

In this context, physics education research plays a pivotal role as the science principles underlying climate change are physics phenomena. The four dimensions of climate literacy (USGCRP 2009) are intertwined and cannot be separated. Hence, climate literacy needs to be fostered in physics education and physics education research needs to explore learning processes (study 1), develop test instruments to diagnose students' competence (study 3), improve teaching approaches (study 1) and provide insights into teachers' (pre- and in-service) perceptions such as attitudes, self-efficacy, willingness, etc. (studies 2 and 4). In presenting four selected studies, we aimed to give insights into how these issues can be addressed in physics education research. For example, Design-Based Research projects can contribute to developing effective teaching and learning sequences addressing one or more dimensions of climate literacy. In study 1, this approach was outlined for the context of the greenhouse effect in lower secondary education. Another approach is to apply successful instruction strategies to the field of climate change education, like the tutorials developed in study 2. These materials build on ideas and insights from physics education research, and their evaluation shows promising results. When it comes to diagnosing learners' (students as well as pre-service teachers) knowledge, skills, and competence, validated and reliable test instruments need to be developed. Study 3 presents an example of such a test instrument, the "CTCC," linking climate change content knowledge with critical thinking skills. To inform interventions targeted at in- and pre-service teachers, the theory of planned behavior provides a valuable theoretical framework to investigate their prerequisites to teaching climate change and climate action as SSI, as shown by study 4 and study 2. Thereby, the interconnectedness of the different studies becomes visible: Studies that provide validated test instruments (study 3) and/or analyses of competences, conceptual understanding, or constructs like attitudes, willingness, self-efficacy, etc. (study 4) can inform the development of new interventions (studies 1 and 2), which, in turn, need to be evaluated (studies 1 and 2).

Several conclusions can be drawn from the presented studies concerning effective teaching in the context of climate literacy. The results of study 1 show that carefully designed visual representations hold great potential in supporting conceptual understanding of the greenhouse effect. However, students have difficulty applying central physics concepts like reflection and absorption of electromagnetic radiation to the context of the greenhouse effect, a very complex phenomenon. Therefore, we conclude that effective physics teaching needs to consider these difficulties. Study 2 provides first insights into the interconnectedness of pre-service teachers' conceptual understanding of and willingness to teach climate change. It also discusses the possible influence of other factors influencing the development of pre-service teachers' willingness to teach climate change, like their attitudes. Future research should use these first results as a starting point for further investigations of these relationships. Study 3 reports promising results in developing a validated and reliable test instrument that connects the content areas of climate change understanding and critical thinking skills. This opens up possibilities for an application in physics classrooms, to diagnose students' critical thinking skills in the context of climate change and base teaching interventions on the results. Additionally, future research can use this test instrument to test the effectiveness of teaching interventions. Study 4 is part of an analysis which factors influence in-service teachers' willingness and intention to teach climate action as SSI, focusing on teachers' topic teaching experience. For supporting the implementation of climate action as SSI in physics classrooms, teachers' perceptions need to be considered. The insights from study 4, for example concerning the relation of teaching experience and perceived ease, can be used to design and evaluate teacher education courses. Collectively, the findings from the studies highlight the need for teaching guidelines and teacher training programs that integrate evidence-based approaches, such as the use of effective visual representations and strategies to enhance teachers' willingness to teach climate action.

To conclude, the four studies presented in this contribution can serve as examples of how the dimensions of climate literacy can be addressed in physics education research, aiming to support effective teaching and learning by focusing on in-service and pre-service teachers' perceptions and conceptual understanding, upper secondary and university students' critical thinking skills in the context of climate change, and lower secondary students' conceptual understanding. By introducing these studies and the issues addressed by them, we intended to give examples of aspects that need to be considered in physics education research to address the dimensions of climate literacy and demonstrate possibilities for how they can be targeted.

References

Ajzen I (2011) Behavioral interventions, social psychology and evaluation. In: Mark M, Donaldson S, Campbell B (eds) Social psychology and evaluation. Guilford Press, New York, pp 74–103

Austrian Federal Ministry of Education, Science and Research (2023) Gesamte Rechtsvorschrift für Lehrpläne – allgemeinbildende höhere Schulen.https://ris.bka.gv.at/geltendefassung.wxe?abfrage=bundesnormen%20&gesetzesnummer=10008568&ShowPrintPreview=True. Accessed 19 Jan 2024

Behrendt S (2023) lm.beta: add standardized regression coefficients to linear-model-objects (Version 1.7–2) [Computer Software] https://CRAN.R-project.org/package=lm.beta. Accessed 21 May 2024

Bond TG, Fox CM (2015) Applying the Rasch model: fundamental measurement in the human sciences, 3rd edn. Routledge, New York

Boone WJ, Staver JR, Yale MS (2014) Rasch analysis in the human sciences. Springer, Dordrecht

Borgerding LA, Heisler JL, Beaver BC, Prince AO (2024) Secondary science teachers' views and approaches for teaching for climate justice and action. J Sci Teacher Educ 35:320–342. https://doi.org/10.1080/1046560X.2023.2265715

Chang CH (2022) Climate change education. Knowing, doing and being. Routlege, London

Chen L, Xiao S (2021) Perceptions, challenges and coping strategies of science teachers in teaching socioscientific issues: a systematic review. Educ Res Rev 32:100377. https://doi.org/10.1016/j.edurev.2020.100377

Eyring V et al (2023) Human influence on the climate system. In: Intergovernmental Panel on Climate Change (ed) Climate change 2021. The physical science basis: Working Group I contribution. Cambridge University Press, Cambridge, pp 423–552

Fasching M, Schubatzky T, Hopf M (2025) Predictors of in-service teachers' intention and willingness to teach climate action in physics classrooms. J Sci Teacher Educ 36(6):831–856. https://doi.org/10.1080/1046560X.2025.2452735

Förtsch C, Werner S, von Kotzebue L, Neuhaus BJ (2016) Effects of biology teachers' professional knowledge and cognitive activation on students' achievement. Int J Sci Educ 38(17):2642–2666. https://doi.org/10.1080/09500693.2016.1257170

Fraude C, Bruhn T, Stasiak D, Wamsler C, Mar K, Schäpke N, Schroeder H and Lawrence M (2021) Creating space for reflection and dialogue. Examples of new modes of communication for empowering climate action. GAIA 30:174–180. https://doi.org/10.14512/gaia.30.3.9

Halpern DF (1998) Teaching critical thinking for transfer across domains: disposition, skills, structure training, and metacognitive monitoring. Am Psychol 53:449–455

Heuckmann B, Hammann M, Asshoff R (2020) Identifying predictors of teachers' intention and willingness to teach about cancer by using direct and belief-based measures in the context of the theory of planned behaviour. Int J Sci Educ 42:547–575. https://doi.org/10.1080/09500693.2020.1717671

Kabapınar F (2005) Effectiveness of teaching via concept cartoons from the point of view of constructivist approach. Educ Sci: Theory Pract 5:135–146. https://doi.org/10.12691/education-9-9-5
Kuckartz U, Rädiker S (2022) Qualitative Inhaltsanalyse. Methoden, Praxis, Computerunterstützung. Grundlagentexte Methoden. Beltz Juventa, Weinheim
Kulgemeyer C (2018) Towards a framework for effective instructional explanations in science teaching. Stud Sci Educ 54(2):109–139. https://doi.org/10.1080/03057267.2018.1598054
Kulgemeyer C, Riese J (2018) From professional knowledge to professional performance: the impact of CK and PCK on teaching quality in explaining situations. J Res Sci Teach 55:1393–1418. https://doi.org/10.1002/tea.21457
Li CJ, Monroe MC, Oxarart A, Ritchie T (2021) Building teachers' self-efficacy in teaching about climate change through educative curriculum and professional development. Appl Environ Educ Commun 20:34–48. https://doi.org/10.1080/1533015X.2019.1617806
Liu S, Roehrig GH, Bhattacharya D, Varma K (2015) In-service teachers' attitudes, knowledge and classroom teaching of global climate change. Sci. Educ 24:12–22
Lovett P, Shah M (ed) (2007) Thinking with data. In: The 33rd annual carnegie symposium on cognition. Lawrence Erlbaum Associates, New Jersey
Mayer RE, Moreno R (2003) Nine ways to reduce cognitive load in multimedia learning. Educ Psychol 38:43–52. https://doi.org/10.1207/S15326985EP3801_6
McDermott LC, Shaffer PS (2002) Tutorials in introductory physics. University of Washington
Obczovsky M, Bernsteiner A, Haagen-Schützenhöfer C, Schubatzky T (2024) Systematizing decisions in design-based research: from theory to design. Sci Educ 109:523–536. https://doi.org/10.1002/sce.21915
Plutzer E, McCaffrey M, Hannah AL, Rosenau J, Berbeco M, Reid AH (2016) Climate confusion among U.S. teachers. Science 351:664–665. https://doi.org/10.1126/science.aab3907
Pörtner HO et al (ed) (2022) Climate Change 2022: impacts, adaptation, and vulnerability. Contribution of Working Group II to the sixth assessment report of the intergovernmental panel on climate change. Cambridge University Press, Cambridge, New York
R Core Team (2023) R: a language and environment for statistical computing (Version 4.3.1) [Computer software]. https://www.R-project.org/. Accessed 14 May 2024
Rasch G (1960) Probabilistic models for some intelligence and attainment tests. Danmarks Paedagogiske Institut, Copenhagen
Robitzsch A, Kiefer T, Wu M (2022) TAM: test analysis modules: R package (Version 4.1–4) [Computer software]. https://CRAN.R-project.org/package=TAM
Rosseel Y (2012) Lavaan: an R package for structural equation modeling [Computer Software]. J Stat Soft 48:1–36
Sadler PM, Sonnert G, Coyle HP, Cook-Smith N, Miller JL (2013) The influence of teachers' knowledge on student learning in middle school physical science classrooms. Am Educ Res J 50:1020–1049. https://doi.org/10.3102/0002831213477680
Sadler TD (2011) Socio-scientific issues in the classroom: teaching, learning and research. Springer, Netherlands, Dordrecht
Sandoval WA, Bell P (2004) Design-based research methods for studying learning in context: introduction. Educ Psychol 39(4):199–201. https://doi.org/10.1207/s15326985ep3904_1
Schubatzky T, Wackermann R, Wöhlke C, Haagen-Schützenhöfer C (2024) How well do German A-level graduates understand the scientific underpinnings of climate change? Sustainability 16:7264. https://doi.org/10.3390/su16177264
Schubatzky T, Wackermann R, Wöhlke C, Haagen-Schützenhöfer C, Jedamski M, Lindemann H, Cardinal K (2023) Entwicklung des Concept-Inventory CCCI-422 zu den naturwissenschaftlichen Grundlagen des Klimawandels. Zeitschrift für Didaktik der Naturwissenschaften 29:159. https://doi.org/10.1007/s40573-023-00159-8
Shepardson DP, Roychoudhury A, Hirsch AS (2017) Using conceptual and physical models to develop students' mental models of the greenhouse effect. In: Shepardson DP, Roychoudhury

A, Hirsch AS (eds) Teaching and learning about climate change—a framework for educators. Routledge, New York
Tiruneh DT, De Cock M, Weldeslassie AG, Elen J, Janssen R (2017) Measuring critical thinking in physics: development and validation of a critical thinking test in electricity and magnetism. Int J Sci Math Educ 15:663–682. https://doi.org/10.1007/s10763-016-9723-0
Treen K, Williams H and O'Neill S (2020) Online misinformation about climate change. WIREs Clim Change 11. https://doi.org/10.1002/wcc.665
UN General Assembly (2015) Transforming our world: the 2030 agenda for sustainable development. https://www.un.org/en/development/desa/population/migration/generalassembly/docs/globalcompact/A_RES_70_1_E.pdf. Accessed 14 May 2024
USGCRP (2009) Climate literacy: the essential principles of climate science
Wiesner H, Wodzinski R (1996) Akzeptanzbefragungen als Methode zur Untersuchung von Lernschwierigkeiten. In Duit R (ed) Lernen in den Naturwissenschaften. Beiträge zu einem Workshop an der Pädagogischen Hochschule Ludwigsburg. IPN, pp 250–274
Wildbichler S, Schubatzky T, Haagen-Schützenhöfer C (2024) Students' ideas about the scientific underpinnings of climate change: a systematic review of the literature. Stud Sci Educ 61(1):117–169. https://doi.org/10.1080/03057267.2024.2395206
Wildbichler S, Wackermann R, De Cock M, Ivanjek L, Micoloi M, Pospiech G, Schubatzky T (2025) Tutorials in climate change. Phys Teach 63:414–417. https://doi.org/10.1119/5.0224098
Woo SE, Jebb AT, Tay L, Parrigon S (2018) Putting the "person" in the center. Organ Res Methods 21(4):814–845. https://doi.org/10.1177/1094428117752467

Rethinking Agency in the Context of Climate Change: The ENCOMPASS Three-Lens Perspective

Giulia Tasquier, Francesca Pongiglione, and Elena Claire Ricci

Abstract Younger generations are showing an unprecedented willingness to address the climate crisis. However, a concurrent willingness to act tends to fall short in the absence of adequate information on both the causal mechanisms that connect human actions to climate change and the behaviours that are most effective in contributing to emission cuts. This paper highlights the role of science education in addressing these shortcomings and fostering transformative agency. By integrating insights from philosophy and economic studies, science education can empower students with the knowledge, ethical grounding and practical skills necessary to act as agents of change. We propose a multi-disciplinary perspective for science education that emphasises interdisciplinarity and transformative competencies, preparing learners to navigate the complexities of climate change and sustainability.

Keywords Science education · Climate change · Agency · Philosophy · Economics · Sustainability

1 Introduction

Climate change represents an unprecedented global challenge, profoundly impacting ecosystems, economies and societies. Over recent decades, the urgency of mitigating its effects has gained widespread recognition (IPCC, 2022). However, being aware of climate change does not automatically imply being practically engaged with its

G. Tasquier (✉)
Department of Physics and Astronomy "A. Righi", Alam Mater Studiorum, University of Bologna, Bologna, Italy
e-mail: giulia.tasquier2@unibo.it

F. Pongiglione
Faculty of Philosophy, Vita-Salute S. Raffaele University, Milan, Italy

E. C. Ricci
Department of Management, University of Verona, Verona, Italy

D. Sokołowska and E. McLoughlin (eds.), *Embracing Changes Together in Physics Education*, Challenges in Physics Education,
https://doi.org/10.1007/978-3-032-27385-7_16

mitigation. European surveys among the general population reveal pervasive misconceptions about the actions required to address climate change effectively. Many individuals focus on low-impact measures, such as reducing plastic consumption, while neglecting more impactful strategies like advocating for systemic policy changes or altering energy use patterns (Special Eurobarometer, 2023).

The role of education in bridging this gap cannot be overstated. Science education, in particular, is uniquely positioned to equip learners with the knowledge, skills and values necessary to understand and address the complexities of climate change. Yet, traditional approaches to science education often emphasise factual knowledge over action competencies, failing to foster the transformative agency required to navigate the socio-ecological challenges of the Anthropocene (White et al. 2023).

Despite ongoing efforts by international organisations to promote scientific literacy, nearly half of the world's school curricula lack explicit references to climate change (UNESCO, 2021). This omission undermines students' ability to critically evaluate climate science, engage in ethical reasoning and participate in effective action. Addressing this educational gap is critical for empowering future generations to act as agents of change. Recent frameworks, such as GreenComp and the PISA 2025 science framework, underscore the importance of embedding interdisciplinary, action-oriented learning in science education (Bianchi et al. 2022; White et al. 2023).

The limitations of traditional science education highlight the need for a reimagined approach that integrates insights from multiple disciplines. Philosophy and economics offer complementary perspectives that enrich science education's potential to address climate challenges. Philosophy provides a foundation for exploring ethical considerations and moral responsibilities, while economic science contributes with more pragmatic tools for understanding decision-making processes and systemic trade-offs. Together, these disciplines can enhance science education by fostering critical, ethical and systemic thinking.

This paper argues for an integrated approach that combines the strengths of science education, philosophy and economics to both unpack and deepen the concept of agency and cultivate transformative agency in educational contexts. By embedding interdisciplinary competencies within educational practices, this approach aims to empower students to critically engage with climate issues, envision sustainable futures and take meaningful action. The main objective of this work is to theoretically explore how science education can be restructured to foster transformative agency in addressing climate change. Specifically, the research seeks to answer the following question: *How can a multi-disciplinary and multi-dimensional framework integrating science education, philosophy and economics enhance students' ability to act as agents of change in the context of climate change?*

This paper is primarily a conceptual contribution, developed through a synthesis of existing literature in science education, philosophy and economics. While it does not present empirical data, it lays the groundwork for future empirical research, which will test and operationalise the proposed three-lens framework in real classroom contexts.

The paper is structured as follows: Sect. 2 explores the concept of agency within science education and the role of science education in fostering agency, emphasising its transformative potential. Section 3 discusses the contributions of philosophy and economics, highlighting their relevance to climate education. Section 4 proposes a multi-disciplinary and multi-dimensional framework for integrating these perspectives into science education. Finally, Sect. 5 concludes by reflecting on the implications of this approach for educational practices and policy.

2 Agency within Science Education Research Area

Science education plays a critical role in equipping learners with the competencies needed to address climate change. As global challenges grow more complex, the demand for education systems to move beyond traditional knowledge transfer becomes increasingly urgent. The evolving definition of scientific literacy provides a pathway for such transformation. Vision III, as discussed by Sjöström and Eilks (2018), extends the concept of literacy from foundational scientific knowledge (Vision I) and reasoning in socio-scientific issues (Vision II) to include transformative dimensions that emphasise actionable agency.

Scientific literacy encompasses more than the mere understanding of facts. It requires students to critically evaluate scientific information, engage in systems thinking and make informed decisions. The PISA 2025 science framework identifies transformative agency as central to equipping learners for the uncertainties of the future. Agency, in this context, is defined as the ability to act purposefully and effectively in response to evolving socio-ecological challenges (White et al. 2023).

However, the current state of science education falls short in fostering these competencies. Traditional curricula prioritise knowledge acquisition over the application of knowledge to real-world challenges. As a result, students are often unprepared to navigate the complexities of climate change, limiting their ability to act as informed and responsible agents of change (Tasquier et al. 2022, 2025).

The evolution of the concept of agency within science education highlights a shift from foundational knowledge (Vision I) to reasoning within socio-scientific issues (Vision II) and, ultimately, to actionable, transformative agency (Vision III). Vision I emphasised equipping students with the scientific principles underlying phenomena, fostering foundational literacy in isolated contexts. Vision II expanded this scope, integrating socio-scientific issues to promote critical thinking and the application of science to societal challenges. However, despite its contributions, Vision II often limited agency by failing to emphasise collective and systemic change, instead focusing primarily on individual cognitive development. Vision III moves beyond these constraints by advocating for competencies that empower learners to envision sustainable futures and act on complex socio-ecological problems (Sjöström & Eilks, 2018).

Despite its potential, science education faces several challenges in fostering agency. One major issue is the persistent focus on individual action rather than systemic change. While personal behavioural shifts are important, they are insufficient to address the scale of the climate crisis. Educators must therefore broaden their focus to include collective and political dimensions of action (Otto et al. 2020).

Additionally, other studies have highlighted the need for interdisciplinary approaches in science curricula to bridge the gap between scientific knowledge, ethical reflection and systemic economic understanding (Tasquier et al. 2022). Without such integration, students are often unable to contextualise their learning within broader frameworks of societal and global change, thereby limiting their ability to develop transformative agency. This gap underscores the need for a more holistic approach that incorporates insights from philosophy and economics. Philosophy contributes a deeper understanding of moral values and responsibilities, while economic science offers tools for evaluating trade-offs and systemic impacts. Together, these perspectives provide a richer framework for fostering transformative agency.

3 The Role of Philosophy and Economics in Fulfilling the Concept of Agency

Integrating philosophy and economics into science education offers a profound opportunity to address the limitations of traditional curricula and enhance the concept of agency. Philosophy contributes a critical lens through which students can examine the ethical dimensions of climate action. It enables learners to engage with moral reasoning, explore the values underpinning decisions and reflect on their responsibilities as global citizens. For instance, understanding the demands of intergenerational justice or the ethical considerations surrounding resource distribution adds depth to students' understanding of climate issues. These philosophical perspectives empower students to not only grasp the scientific principles of climate change but also connect them with broader societal and ethical implications.

Economics, on the other hand, provides a pragmatic framework for analysing decision-making processes and systemic trade-offs. By introducing economic models, students can explore the complexities of balancing short-term costs with long-term benefits, understanding incentives and evaluating policy effectiveness. Economic science sheds light on the systemic barriers to sustainable action, such as market failures, and individual and societal constraints. For example, exploring the economic principles of carbon pricing or analysing the cost-benefit dynamics of renewable energy investments equips students with practical tools to assess real-world challenges. This pragmatic approach complements the ethical insights provided by philosophy, grounding moral considerations in actionable strategies.

Together, philosophy and economic science bridge the gap between knowledge and action. They illuminate the interplay between individual choices and systemic outcomes, emphasising the importance of both ethical reflection and practical decision-making. By integrating these disciplines, science education can foster a more holistic understanding of agency. Students learn to navigate the tensions between personal values and societal needs, critically evaluate trade-offs and envision pathways towards sustainable futures. This approach aligns with contemporary educational frameworks, such as GreenComp, which advocate for systems' thinking- and future-oriented competencies (Bianchi et al. 2022).

Furthermore, the integration of philosophy and economics enriches the pedagogical landscape by encouraging dialogue and critical inquiry. Philosophical discussions about the nature of responsibility, equity and justice intersect with economic analyses of feasibility and efficiency, creating a multi-dimensional learning experience. This dynamic interplay fosters a deeper sense of agency, equipping students to address the complexities of climate change with both conviction and competence. By embedding these perspectives into science education, we support the importance of cultivating learners who are not only scientifically literate but also ethically grounded and pragmatically skilled, ready to act as transformative agents of change.

4 The ENCOMPASS Three-Lens Perspective

4.1 A Multi-Disciplinary and Multi-Dimensional Approach to Agency

Integrating science education with philosophy and economics is not merely an additive process but a transformative one. Each discipline brings unique insights to the concept of agency, creating a multi-dimensional framework that has the potential to empower people to act effectively in addressing climate challenges. Our conceptualisation resonates with recent debates on human agency in the Anthropocene (Otto et al. 2020; Kranz et al. 2025) and builds on Vision III of scientific literacy as outlined by Sjöström and Eilks (Sjöström & Eilks, 2018), which highlights the transformative potential of science education for sustainability. By integrating philosophy and economics, our framework deepens this perspective, operationalising transformative agency through ethical and systemic dimensions. Science education provides the foundational knowledge necessary for understanding the mechanisms and impacts of climate change. However, it often stops short of equipping learners to translate this knowledge into meaningful action. Philosophy and economics address this gap, offering complementary perspectives that enhance the depth and breadth of educational practices.

Philosophical reflection is essential to understand why people are being asked to take action on climate change. A purely scientific explanation cannot provide these reasons, which are uniquely ethical. There is a need to address the suffering that this phenomenon is already causing and that will worsen over time, and to reflect on one's duties. As observed by several studies in environmental ethics, it is by no means straightforward to theorise the existence of an individual duty to engage in the fight against climate change (Fragnière, 2016). To do so requires rethinking traditional moral theories by considering some of the most problematic issues, including the global dimension within which human actions unfold, the lack of direct causality between actions and consequences and the absence of negative intentions behind actions that prove harmful to the environment. Without such an inquiry, it is impossible to provide new generations with sufficient elements to motivate the call for their involvement in climate action, as well as in other global issues.

Economic science complements this ethical lens by providing a practical understanding of decision-making processes and systemic trade-offs. It equips people with tools to analyse the feasibility and efficiency of various climate solutions, bridging the gap between abstract principles and real-world applications. By examining market failures (for example, due to externalities or information asymmetry) and promoting thinking in terms of differential costs and benefits of different (individual or collective) strategies, economics helps learners navigate the complexities of policymaking and resource allocation. At a more individual level, behavioural economics can shed light on preferences and on how these are formed and evolve over time, following both rational and more impulsive mechanisms (Kahneman, 2011), highlighting barriers and enablers of climate action. This pragmatic perspective ensures that individuals can assess the viability of their actions, on one hand, reflecting on the complexity of their own decision-making processes as citizens and consumers (Thaler & Sunstein, 2008) and, on the other hand, evaluating the effects of their action within broader socio-economic systems (Dietz, Ostrom & Stern, 2009). This enables them to better understand the implications of policies and their potential effectiveness, and thus advocate for or against these in a more conscious manner.

Figure 1 illustrates the integration of these three disciplines within the framework of agency, highlighting their unique contributions and synergies.

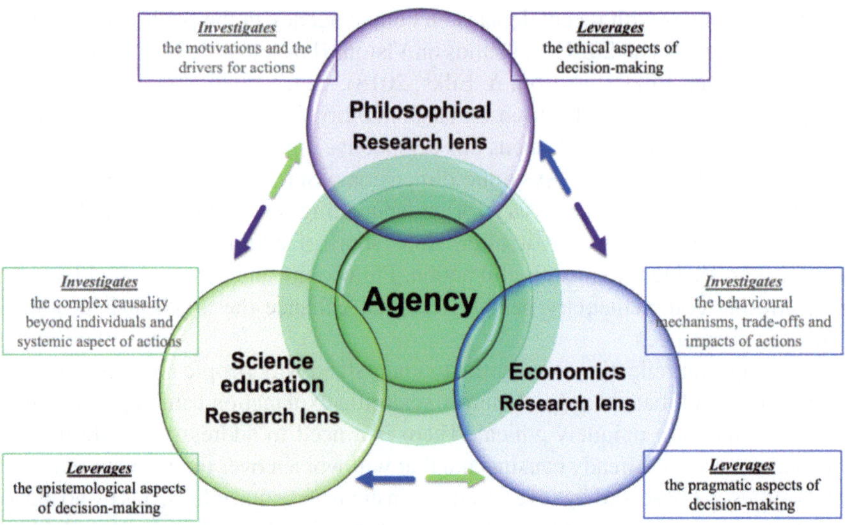

Fig. 1 Features of the three-lens perspective

This model underscores the importance of interdisciplinary collaboration in fostering transformative agency. By leveraging the strengths of science education, philosophy and economics, we can create a comprehensive approach that prepares learners to navigate the complexities of a rapidly changing world and contribute to a sustainable future (Tasquier et al. 2022; 20,225).

The integration of these disciplines within science education creates a holistic approach to agency, characterised by three interconnected dimensions. Firstly, the ***epistemic-driven*** dimension of agency emphasises the critical evaluation of scientific knowledge, encouraging students to question assumptions, synthesise information and develop evidence-based solutions. Secondly, the ***ethical-reflective*** dimension of agency fosters a value-oriented approach to decision-making, enabling learners to navigate moral dilemmas and align their actions with a broader concept of moral responsibility. Finally, the ***systemic-pragmatic*** dimension of agency focuses on the strategic and operational aspects of climate action, preparing students to implement solutions that are both effective and equitable. Together, these dimensions form a cohesive framework that transcends the limitations of traditional curricula, empowering students to address the multifaceted challenges of the Anthropocene.

4.2 Three Interrelated Dimensions of Agency

This ENCOMPASS three-lens perspective offers a structured approach to integrate three key dimensions of agency—epistemic-driven, ethical-reflective and systemic-pragmatic, usually kept separate. At the heart of this framework is the recognition that agency is inherently situated within multi-level causal structures, requiring individuals to navigate epistemic, ethical and systemic dimensions of the complex reality. Science education engages with the epistemic-driven dimension, allowing learners to grapple with how knowledge is constructed and applied in complex systems. Philosophy introduces the ethical-reflective dimension, demanding reflection on the moral weight of actions, responsibility and justice within human-environment interactions. Economics, in turn, provides an understanding of the systemic-pragmatic dimension, equipping students to analyse the mechanisms through which individual choices aggregate to shape collective outcomes. Figure 2 shows the three interrelated dimensions of agency in relation to the different disciplinary fields.

Specifically, science education serves as the foundation of epistemic agency, offering students forms of reasonings and methods to navigate the complexities of scientific knowledge production, uncertainty and causality. The science of climate change itself is deeply embedded in questions of causal relationships—understanding feedback loops, probabilistic models and non-linear interactions that shape planetary systems. The challenge in science education is not simply to convey climate facts but to cultivate an understanding of how scientific knowledge is constructed and revised in response to emerging evidence. Epistemic complexity becomes particularly salient in confronting multi-factorial causality in climate systems, where no single variable can account for environmental change. Greenhouse gas emissions, biodiversity loss,

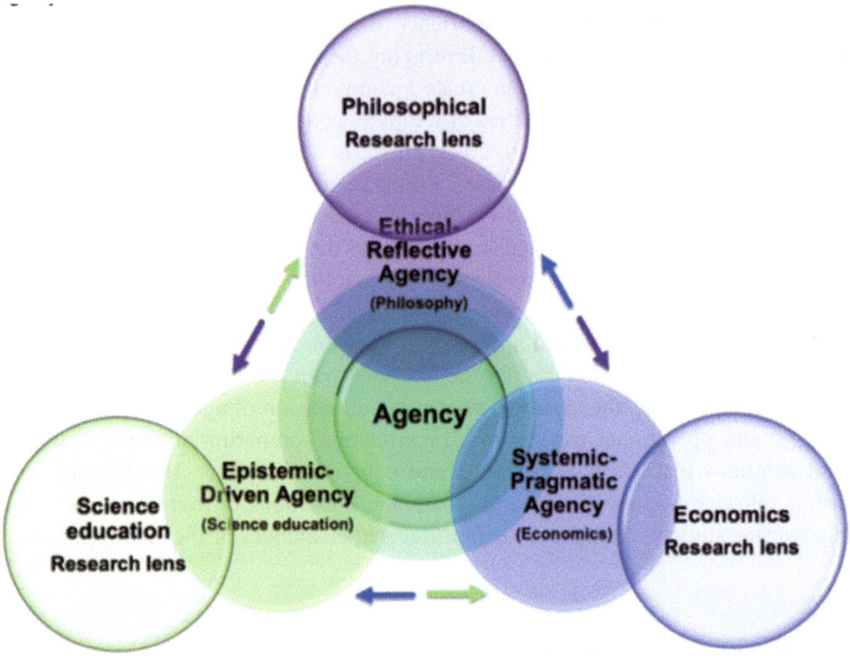

Fig. 2 Three interrelated dimensions of agency

oceanic circulation shifts and land-use transformations interact in ways that resist simplistic causal explanations. Engaging with this complexity requires students to move beyond deterministic thinking, fostering an ability to reason through probabilistic models, systems thinking and the limits of predictability in climate science (Sterman, 2000). Without developing epistemic agency, students can risk either overconfidence in isolated solutions or paralysis in the face of uncertainty—both of which undermine agency. However, this epistemic dimension needs to be integrated with broader competencies to ensure that knowledge is actionable. Indeed, while empirical data and climate models provide insights into future scenarios, the translation of knowledge into decision-making requires additional competencies. This epistemic awareness redefines agency, developing the cognitive flexibility to navigate complex, contested and evolving knowledge landscapes.

The philosophical lens extends agency beyond scientific knowledge by fostering moral reflection on responsibility, justice and collective action. Philosophy critically examines the motivations and consequences of climate-related decisions, ensuring that agency is not only informed by science but also driven by values. While science education provides students with the forms of reasonings to navigate the complexity of uncertainty and causality in climate systems, philosophy asks them to reflect on the implications of acting (or failing to act) within these systems. The ethical dimensions of climate change education cannot be reduced to abstract discussions of right and wrong; rather, they require an engagement with the moral weight of causation, considering who is responsible, who is affected and how different forms of agency contribute to (or mitigate) harm. One of the central ethical challenges in climate change education is the dispersion of responsibility across temporal, spatial and political scales. Unlike direct harm scenarios, climate change involves diffused causal chains, where individual actions accumulate into systemic consequences that disproportionately affect marginalised communities and future generations. Philosophical inquiry enables students to confront these dilemmas by engaging with questions of intergenerational justice, ecological responsibility and collective action ethics (Cripps, 2013; Gardiner, 2011). This type of reasoning within philosophy helps to move beyond an individualistic conception of responsibility. Furthermore, philosophy interrogates the tensions between personal and collective agency, exploring the ethical implications of action at different scales. Should individuals bear responsibility for reducing their carbon footprints, or does meaningful action require systemic transformation? How should ethical considerations shape climate policies, and what justifications exist for prioritising certain actions over others? These questions push us to recognise that ethical agency is not merely about acting in accordance with personal values but also involves engaging in critical deliberation about the justifications, trade-offs and consequences of different choices. Ethical-reflective agency is at the base of the development of moral reasoning competencies that equip people to navigate value conflicts and decision-making under uncertainty. Without developing ethical-reflective agency, the educational process risks reducing climate action to technical solutions or personal lifestyle choices, disconnected from broader social and moral contexts. Students may lack the capacity to navigate the value conflicts and uncertainties that climate decision-making entails.

The economic lens introduces a perspective that is both pragmatic and systemic, illuminating how individual and collective actions interact within complex institutional and financial structures. Differently from the epistemic-driven dimension of science or the ethical-reflective dimension of philosophy, the systemic-pragmatic dimension of economics emphasises causality within human-designed systems, where choices are shaped by incentives, constraints and policy mechanisms. While science education helps students understand the dynamics of the climate system, and philosophy deepens this through ethical reflection, economics expands agency by situating decision-making within real-world structures—highlighting how choices are shaped by incentives, constraints and institutional frameworks. One of the fundamental challenges in climate education is recognising that agency is not exercised in a vacuum but within existing economic and political structures. Understanding economic causality allows students to explore why certain climate policies succeed or fail, how systemic incentives shape behaviour, and what constraints limit the scalability of solutions. This dimension of agency moves students beyond idealistic commitments to climate action, equipping them to navigate the realities of policy implementation, financial mechanisms and institutional constraints. Economic literacy is particularly crucial in climate education because many high-impact decisions involve navigating trade-offs and resource allocation. Addressing climate change requires weighing immediate costs against long-term benefits, balancing economic growth with sustainability and understanding the implications of policy interventions such as carbon pricing, subsidies and regulatory frameworks. Without developing systemic-pragmatic agency, students may remain limited to ethical intention or scientific understanding, so climate action risks becoming either idealistic or impractical. Systemic-pragmatic agency ensures that learners can navigate the real-world tensions by anchoring agency not only in values, but in viable pathways for transformation. Furthermore, economics provides insight into how agency ranges from individual choices to collective transformations. It highlights the mechanisms through which policy, market dynamics and institutional change influence climate outcomes, demonstrating that effective climate action requires not only individual ethical agency but also engagement with broader economic structures and political processes. Integrating economics into science education enables learners to navigate the practical dimensions of climate action, ensuring that agency is both ethically guided and pragmatically viable.

Interweaving epistemic-driven, ethical-reflective, and systemic-pragmatic dimensions of agency—respectively grounded in science education, philosophy and economics—constructs a multi-disciplinary and multi-dimensional approach that both unpacks and fulfils the complex space of the concept of agency. This approach has several educational potentialities. For instance, science education provides the epistemic tools to understand how climate systems function and how knowledge interacts with decision-making. Philosophy fosters ethical reflection on responsibility, justice and the moral weight of action. Economic science situates agency within institutional and systemic structures, offering tools to navigate constraints, trade-offs and mechanisms of large-scale change. Together, these disciplines ensure that agency in climate education is neither naively individualistic nor overwhelmed by systemic inertia. Instead, it becomes an informed, reflective and action-oriented capacity, one that enables students to move between different scales of agency, from personal commitment to collective transformation.

4.3 Building Educational Design Principles

This ENCOMPASS three-lens perspective offers a structured approach to integrate three key dimensions of agency—epistemic-driven, ethical-reflective and systemic-pragmatic—into educational design. This framework shifts from the theoretical elaboration of the framework above to practical applications, turning the lenses into actionable educational principles that guide the design and implementation of curricula. Rather than isolating these dimensions within separate educational strategies, the challenge of educational design lies in embedding them simultaneously into learning environments so that students develop the capacity to navigate scientific, ethical and systemic complexities in an integrated way. The following design principles articulate how this triadic framework generates transversal design principles that integrate epistemic-driven, ethical-reflective and systemic-pragmatic agency in a coherent and interdependent manner.

Design Principle 1 (DP1): Moving from Abstraction to Situated Real-World Complex Challenges.

To cultivate agency, it is needed to generate a shift from abstract scenarios to meaningful engagement with real-world challenges—essential for cultivating relevance and agency-oriented thinking. A transversal approach requires that climate learning is embedded within context-rich, socio-ecological settings where students engage with uncertain, contested and evolving knowledge. Epistemic-driven agency is fostered by requiring students to work with dynamic, incomplete and contested scientific knowledge—for instance, by engaging with real-world climate models, assessing the uncertainties in emissions projections or questioning how different scientific disciplines construct knowledge about climate change. This prevents students from viewing science as a static repository of facts and instead encourages them to engage with knowledge as an evolving and contested process (Tasquier & Pongiglione, 2017). Ethical-reflective agency emerges when students critically examine how scientific uncertainty intersects with values and decision-making. Systemic-pragmatic agency is developed when students engage in decision-making exercises that require balancing uncertainty with action.

Climate change is not a problem with clear-cut answers. It involves dynamic systems, uncertain projections and contested knowledge. This principle aims to mirror this complexity by encouraging students to navigate ambiguity, explore questions with no single correct answer and understand that knowledge itself is under construction.

Design Principle 2 (DP2): Embedding Reflexive and Multi-perspective Deliberation.

To develop transformative agency, it is required to embed diverse perspectives, values and structures that shape climate action. This principle ensures that students learn to deliberate across multiple knowledge systems, ethical frameworks and economic constraints, fostering the ability to reason critically, negotiate tensions and justify decisions in complex contexts. Epistemic-driven agency is cultivated by requiring students to engage with diverse forms of knowledge production—from climate science and diverse form of ecological knowledge to economic models and policy evaluations. This pluralistic approach helps students understand that no single discipline holds all the answers and that complex problems require knowledge integration and negotiation. Ethical-reflective agency develops when students are asked to justify decisions across multiple ethical frameworks. For instance, they might compare deontological, utilitarian and justice-based arguments for climate action, or debate how different cultural traditions conceptualise human responsibilities to nature. This deliberative approach fosters students' ability to engage in reasoned, value-based discourse, preparing them to contribute meaningfully to democratic debates on climate policy (Thaler, 2008). Systemic-pragmatic agency emerges through multi-stakeholder deliberation exercises where students must balance competing interests, trade-offs and feasibility constraints. Whether debating energy transitions, carbon pricing or environmental regulations, students learn that climate action is not just a matter of knowledge or ethics but also of navigating political and economic realities.

Climate change decision-making is shaped by competing values, knowledge pluralism and structural constraints. This principle aims to reflect that complexity by enabling students to navigate tensions across perspectives, justify choices and participate in decision-making where no solution is neutral.

Design Principle 3 (DP3): Fostering Strategic Action Through Applied and Adaptive Learning.

To cultivate agency, it is necessary to address the gap between knowing about climate change and being able to act meaningfully within real-world constraints. This principle calls for learning experiences that are applied, strategic and adaptive—enabling students to engage with the political, institutional and economic dimensions of climate action in ways that are both context-sensitive and impact-oriented. Epistemic-driven agency is strengthened when students learn to critically assess different pathways for action, recognising the limits of purely scientific interventions. This includes evaluating technological solutions, policy innovations and economic instruments, ensuring that they can apply scientific knowledge in ways that are context-sensitive and adaptive (Sterman, 2000). Ethical-reflective agency is reinforced when students engage in consequentialist reasoning about the impact of different interventions. For example, students might examine how climate policies distribute benefits and burdens across different social groups, ensuring that their decisions account for intergenerational and global justice considerations (Gardiner, 2011). Systemic-pragmatic agency is realised when students engage in applied, experiential learning that mirrors real-world decision-making processes. This includes developing and testing climate action plans, engaging in community-based research and participating in policy advocacy initiatives. Learning through adaptive action cycles ensures that students develop the capacity to assess, refine and iterate their interventions in response to real-world feedback.

Climate action is shaped by dynamic constraints and evolving opportunities; this principle aims to empower students to transform knowledge into strategic agency, capable of navigating real-world challenges with adaptability, accountability and impact. The integration of epistemic-driven, ethical-reflective and systemic-pragmatic agency within climate education presents both opportunities and challenges. While this three-lens framework offers a structured way to embed complexity into science education, its implementation requires a careful balance between disciplinary depth and interdisciplinary integration. The design principles outlined here emphasise situated inquiry, deliberative reasoning and strategic action, ensuring that students engage with climate change as both a scientific and a socio-political challenge.

5 Reflections on the Role of Social Media and Science Education

The interdisciplinary integration of science education, philosophy and economics redefines the concept of agency, offering a transformative approach to climate education. This approach directly addresses the research question by showing how these three disciplines can synergistically enhance students' ability to act as agents of change. By embedding these dimensions into science education, we move beyond traditional frameworks that emphasise knowledge acquisition, enabling students to connect abstract concepts with real-world applications. This integration fosters a deeper understanding of the complexities of climate change, equipping learners to address both its causes and consequences effectively (Tasquier & Pongiglione, 2017).

The impact of this approach on science education research is profound. It challenges existing paradigms by introducing a multi-dimensional framework that aligns with global competencies such as those outlined in GreenComp (Bianchi et al. 2022). This contribution advances the field by emphasising the transformative potential of interdisciplinary education. By connecting scientific literacy with ethical reflection and pragmatic action, the proposed framework supports the development of adaptive, future-ready learners who can navigate the uncertainties of the Anthropocene (White et al. 2023).

Furthermore, this approach promotes a shift from individualistic to systemic perspectives on climate action. By encouraging students to consider the interconnectedness of social, economic and environmental systems, it fosters a more comprehensive understanding of sustainability. This perspective aligns with the pressing need for education systems to cultivate collective agency (Kranz et al. 2025) and foster collaboration across disciplines (Knain et al. 2025). Future research will operationalise this framework through empirical studies and collaborative design with teachers, aiming to translate the three-lens perspective into concrete pedagogical practices.

In conclusion, the integration of philosophy and economic studies with science education offers a robust framework for expanding the concept of agency and cultivating transformative agency in educational contexts. This interdisciplinary approach equips students with the skills, knowledge and values needed to navigate the complexities of climate change and contribute to a sustainable future. By redefining the role of education in the Anthropocene, this work provides a blueprint for reimagining science education as a leverage for global change (Tasquier et al. 2025).

References

Bianchi G, Pisiotis U, Cabrera Giraldez M (2022) GreenComp—the European sustainability competence framework. In: Bacigalupo M, Punie Y (eds) EUR30955EN. Publications Office of the European Union, Luxembourg

Cripps E (2013) Climate change and the moral agent: individual duties in an interdependent world. Oxford University Press

Dietz T, Ostrom E, Stern PC (2009) The struggle to govern the commons. Science 302(5652):1907–1912

Fragnière A (2016) Climate change and individual duties. WIREs Clim Change 7:798–814

Gardiner SM (2011) A perfect moral storm: the ethical tragedy of climate change. Oxford University Press

IPCC (2022) Climate change 2022: mitigation of climate change. Contribution of working group III to the sixth assessment report of the intergovernmental panel on climate change

Kahneman D (2011) Thinking, fast and slow. Farrar, Straus and Giroux, New York

Knain E, Tasquier G, Jornet A, Rokenes H, Lee Mueller M (2025) School boundaries coming alive: transforming education towards sustainability through open schooling initiatives. Acta Psychol 258:105192. https://doi.org/10.1016/j.actpsy.2025.105192

Kranz J, Breitenmoser P, Laherto A, Krug A, Schwichow M, Tasquier G (2025) Science education for collective agency in the climate crisis: a social identity approach. Res Sci Educ. https://doi.org/10.1007/s11165-025-10282-w

Otto IM, Wiedermann M, Cremades R, Donges JF, Auer C, Lucht W (2020) Human agency in the Anthropocene. Ecol Econ 167:106463

Sjöström J, Eilks I (2018) Reconsidering different visions of scientific literacy and science education based on the concept of Bildung. In: Dori YJ, Mevarech Z, Baker D (eds) Cognition, metacognition, and culture in STEM education. Springer, pp 65–88

Special Eurobarometer SP538: Climate change (v1.00). (2023). [Data set]. European Commission, Directorate-General for Communication

Sterman J (2000) Business dynamics: systems thinking and modeling for a complex world. Irwin/McGraw-Hill

Tasquier G, Knain E, Røkenes H, Jornet A (2025) Exploring narratives of change for imagining the future and fostering agency toward sustainability. In: Cakmakci G, Tasar MF (eds) Connecting science education with cultural heritage. Contributions from science education research, vol 15. Springer, Cham. https://doi.org/10.1007/978-3-031-90944-3_22

Tasquier G, Knain E, Jornet A (2022) Scientific literacies for change making: equipping the young to tackle current societal challenges. Front Educ 7:689329

Tasquier G, Pongiglione F (2017) The influence of causal knowledge on the willingness to change attitude toward climate change: results from an empirical study. Int J Sci Educ 39(13):1846–1868. https://doi.org/10.1080/09500693.2017.1355078

Thaler R, Sunstein C (2008) Nudge: the gentle power of choice architecture. Yale University Press, New Haven, CT

UNESCO (2021) Getting every school climate-ready: how countries are integrating climate change issues in education. UNESCO report

White PJ, Ardoin NA, Eames C, Monroe MC (2023) Agency in the Anthropocene: supporting document to the PISA 2025 science framework. OECD Educ Work Pap 297

STEM Researchers' Practices for Public Trust in Science Enhancement

Athanasia Kokolaki, Eleni Botzaki, Emily Michailidi, and Dimitris Stavrou

Abstract The present study explores the current practices employed by STEM researchers when communicating climate change to the general public, as well as how these practices are formulated after a professional development program focused on strategies for enhancing public engagement with climate action. The study took place in the context of the European project "STAGE." Eighteen researchers working either in national university institutions and research centers or in science centers and museums participated in the research. The results indicate that the predominant science communication practices employed by STEM researchers include conducting school visits to research centers and delivering public speeches to disseminate their research findings. Additionally, the results highlight researchers' intention to enrich their current outreach activities not only with interactive materials but also with the negotiation of the social aspects of science.

Keywords Science communication · Public engagement · Climate change · STEM researchers · Professional development

---

A. Kokolaki (✉) · E. Botzaki · E. Michailidi · D. Stavrou
Department of Primary Education, University of Crete, Rethymno, Greece
e-mail: akokolaki@edc.uoc.gr

E. Botzaki
e-mail: e.botzaki@edc.uoc.gr

E. Michailidi
e-mail: e.michailidi@uoc.gr

D. Stavrou
e-mail: dstavrou@uoc.gr

D. Sokołowska and E. McLoughlin (eds.), *Embracing Changes Together in Physics Education*, Challenges in Physics Education,
https://doi.org/10.1007/978-3-032-27385-7_17

1 Introduction

1.1 Trust in Science

Climate change, sustainability, nanomaterials, and energy choices are all examples of contemporary socioscientific issues that represent an area of science which is controversial, incorporates issues that usually raise debates among experts, politicians, citizens, etc., and highlights the need for non-experts' involvement in decision-making and action-taking processes either individually or collectively (Dillon and Avraamidou 2020). The inherent uncertainty of such issues, as well as the involvement of many actors in the corresponding debates, makes the dissemination of relevant misinformation through public media widespread. However, the abundance of available information, particularly online, makes it challenging for non-experts to process and evaluate such information while the acquisition of the specialized knowledge and skills needed to address such complex socioscientific issues seems unrealistic for the general public (Osborne et al. 2022). Therefore, enhancing public trust in science is essential to enable non-experts to engage in informed and reasoned discourse about science and technology, based on fundamental knowledge on scientific content and processes (Hendriks et al. 2016; Krüger et al. 2022).

Studies regarding public trust in science levels are limited. However, existing research indicates that the mean value of trust in science among secondary school students when it comes to current socioscientific issues is similar to that of university students, an indication that trust in science might be fostered to a certain degree during school education (Krüger et al. 2022). Research highlights various factors contributing to the public's distrust toward science, including a. the level of scientific literacy; b. their sociocultural background; c. their religious beliefs as well as d. their political ideology (Hendriks et al. 2016; Krüger et al. 2022). Focusing on the level of scientific literacy, fostering trust in science in school education entails the development of students' skills in evaluating the information they encounter, as well as assessing the scientific consensus concerning the issue under examination (Osborne et al. 2022). These skills are supposed to be cultivated through the understanding of the social aspects of science (Osborne et al. 2022; Osborne and Pimentel 2023). The Family Resemblance Approach (FRA) framework (Erduran and Dagher 2014), which provides a holistic understanding of science, seems suitable for the negotiation of such aspects given that FRA analyzes: (i) the methods and practices that underpin scientific investigations; (ii) the norms and values that scientists employ in their work; (iii) how scientists engage in professional settings; (iv) the social mechanisms through which scientists review, evaluate, and validate scientific knowledge; (v) the underlying financial and political dimensions of science and (vi) the inherent uncertainty of science-in-the-making by dismissing the idealized and unrealistic image of an unerring and absolutely certain science (Erduran and Dagher 2014). However, the aforementioned aspects and scientists' lived experiences are far beyond the aspects of the nature of science that students usually explore during their school life (Fensham 2014).

1.2 Climate Change and Public Trust in Science

In recent years, there has been a growing international acceptance that climate change is one of the most urgent and challenging socioscientific issues, due to its environmental, political, economic, and social implications (IPCC 2023; Leiserowitz et al. 2021). However, while the scientific community has reached a consensus regarding the scientific aspects of climate change and the role of anthropogenic activities in exacerbating the phenomenon, various controversial, yet convincing, information relating both the causes and the extent of climate change continues to be widely disseminated through mass media and social media (IPCC 2023; Leiserowitz et al. 2021). Common misconceptions about climate change, often communicated through media, include: (i) questioning the anthropogenic nature of climate change; (ii) the economic and political implications of the proposed response and adaptation measures; (iii) the perception of climate change as an irrelevant and distant issue; and (iv) the perception that the response to climate change is only a governmental and industrial responsibility (Fleming et al. 2021; Whitmarsh et al. 2013). These misconceptions seem to reduce individuals' motivation and ability to make informed decisions and to take individual and/or collective actions toward climate crisis and the creation of a sustainable society (Whitmarsh et al. 2013).

Recent research further highlights that this kind of misleading information is usually disseminated through specific deceptive tactics employed, such as the use of emotional language, the explicit questioning of scientists and scientific institutions, the emphasis on the inherent uncertainty of science-in-the-making, and the deliberate repetition of myths (Lewandowsky et al. 2020). The abundance of such information, particularly through online and social media environments, makes it increasingly challenging for non-expert audiences to recognize and interpret these deceptive tactics especially when they lack an understanding of the nature of science.

Within this context, fostering trust in science is closely related to making visible the epistemic and social processes through which scientific knowledge is gradually established, especially in contexts characterized by uncertainty and debate. In this regard, the role of STEM researchers becomes particularly important. STEM researchers are supposed to be the agents-communicators of contemporary scientific and technological issues in society and are thus considered able to enhance public engagement with science by facilitating interaction with scientific practices and increasing motivation and enthusiasm for science (Koswatta et al. 2022; Baram-Tsabari and Lewenstein 2017). However, merely enriching science communication activities with additional scientific data is not considered sufficient to enhance public engagement with science (Richards and Carruthers Den Hoed 2018). Moreover, communicating scientific findings with non-expert audiences is not typically included in the training that most STEM researchers receive (Baram-Tsabari and Lewenstein 2017), making it more difficult for them to respond effectively to the needs and interests of general audiences. Therefore, the need to equip STEM researchers with knowledge, skills, and practices necessary to increase public trust in science and public engagement with science arises. A first step in this direction is to explore how STEM researchers intend to communicate contemporary scientific issues, such as climate change, to non-experts. Thus, the research question guiding the present study is: *How do STEM researchers propose to adapt their science communication strategies to build public trust in science?*

2 Methodology

2.1 Context of the Study

The present study took place in the context of the European Erasmus+ project "STAGE" which aimed at training STEM researchers and scientists on science communication practices in order to foster public engagement with science. The project was the result of a partnership among four universities and one research and technological center from five European countries, namely the Netherlands, Cyprus, Italy, the Czech Republic, and Greece.

In the context of "STAGE," a professional development program for STEM researchers to become familiar with effective communication practices regarding climate change was developed. More specifically, the professional development program is comprised of six modules that focus on STEM researchers' familiarization with a. the scientific aspects of climate change that are considered crucial for science communication activities (module 1); b. the social, ethical, and cultural dimensions of climate change (module 2); c. the public's misconceptions about climate change, as well as the factors contributing to public distrust toward science (module 3); d. the development of STEM researchers' written and oral skills for science communication (module 4); e. the negotiation of sustainability issues (module 5); and finally f. STEM researchers' understanding of the role and value of public engagement in science for their career (module 6). Additionally, a toolkit including both theoretical resources and examples of good practices, 20 infographics, and an open e-learning platform intended for use by STEM researchers and scientists has also been developed (https://stage.cardetprojects.com/).

The research team from the University of Crete developed the module 3 of the professional development program to elaborate with STEM researchers on the public's common misconceptions about climate change, as well as on the factors contributing to public distrust toward science. Particularly, the materials are structured in three parts. The first part refers to misinformation regarding climate change; the second part concerns practices for fostering trust in science; and the third part supports STEM researchers' reflection on their current science communication practices aimed at enhancing the public's trust in science.

Initially, the first activity of the module asks STEM researchers to analyze comics and discuss, in groups, common climate change misinformation conveyed through them. Additionally, they are encouraged to find and present other images and comics available online that may also spread misinformation about climate change.

1. Discuss with your group and recognize the features of common headlines & social media posts regarding climate change.

FEATURE	Post 1	Post 2	Post 3	Headline 1	Headline 2
"Continued influence effect"					
Sense of familiarity by repeated info *("Illusory truth effect")*					
Use of emotional language					
Use of "just asking questions" strategy					
Questioning scientific consensus					
Highlighting scientific uncertainty					
Undermining scientists' credibility					
Provision of pseudoscientific alternatives					

Fig. 1 Criteria for analyzing deceptive tactics of misinformation

In the second activity, STEM researchers are called to analyze online articles and social media posts that convey misinformation on climate change, and to discuss in their groups the deceptive tactics employed, such as the use of emotional language and emphasis on scientific uncertainty (Fig. 1) (Lewandowsky et al. 2020).

In the third activity, STEM researchers become familiar with a common practice for debunking misinformation, which can be used during their science communication activities, and they are asked to implement this practice in order to develop a refutation of a specific myth regarding climate change (Fig. 2) (Lewandowsky et al. 2020).

In the fourth activity, STEM researchers discuss and brainstorm on the skills the public should be equipped with in order to respond to current socioscientific issues like climate change. Afterward, they become familiar with the social aspects of science through the FRA framework and wheel (Erduran and Dagher 2014). Finally, they work in groups to identify and discuss the social aspects of science they consider important for engaging with the public (Fig. 3).

Finally, in the fifth activity, STEM researchers are asked to describe how they would redesign their current science communication activities in order to foster the public's trust in science.

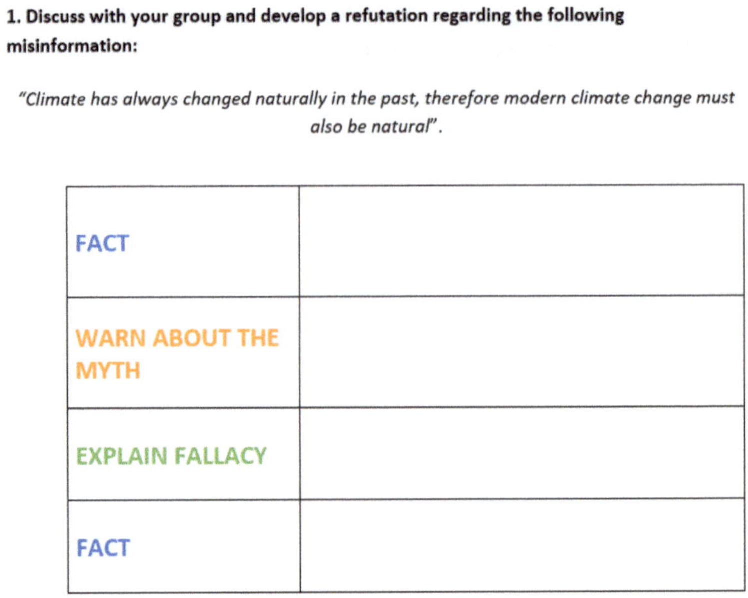
1. Discuss with your group and develop a refutation regarding the following misinformation:

"Climate has always changed naturally in the past, therefore modern climate change must also be natural".

FACT	
WARN ABOUT THE MYTH	
EXPLAIN FALLACY	
FACT	

Fig. 2 Activity on a common debunking practice

Fig. 3 Activity on the negotiation of the social aspects of science

2.2 *Participants*

As far as the empirical study, 18 researchers who work either in university institutions and research centers in Greece or in science and technology centers participated in the research. The implementation of the materials included two synchronous sessions during which participants elaborated on the public's misconceptions about climate change, practices for debunking misinformation, and the negotiation of the social aspects of science, as well as asynchronous planned reading of resources that are available through the "STAGE" toolkit and infographics. Finally, participants were asked to asynchronously provide their proposals for redesigning their current science communication activities in order to support the public in building their trust in science.

2.3 *Data Collection and Analysis*

Data was collected through questionnaires before and after the implementation of the materials with the 18 STEM researchers. Specifically, the initial questionnaire was focused on participants' current science communication practices, while the final questionnaire was focused on the redesign of their current practices.

For the data analysis, qualitative methods of content analysis were used (Bryman 2017). Data analysis adopted a mixed top-down and bottom-up approach. The coding process started with relevant literature and theoretical frameworks on public trust in science and the nature of science in mind, particularly the Family Resemblance Approach (Erduran and Dagher 2014), which informed the initial analytical focus. Participants' responses from the pre- and post-questionnaires were then read to identify meaning units related to their current and redesigned science communication practices. These meaning units were coded, and during this process additional patterns emerged from the data, leading to the refinement and expansion of the initial categories. The systematic categorization focused on participants' redesign intentions following the professional development program and resulted in four main categories: a. the negotiation of the social aspects of science; b. the negotiation of the cognitive and epistemic aspects of science; c. the negotiation of strategies for evaluating information and resources as well as d. the use of appropriate interactive materials during science communication activities (Table 1).

Table 1 Categorization of STEM researchers' suggestions regarding the redesign of their current science communication activities

Categories	Criteria
Social aspects of science	- professional activities
	- social certification and dissemination
Cognitive and epistemic aspects of science	- aims and values of science
	- scientific methods
	- authentic scientific practices
	- use of real data
Information evaluation strategies	- check the credibility of the source
	- check the expertise of the source
Use of interactive materials	- video, animations, images, etc.

To elaborate on the coding process, references highlighting collaboration, critique, and the gradual development of consensus in scientific work were coded under the category "Social aspects of science." For example, references such as *"the public should understand that scientific knowledge does not come from isolated individuals, but from collaboration, disagreements, and gradual consensus among scientists during conferences, research projects and papers"* were categorized accordingly. Similarly, references to the enactment of authentic scientific practices during outreach activities were coded under the category "Cognitive and epistemic aspects of science." Specifically, statements describing the performance of live experiments to demonstrate how scientific research is conducted or to reinforce public trust in scientific methodology were included within this category.

3 Results

As far as the results of the study, before the implementation, participants described relatively general practices such as the organization of school visits to the labs and research centers in which they work, their participation in STEM education research programs and in researchers' night activities as well as the provision of public talks. For example, one participant stated: *"I usually talk to secondary school students about my research work when they visit our institute as part of school visits; however, sometimes I feel that most students cannot follow me."* Another early-career researcher noted: *"My main experience is in Researchers' Night activities, and I really like this contact with society, but I would like to learn more in order to make these activities more attractive."* Additionally, participants mentioned their involvement in STEM education projects for supporting schools and STEM education groups, as illustrated by the following statement: *"Schools or STEM education groups often contact us to support their projects and provide expertise in scientific content."* These responses suggest that participants have a positive attitude toward public engagement activities while simultaneously recognizing the need to further develop their science communication skills in order to better address the needs and interests of non-expert audiences.

After the implementation of the professional development program, STEM researchers were able to articulate more specific and concrete plans for redesigning their science communication activities (Table 2). Initially, they emphasized the need for revising and enriching their materials with interactive elements, such as videos and infographics. For instance, one participant mentioned: *"I feel that I need to update my talks and presentations by incorporating more graphs, pictures, animations, and multimedia elements."* Several participants also explicitly linked the redesign of materials with the increased audience engagement, noting that *"Since my expertise lies in chemistry, it is evident that my future outreach efforts should integrate activities that engage the audience."*

In addition to changes in the type of materials, STEM researchers also focused on the content of their science communication activities. Specifically, they proposed to

Table 2 Results regarding STEM researchers' suggestions about the redesign of their current science communication activities

Categories	Frequency
Social aspects of science	8
Cognitive and epistemic aspects of science	11
Information evaluation strategies	5
Use of interactive materials	12

include, during their public events, discussions and reflections on the cognitive and epistemological aspects of the scientific process in order to familiarize the public with authentic scientific practices and engage them with real data. For example, one participant noted: *"Providing the audience with real data, showing and explaining how a research is conducted may be very helpful for my future events"* while another one mentioned *"My future activities should focus on helping my audience develop tools to challenge their viewpoints and evaluate their sources"*. The performance of live experiments was also repeatedly mentioned as a key strategy, with participants noting their intention to conduct such activities. For example, one participant stated: *"I should think of performing live experiments in order to make it more understandable to visitors how scientific research is performed"* while another one said: *"Performing live experiments for the public may reinforce their trust in scientific methodology"*.

Finally, several participants highlighted the need to negotiate the social aspects of the scientific process with the public. Specifically, they mentioned both the collaboration among scientists and the communication of their research findings to the wider community through conferences and articles in scientific journals as fundamental practices with which the public should be familiarized in order to understand the procedures of evaluation, validation, and gradual establishment of the new scientific knowledge (Table 2). This focus was further reflected in intentions to modify existing activities, as one participant stated: *"Participating in the workshops made me realize that I need to revise my talks and materials to include the aspects discussed during the training program...For example by focusing on our authentic experiences as researchers and on experiments conducted through collaboration with other research groups abroad, as well as on how we collectively interpret data after the experiments."*

These findings indicate a shift toward more reflective and participatory science communication practices after STEM researchers' participation in the professional development program.

4 Discussion and Conclusions

The findings of the present study indicate that STEM researchers intend to change their science communication practices after their participation in a training program. Prior to the implementation of the professional development program, outreach activities were mainly content-oriented, such as public lectures and school visits to research institutes. Although these practices reflect STEM researchers' positive disposition toward public engagement, they were often described as less effective in engaging the public with science. This finding aligns with previous research indicating that STEM researchers frequently engage in outreach activities without having received systematic training in science communication (Baram-Tsabari and Lewenstein 2017).

Following the implementation of the training program, STEM researchers articulated more specific and reflective plans for redesigning their science communication activities, moving toward approaches that emphasize interaction and audience engagement. In particular, emphasis was given to the enrichment of their science communication activities with cognitive and epistemic aspects of science such as the engagement with authentic scientific practices, the integration of real data, and the inclusion of live experiments. This finding is particularly promising, given that fostering public trust in science requires non-experts' engagement with the process of scientific knowledge production (Osborne et al. 2022; Osborne and Pimentel 2023).

Furthermore, STEM researchers emphasized the importance of communicating the social aspects of science in order to support non-experts in the development of a more holistic image of science (Erduran and Dagher, 2014) and to challenge the perception of science as a solitary activity. Specifically, STEM researchers identified the need to familiarize non-experts with the collaborative nature of scientific work, the inherent uncertainty of science-in-the-making as well as the role of consensus in the validation and establishment of new scientific knowledge. Addressing these aspects may also support non-experts in tackling misinformation, as it may enable them to recognize deceptive tactics that tend to question the scientific enterprise (Lewandowsky et al. 2020).

The findings highlight the potential of professional development programs to support STEM researchers in rethinking their roles as science communicators. Structured opportunities to reflect on public misconceptions, deceptive misinformation tactics, and the cognitive-epistemic and social aspects of science appear to enable STEM researchers to move from content-oriented outreach activities toward more participatory communication practices. Such programs may therefore play a critical role in strengthening public engagement with complex socioscientific issues such as climate change. However, the present study focuses on STEM researchers' intentions, and thus future research could examine how these intentions are enacted in practice and through which approaches. Currently, these materials are being revised for use in pre-service primary teacher education programs, with the aim of fostering trust in science among future primary teachers.

Acknowledgments The research work is supported by "STAGE" project which was funded by the Erasmus+ Programme of the European Union. Grant agreement no. 2021-2-NL01-KA220-HED-000048944.

References

Baram-Tsabari A, Lewenstein BV (2017) Science communication training: what are we trying to teach? Int J Sci Educ Part B 7(3): 285–300.Oxford University Press doi:https://doi.org/10.1080/21548455.2017.1303756

Bryman A (2017) Social research methods (5th ed.)., Oxford. ISBN 978-0-19-968945-3

Dillon J, Avraamidou L (2020) Towards a viable response to COVID-19 from the science educationcommunity. J Activist Sci Technol Educ 11(2):1–6. https://doi.org/10.33137/jaste.v11i2.34531

Erduran S, Dagher ZR (2014) Reconceptualizing nature of science for science education. Springer, Netherlands, pp 1–18. https://doi.org/10.1007/978-94-017-9057-4_1

Fensham PJ (2014) Scepticism and trust: two counterpoint essentials in science education for complex socio-scientific issues. Cult Stud Sci Educ 9:649–661. https://doi.org/10.1007/s11422-013-9560-1

Fleming W, Hayes AL, Crosman KM, Bostrom A (2021) Indiscriminate, irrelevant, and sometimes wrong: causal misconceptions about climate change. Risk Anal 41(1):157–178. https://doi.org/10.1111/risa.13587

Hendriks F, Kienhues D, Bromme R (2016) Trust in science and the science of trust. In: Blöbaum B (ed) Trust and communication in a digitized world. Springer International Publishing, pp 143–159. https://doi.org/10.1007/978-3-319-28059-2_8

IPCC (2023) Climate change 2023: synthesis report. Contribution of working groups I, II and III to the sixth assessment report of the intergovernmental panel on climate change. Lee H, Romero J (eds). IPCC. https://doi.org/10.59327/IPCC/AR6-9789291691647

Koswatta TJ, Parrella JA, Leggette HR, Ramasubramanian S, Rutherford T (2022) Improving public science communication: a case study of scientists' needs when communicating beyond the academy. Int J Sci Educ Part B 12(2):174–179. https://doi.org/10.1080/21548455.2022.2055191

Krüger JT, Höffler TN, Parchmann I (2022) Trust in science and scientists among secondary school students in two out-of-school learning activities. Int J Sci Educ Part B 12(2):111–125. https://doi.org/10.1080/21548455.2022.2045380

Leiserowitz A, Roser-Renouf C, Marlon J, Maibach E (2021) Global warming's six americas: a review and recommendations for climate change communication. Curr Opin Behav Sci 42:97–103. https://doi.org/10.1016/j.cobeha.2021.04.007

Lewandowsky S, Cook J, Ecker UKH, Albarracín D, Amazeen MA, Kendeou P, Lombardi D, Newman EJ, Pennycook G, Porter E, Rand DG, Rapp DN, Reifler J, Roozenbeek J, Schmid P, Seifert CM, Sinatra GM, Swire-Thompson B, van der Linden S, Vraga EK, Wood TJ, Zaragoza MS (2020) The debunking handbook 2020. https://skepticalscience.com/debunking-handbook-2020-downloads-translations.html

Osborne J, Pimentel D (2023) Science education in an age of misinformation. Sci Educ 107(3):553–571. https://doi.org/10.1002/sce.21790

Osborne J et al (2022) Science education in an age of misinformation. Stanford University, Stanford, CA. https://coilink.org/20.500.12592/q8zfpd

Richards GW, Carruthers Den Hoed R (2018) Seven strategies of climate change science communication for policy change: combining academic theory with practical evidence from science–policy partnerships in Canada. In: Handbook of climate change communication, practice of climate change communication, vol 2, pp 147–160. https://doi.org/10.1007/978-3-319-70066-311

Whitmarsh L, O'Neill S, Lorenzoni I (2013) Public engagement with climate change: what do we know and where do we go from here? Int J Media Cult Politics 9(1):7–25. https://doi.org/10.1386/macp.9.1.7_1

Supporting Physics Teacher Professional Learning

Teacher Professional Learning in Transformative Times: Three Collaborative Visions for Physics Education—From Cognitive Science to Artificial Intelligence

Magdalena Kersting

Abstract Physics education today unfolds against a backdrop of rapid societal transformations. Climate change, geopolitical crises and quickly evolving technologies outstrip the pace of curriculum development, calling for a re-evaluation of educational practices. In line with the overarching conference theme 'Embracing Changes Together', this paper presents three collaborative visions for the future of physics education that attempt such a re-evaluation. My common anchor for these visions is the focus on teacher professional learning, since preparing and supporting physics teachers is key to transforming educational practices. The first vision promotes interdisciplinary collaboration, merging cognitive science with physics education. The Science Education Network for Supporting Embodied Sense-Making (SENSES) serves as an example of high-school teachers co-creating instructional resources inspired by embodied cognition principles to make physics more accessible to a broader range of students. The second vision advocates for collaborative inquiries between teachers and research physicists. The Danish cross-institutional STEM teaching master programme (KASTEM) illustrates how such collaboration can empower teachers to work alongside climate and space physicists and co-develop inquiry-based lesson plans that bridge cutting-edge science with primary and middle school curricula. The third vision embraces the collaborative potential of generative artificial intelligence (GenAI). By treating GenAI not as a mere tool but as a non-human co-creator and collaborator in the learning process, we can explore synergies between embodied minds and disembodied machines, thereby shifting our understanding of the role of the teacher—and what physics teaching and learning mean at a deeper level. To conclude, these visions call on the physics education community to embrace change collaboratively, pooling insights across fields, professions and technologies to support teachers and advance learning.

M. Kersting (✉)
Department of Science Education, University of Copenhagen, Copenhagen, Denmark
e-mail: science@magdalenakersting.com

D. Sokołowska and E. McLoughlin (eds.), *Embracing Changes Together in Physics Education*, Challenges in Physics Education,
https://doi.org/10.1007/978-3-032-27385-7_18

Keywords Teacher professional learning · Physics education · Embodied cognition · Generative artificial intelligence · Collaborative inquiry

1 Introduction

Physics education today unfolds against a backdrop of rapid societal transformations. Climate change, geopolitical crises and quickly evolving technologies challenge established educational frameworks and outstrip the pace of curriculum development, calling for a re-evaluation of educational practices. In response, Dagmara Sokolowska, the chair of the 4th World Conference of Physics Education 2024 in Krakow, encapsulated her hope for a possible path forward by stating:

The conference theme, Embracing Changes Together, reflects the emergent need to join all educational forces to address—through a collaborative effort—the challenges confronting physics education emerging from the rapidly changing world.

Inspired by this theme, for me, the crucial question then becomes: How can we embrace change collaboratively, pooling insights across fields, professions and technologies to support teachers and advance learning? This question will guide me in outlining three visions of how such collaboration may take form; three collaborative visions for the future of physics education: first, collaborating across disciplines; second, collaborative inquiries between teachers and research physicists; and third, collaborating with artificial intelligence. Each of these collaborative visions addresses a distinct yet interconnected challenge that contemporary physics education must navigate:

- *Instructional Challenge*: Broadening student success in and access to physics education.
- *Curricular Challenge:* Ensuring curriculum relevance by integrating contemporary scientific advancements.
- *Technological Challenge:* Reconceptualising teaching, learning and understanding physics in the era of GenAI.

In the following sections, I elaborate on each vision, link them to teacher professional learning and ground them in successful examples from my own work, currently in the Danish context. My focus on teachers and teacher professional learning[1] is a deliberate one and stems from robust evidence that preparing and supporting teachers is arguably the most effective lever for transforming educational practices. Indeed, numerous studies have demonstrated that teaching quality is the most significant in-school factor affecting student achievement (Hattie 2009; OECD 2005). More specifically, teachers' practices and the 'quality of their teaching as perceived by the students' (Hattie 2009, p. 34) matter, and we know of the significance

[1] Following Farmer (2024), I use the term *teacher professional learning* to denote activities undertaken by teachers, while *teacher professional development* refers to the professional growth resulting from such activities.

of cognitive activation (including subject content coverage), instructional clarity, classroom management and teacher-student interactions as key factors influencing student outcomes (Klette et al. 2017). Of course, external factors such as socioeconomic background, school resources, curriculum quality and parental involvement also play significant roles in student achievement (Hattie 2009; OECD 2005). However, as physicists and physics educators, I believe teachers represent our most direct point of leverage for systemic improvement. Many of us either educate pre-service teachers, provide professional learning opportunities for in-service teachers or collaborate closely with teachers in other formal and informal learning spaces. This existing involvement is an encouraging starting point because ongoing teacher training and professional learning have been shown to enhance teaching quality (Darling-Hammond et al. 2017; European Commission 2015; Farmer 2024). Thus, while we all operate in different contexts and educational systems, teachers represent one common and critical factor in directing our efforts meaningfully towards educational impact.

1.1 The Distinct Position of Physics Education Research

Before looking more closely at the three collaborative visions, I want to acknowledge the unique position of physics education research (PER) and its distinct potential to respond to the challenges of our times. Of course, we all think that physics, with its strong connection between fundamental principles and applications, can equip learners with the knowledge, will and skills to navigate the complexities of our times (Blandford and Thorne 2020). However, I argue that PER is equally important in addressing contemporary challenges, not the least because PER is, by its very nature, cross-disciplinary: it draws on epistemologies from both natural and social sciences. This epistemological flexibility constitutes what I consider our disciplinary 'superpower'. As physics education researchers, we are not confined to the methods and theories of a single discipline but instead integrate approaches and ways of knowing from different fields. We develop fluency in understanding, employing and combining different epistemic standards—from the reductionist approaches that isolate variables in physical systems to the all-encompassing perspectives on learners that acknowledge the intricate interplay of parameters at every possible level, from the neurobiological bases of cognition to the individual psychological level of motivation, prior knowledge and conceptual development, and the collective sociocultural dimension of classroom dynamics, identity formation, and cultural practices that shape what counts as knowledge and how it is constructed.

This cross-disciplinary positioning, this inhabiting of a methodological middle ground, grants us both an opportunity and a responsibility to view educational problems holistically. It enables us to recognise patterns across seemingly disparate domains and pool different methodologies to address the complex realities of physics classrooms. More importantly, it positions us as intellectual mediators, translating ideas from one field to another, particularly when working with teachers who must navigate both the disciplinary content of physics and the pedagogical challenges of conveying that content effectively.

The three visions presented in this chapter exemplify this mediating role—whether bridging cognitive science with classroom practice, connecting research physics with instructional implementation or navigating the interaction between human cognition and artificial intelligence. Each vision leverages the capacity of PER to work across boundaries, drawing on diverse epistemological traditions to address the instructional, curricular and technological challenges facing contemporary physics education.

2 Vision I—Collaborating Across Disciplines

2.1 The Instructional Challenge: Broadening Student Success in and Access to Physics Education

The cross-disciplinary nature of PER that I have just outlined positions us well to address the first major challenge: broadening student success in and access to physics education, thereby enhancing scientific literacy. This instructional challenge sits at the intersection of conceptual understanding, educational access and societal needs. Enhancing scientific literacy—the capacity to engage with, comprehend and evaluate scientific issues—has become increasingly urgent in the face of substantial environmental, socioeconomic and geopolitical challenges (OECD 2019). Clearly, physics education can contribute greatly to achieving this goal. Yet despite this potential, physics education continues to face a persistent paradox: at precisely the moment when we most need diverse perspectives in physics, we still fail to include large segments of potential learners, while many who do enter the field struggle with the material and conceptual challenges.

First, among those who do have access to physics education, deep conceptual understanding often remains elusive, with many students failing to reason correctly across contexts and domains, even at the university level (Sands 2014). Research consistently shows that students at all levels develop incorrect ideas about fundamental physics principles that persist despite traditional instruction (Docktor and Mestre 2014; McDermott 2001). These conceptual difficulties transcend demographics and educational settings, reflecting inherent challenges in bridging intuitive understanding with formal physics knowledge. Besides, the disconnect between mathematical formalism and physical meaning remains a stubborn obstacle to meaningful learning in physics (Bouchée et al. 2022; Gifford and Finkelstein 2020; Uhden et al. 2012).

Compounding these conceptual challenges are significant access and inclusion barriers. The evidence for these challenges is substantial and multifaceted. Women remain significantly underrepresented in physics at all levels, from secondary education to faculty positions (Archer et al. 2010; Avraamidou 2020; Bøe et al. 2011; Dare and Roehrig 2016; Kelly 2016; Mujtaba and Reiss 2013). There is also a gender gap regarding achievement and positive attitudes towards physics (Häussler and Hoffmann 2002; Nissen and Shemwell 2016; Reid and Skryabina 2003; Wilson et al. 2016). Rather than challenging this status quo, physics education often reinforces gendered cultural norms through instructional practices, curricular examples and classroom dynamics; indeed, traditional physics education is often disembodied and privileges abstract representations that ignore cultural and bodily experiences, which can alienate students, particularly those from underrepresented groups (Kaur et al. 2020; Kersting et al. 2021b; Lin-Siegler et al. 2016; Ross et al. 2023).

Beyond gender disparities, broader engagement issues persist as well. Student performance in science and mathematics has declined significantly across OECD countries in recent years (OECD 2023), with many students feeling disconnected from physics material that is presented as abstract, unrelatable or divorced from real-world challenges. The field struggles to convey personal relevance (Kapon et al. 2018), resulting in many students failing to develop agency and ownership of physics concepts.

The twin challenges of disciplinary success in and inclusive access to physics are deeply interconnected and resistant to simple solutions. Indeed, in many ways, these challenges underlie the very existence of PER as a field. However, our cross-disciplinary 'superpower' provides a promising path forward through interdisciplinary collaboration, particularly through frameworks that can address both conceptual understanding and access through a unified lens.

2.2 The Vision: Collaborating Across Disciplines

Addressing this instructional challenge requires us to look beyond the borders of traditional physics education. We need to understand not only *how* students learn physics, but also *why* certain groups become disengaged or discouraged. This is where collaboration across disciplines becomes essential.

My first vision proposes an interdisciplinary approach that merges insights from diverse fields with physics education research and, in turn, makes these insights available to teachers. By better understanding the cognitive and sociocultural factors that affect learning, we can improve instructional practices. Indeed, over the last decades, fields as diverse as educational psychology, linguistics, sociology and anthropology (to name a few) have brought significant insights into how students construct knowledge in physics, how they develop their identity as physicists, and how social, cultural and linguistic factors influence their learning experience.[2]

[2] Theories and concepts such as Dual Process Theory (Speirs et al. 2021), Systemic Functional Linguistics (Jeppsson et al. 2025), Communities of Practice (Prefontaine et al. 2021), Situated Learning (Sadler 2009), and Science Capital (Cooper and Berry 2020) have all enriched our understanding of learning processes in physics. Here, we see the power of interdisciplinary collaboration.

Among these interdisciplinary contributions, I see particular promise in embodied cognition, a paradigm from cognitive science that has significant implications for physics education (Kersting et al. 2023; Kersting et al. 2021a). Embodied cognition recognises that thinking and learning are fundamentally grounded in bodily experiences and, consequently, that there are deep connections between sensorimotor actions and abstract forms of cognition (Shapiro 2014). I argue that this paradigm offers pathways to make abstract physics concepts more accessible and meaningful to diverse groups of learners. In particular, embodied cognition can serve three promising functions in physics education.

First, embodied cognition has explanatory power—it helps us understand why certain instructional strategies work better than others by illuminating the underlying cognitive mechanisms. Effective physics learning, across different methods and contexts, often involves making abstract concepts accessible by engaging the body in meaningful ways. This extends beyond purely cognitive aspects to include emotional and affective dimensions of learning physics. In this sense, my vision also aligns with the message of Claudio Fazio, another keynote speaker at the 4th World Conference of Physics Education, who pointed out the importance of affective and emotional dimensions in learning physics.

Second, embodied cognition unifies various research findings in physics education research by revealing a common thread: successful instructional strategies often engage the connection between physical experiences and cognitive processes. Whether we examine the challenge of transitioning from intuitive to formal knowledge, the use of gestures in physics meaning-making, multiple representations of physics concepts, the power of metaphors and analogies, cognitive load considerations or the effectiveness of interactive simulations and virtual reality—all can be viewed through the lens of embodied cognition. This umbrella view allows us to connect seemingly disparate theories and ideas in physics education research, thereby laying the foundations for cumulative research where we can make greater progress more quickly. As such, my first vision also resonates with another keynote speaker's observation, Paula Heron, that PER scholars are now in a better position to connect theories to develop a broader, more coherent, more empirically supported and more powerful framework to guide physics education so that 'how we teach more closely matches how students learn'.

Third, and perhaps most importantly for practice, embodied cognition empowers teachers by providing a new understanding of how existing teaching strategies work. Rather than requiring entirely new methods, it reframes familiar approaches—such as hands-on activities, gestures, physical experiments and metaphors—by revealing how they leverage the natural connection between bodily experiences and cognitive processes. With this deeper insight, teachers can apply these strategies more deliberately and intentionally, optimising their effectiveness and enhancing student learning.

2.2.1 How Embodied Cognition Differs from Previous Approaches

It is important to note that embodied cognition is not simply another iteration of constructivist or hands-on learning approaches. What distinguishes this framework is its strong theoretical grounding across multiple disciplines—cognitive science, neuroscience, philosophy of mind, robotics, artificial intelligence and cognitive linguistics—rather than being merely an educational trend (Shapiro 2014; Shapiro and Spaulding 2021). Embodied cognition offers testable mechanisms that explain why and how physical experiences connect to conceptual understanding, moving beyond the generic notion that 'hands-on learning is good'. It provides a more comprehensive framework that integrates physical, emotional and cognitive aspects of learning rather than treating them as separate domains. This robust, multidisciplinary foundation makes it particularly valuable for physics education research and practice.

2.2.2 Addressing Inclusion Through Embodied Approaches

Beyond enhancing conceptual understanding, embodied cognition also offers promising pathways for addressing the above-mentioned inclusion barriers in physics education. For example, Solomon et al. (2022) demonstrate how embodied approaches in the form of integrating dance into physics learning can serve as a resource for sense-making, giving students access to an expanded vocabulary for understanding physics concepts. Specifically, the study found that Black girls' cultural dance practices are critical 'funds of knowledge' that can be leveraged in physics learning. This approach positions students' cultural and embodied knowledge as valuable in scientific inquiry, countering the traditional disembodied framing of physics. Thus, by incorporating culturally relevant embodied practices, educators can create more inclusive learning environments that validate and engage diverse student experiences.

2.2.3 The Example: SENSES as a Model for Interdisciplinary Collaboration

The Science Education Network for Supporting Embodied Sense-Making (SENSES) exemplifies this vision in practice. Established in early 2024 at the University of Copenhagen, this initiative brings together an interdisciplinary group of researchers and high-school teachers to co-create instructional resources grounded in embodied cognition principles. The network builds partly on my previous work in which I have brought different disciplines into dialogue to explore the implications of embodied cognition for physics education (Kersting et al. 2023). SENSES is the next natural step to include teachers in this collaboration process.

Two integral features of SENSES align particularly strongly with our vision for interdisciplinary collaboration. First, participatory co-design methodologies underpin all workshop activities, ensuring that teachers' practical expertise shapes the development process from the outset rather than being consulted after materials are already conceptualised. For example, the participants create instructional resources addressing challenging physics concepts identified by teachers themselves.

Second, the network employs a reflective practitioner stance based on Schön's notion of reflection in and on action (Schön 1983). These collaborative reflections allow participants to make implicit assumptions explicit, creating opportunities to examine how theoretical constructs from embodied cognition manifest in classroom contexts and how teachers' practical knowledge can inform theory development.

2.2.4 An Embodied Approach in Practice: Congruence in Teaching Astronomical Concepts

One illustrative example of the impact of embodied approaches that we develop in SENSES comes from the work of one of our participating teachers, who decided to explore embodied cognition in his physics classroom for his master's thesis. The teacher investigated whether the congruence between body movements and physics content impacts secondary students' understanding and experience of physics. Here, congruence refers to the degree to which physical movements used in teaching align with the subject content being taught—an important nuance showing that not all movement-based approaches are equally effective.

As part of a teaching unit on 'The Earth as a Planet in the Solar System' with 23 physics students (aged 17–18 years), the SENSES teacher compared two instructional approaches using a mixed-methods approach, including pre- and post-tests, questionnaires, drawings and focus group interviews (Bøgild 2024). Students who experienced the high-congruence approach showed greater improvement in their understanding and reported more positive experiences with the learning activity. This example illustrates how even relatively simple embodied approaches, when theoretically informed, can make a significant difference in students' conceptual understanding and engagement.

2.2.5 A Two-Way Dialogue

While SENSES primarily focuses on integrating cognitive science insights into physics education, we also recognise the importance of a two-way dialogue. Physics education provides particularly rich contexts for testing and refining embodied cognition theories because physics concepts often require bridging intuitive/experiential and formal/mathematical understandings. Many physics concepts extend beyond direct sensory experience, presenting interesting challenges for embodied approaches. By documenting how teachers implement embodied strategies and their effects on student learning, my hope is that SENSES contributes not only to physics education but also to cognitive science, exemplifying the mutual benefits of interdisciplinary collaboration.

While the development and evaluation of SENSES is ongoing, initial teacher feedback has been positive, with participants reporting value in the collaborative approach and the embodied perspective on physics instruction. By bridging cognitive science research and classroom practice through meaningful teacher engagement, SENSES exemplifies the cross-disciplinary collaboration needed to address the instructional challenges facing contemporary physics education, demonstrating how our 'superpower' as physics education researchers can be leveraged to make physics more accessible, relevant and understandable to diverse groups of learners.

3 Vision II—Collaborative Inquiries Between Teachers and Research Physicists

3.1 The Curricular Challenge: Ensuring Curriculum Relevance by Integrating Contemporary Scientific Advancements

My second vision addresses another pertinent challenge, namely, to keep the curriculum relevant and empower teachers to integrate new scientific developments into their practices. Today's scientific landscape is changing rapidly, from gravitational waves to exoplanet discoveries, from quantum computing breakthroughs to climate system tipping points. While these advancements expand our understanding of the universe and reshape our daily lives, they often remain disconnected from classroom practice (Kaur et al. 2024b; Kersting, Blair et al. 2024). Consequently, teachers face the formidable challenge of keeping their lessons, and by extension, the curriculum, relevant. Primary and middle school teachers, who typically have broad pedagogical training but more limited specialised scientific knowledge, find themselves particularly challenged when trying to integrate contemporary scientific developments into their teaching (Balta et al. 2022; Blair et al. 2024; Kaur et al. 2024a).

Traditional professional learning rarely provides teachers with the disciplinary knowledge *and* the pedagogical tools needed to reconstruct novel topics in a way that makes them accessible to students while also linking them to the existing curriculum. Since scientific advancements outstrip the pace of curriculum development, this disconnect creates a widening gap between frontier science and physics education, with consequences that extend far beyond classroom engagement. When students learn physics that seems detached from contemporary scientific developments, it contributes to public misconceptions about the nature and relevance of physics, ultimately affecting scientific literacy on pressing issues where physics understanding is crucial, such as climate change, renewable energy technologies and our digital communication systems.

3.2 The Vision: Collaborative Inquiries Between Teachers and Research Physicists

To address this challenge, we need to create supportive environments where teachers can gain insights into new scientific developments and receive the guidance they need to integrate these insights into their instructional practices. My second vision calls for establishing collaborative inquiries where physics teachers and research physicists work together as equal partners, mutually enriching each other's professional practice through systematically investigating scientific content and pedagogical approaches.

As Blandford and Thorne (2020) have argued, physics teachers should be considered integral members of the broader physics community: 'High-school and community-college science teachers must be acknowledged more publicly for what they do and as belonging to the same profession as research scientists. They must be fully included as members of a community that discovers new elementary particles, develops new types of lasers, and formulates new curricula. Their perspective is crucial as we all grapple with educational content and assessment, outreach, and professional development in a changing world. We research leaders can learn much by listening to high-school teachers. We need to practice inreach as well as outreach'.

This perspective shifts our approach from traditional 'outreach' to inclusive 'inreach', where teachers actively participate in knowledge construction and curriculum innovation alongside researchers. This shift from outreach to inreach represents more than a semantic distinction—it fundamentally reframes the relationship between research and teaching. Drawing inspiration from the concept of practitioner inquiry in teacher professional learning (McLoughlin et al. 2024), my vision positions both teachers and physicists as collaborative inquirers who systematically investigate questions at the intersection of scientific advancement and educational practice. Just as teachers engage students in inquiry-based learning and themselves benefit from practitioner inquiry approaches, the collaboration between teachers and physicists can be structured as a shared inquiry process as well.

These collaborative inquiries mirror the very processes we hope to foster in physics classrooms: asking authentic questions, gathering evidence, developing explanations and communicating findings. In this way, the professional learning of teachers becomes aligned with both the practices of scientific research and the instructional approaches we know benefit students, creating a coherent thread from research physics through teacher learning to student experiences.

Finally, rather than prescribing a single model for such collaborative inquiries, my vision intentionally embraces flexibility, recognising that effective partnerships must be responsive to local educational contexts and available resources. In the next section, I give one example of establishing collaborative inquiries between teachers and research physicists in the Danish context, recognising that there are many other models that realise this vision (Kersting et al. 2024; Serio et al. 2025).

3.3 *The Example: KASTEM as a Model for Collaborative Inquiry*

The Danish cross-institutional STEM teaching master programme (KASTEM) exemplifies how this collaborative vision can be realised in practice. KASTEM targets primary and lower secondary in-service teachers across Denmark, offering a structured framework for teachers to engage with cutting-edge science and translate it into meaningful classroom experiences. The programme consists of 120 ECTS credits, which teachers can complete over 2–4 years, depending on whether they study full-time or part-time alongside their teaching responsibilities. One course in the programme, 'The Empirical Experimental Sciences', explicitly focuses on translating current scientific research into experimental- and empirical-based teaching through collaborative inquiry between teachers, educational researchers and research scientists.

In this course, teachers work alongside climate and space physicists, engaging with authentic scientific questions with societal relevance. The course follows a structure built around three complementary approaches:

- *Inquiry-Based Learning:* Teachers first experience science as learners themselves, participating in inquiry-based activities on astronomy and climate change guided by the Danish 6F inquiry model (an adaptation of the better-known 5E inquiry model: Engage, Explore, Explain, Elaborate and Evaluate, with an added focus on formative feedback (Bybee et al. 2006)). This inquiry stance towards their own learning helps teachers develop a deeper understanding of both the content knowledge and the pedagogical approaches that support authentic scientific inquiry. Notably, principles of embodied cognition are deliberately integrated into these inquiry experiences, creating coherence with the approaches discussed in Vision I and providing teachers with multiple theoretical perspectives to guide their practice.
- *Model of Educational Reconstruction:* Teachers learn to use the model of educational reconstruction both as a scaffolding structure for designing lessons and as a tool for their own professional learning (Duit et al. 2012). The model helps teachers systematically analyse scientific content, consider student perspectives and design lesson plans that align cutting-edge scientific ideas with the existing curriculum in an accessible way.
- *Collaborative Lesson Design and Implementation*: With ongoing guidance from both educational researchers and physicists, teachers design, implement and reflect on inquiry-based lesson plans for their own classrooms. These lesson plans bridge new scientific developments with primary and middle school curricula, making contemporary scientific understanding accessible and relevant to young learners.

KASTEM illustrates a culture of collaborative inquiry where all participants contribute valuable knowledge: scientists bring their disciplinary expertise and frontier knowledge, educational researchers contribute theoretical frameworks and evidence-based instructional approaches, and teachers bring their practical classroom experiences and understanding of their students' needs and capabilities. For example, in recent offerings of the course, teachers worked with an MIT astronomer who shared never-before-seen James Webb Space Telescope images, collaborated with a cosmologist involved in the European Space Agency's Euclid mission and simulated the experimental observation of exoplanets with two astro physicists. In parallel, the students engaged with climate scientists specialising in ice cores and tipping points in Earth's climate system.

The breadth of topics that emerge from these collaborative inquiries also demonstrates how this approach naturally integrates physics into a broader STEM perspective—a key perspective in the educational system in Denmark. In my most recent iteration of the course, teachers chose to educationally reconstruct and develop lesson plans on diverse topics including 'Creating Algae Petrol from Wastewater', 'Searching for Life in Space', 'Weather and Climate—Precipitation and Snow', 'Birth of a Solar System' and 'The Inland Ice and Its Formation', among others. This range reflects both the contemporary scientific interests of the participating teachers and the interdisciplinary nature of modern scientific challenges.

Moreover, the outcomes of these collaborations extend beyond the course and the individual teachers' classrooms. By documenting their processes of educational reconstruction and lesson implementation, teachers in the course create a growing repository of resources that can benefit the broader educational community in Denmark. Besides, the scientists involved report gaining new perspectives on their own research and its communication, exemplifying the mutual benefits of such partnerships.

In conclusion, KASTEM demonstrates that when we create spaces for genuine collaboration between teachers and physicists, we can foster a more integrated community of practice where the boundaries between physics research and physics education become productively blurred. This approach empowers teachers as co-creators rather than mere curriculum implementers, ultimately enhancing students' engagement with physics as a dynamic and relevant field of inquiry.

4 Vision III—Collaborating with Artificial Intelligence

4.1 The Technological Challenge: Reconceptualising Teaching, Learning and Understanding Physics in the Era of GenAI

My third vision addresses an obvious technological challenge with great epistemological implications, namely, that we need to reconsider what it means to learn, teach and know in a world where GenAI is an integral part of everyday life. GenAI technologies, which create novel content in response to prompts, are reshaping numerous domains, including physics and physics education. Built on artificial neural networks that mimic functions of the human brain, these systems replicate 'the higher-order thinking that constitutes the foundation of human learning' (International Commission on the Futures of Education 2023) and have transformative implications for knowledge construction, teaching approaches, learning processes and assessment practices.

This technological transformation introduces a multifaceted challenge. At its core lies an epistemological tension: as GenAI increasingly generates scientific explanations and simulates physical phenomena, questions arise about scientific authority and evidence (Cheung et al. 2024; Erduran 2023; Erduran and Levrini 2024; Tang and Cooper 2024). What happens to the epistemic foundations of physics when explanations and derivations can be generated algorithmically rather than through physical experimentation and human reasoning? How do we establish new epistemic standards that acknowledge the capabilities of AI while preserving the empirical foundations of physics?

Furthermore, the integration of GenAI introduces what we might call an embodied-disembodied challenge. As already explained, human cognition is inextricably linked to sensory and motor experiences, with learning and understanding deeply grounded in bodily interactions with the physical world (Kersting et al. 2021a). This is also true for the scientific method, which is centred around the investigation and manipulation of the material world. Consequently, the embodied nature of cognition is particularly relevant in physics, where conceptual understanding often emerges through hands-on experimentation, physical demonstrations and sensorimotor engagement with phenomena. In stark contrast, GenAI operates without sensory or material grounding, leading to an obvious tension: physics students interact with systems that are neither aligned with human cognition nor the subject matter at hand (Di Paolo et al. 2024).

This embodied-disembodied challenge becomes particularly evident in domains like classical mechanics, where developing physical intuition relies heavily on bodily experience. Consider a student learning about angular momentum conservation. Human understanding typically develops through physical interactions—spinning on a rotating chair while extending and retracting arms or manipulating a bicycle wheel gyroscope. These embodied experiences create cognitive grounding for the abstract mathematical principles. GenAI, in contrast, operates without sensory or material grounding. While GenAI can flawlessly manipulate the mathematical formalisms and generate seemingly accurate explanations of conservation laws, it lacks the embodied foundation that makes these concepts meaningful to human learners.

For teachers, these challenges are particularly acute. Standing at the forefront of these fast technological advancements, they must navigate evolving instructional practices where traditional concepts of teaching and learning increasingly intertwine with AI capabilities. We already see this intertwining in practice, with GenAI supporting students as personalised feedback partners, challenging their argumentation in physics (Bitzenbauer 2023), and aiding teachers in assessment and content generation (Kaldaras and Haudek 2022; Zhai et al. 2022). As these applications multiply, we can anticipate a fundamental shift in the role and responsibilities of physics teachers and, by extension, what it means to teach and learn physics at a deeper level.

4.2 The Vision: Collaborating with Artificial Intelligence

In response, my third vision embraces the collaborative potential of generative artificial intelligence. Rather than viewing GenAI as merely a tool to enhance or accelerate existing educational practices, this vision reconceptualises it as a non-human collaborator in the learning process. By recognising GenAI as a 'teammate' (Dellermann et al. 2019), an 'interaction partner' (Cress and Kimmerle 2023) or a 'non-human actor' (Tang and Cooper 2024) in learning processes, we can explore productive synergies between embodied human cognition and disembodied machine intelligence.

In proposing this collaborative relationship, I want to shift the conversation away from somewhat abstract concerns about what GenAI can 'truly' or 'genuinely' understand and do (Kortemeyer 2023; Kortemeyer et al. 2025; Polverini and Gregorcic 2024b, 2024a) towards their practical applications in physics education. This shift is supported by two key arguments: First, a functional argument: even if AI-generated physics explanations differ fundamentally from human understanding in their origins and mechanisms, the functional output can be similar enough to support meaningful learning interactions. While GenAI lacks embodied understanding of physical concepts, it can nonetheless process and respond to physics questions in ways that mirror important aspects of human dialogue about physics. From an educational perspective, these functional similarities enable productive exchanges that can enhance human learning, regardless of whether the AI 'truly understands' the physics it discusses.

Second, a pragmatic argument: The central aim of physics education is to educate humans, not AI systems. Therefore, the critical question is not what GenAI can or cannot do compared to humans but rather how we can leverage GenAI to help students think and learn physics more effectively. This pragmatic orientation keeps our focus where it belongs—on enhancing human understanding of physics through whatever means prove most effective.

This collaborative vision acknowledges the complementary strengths and limitations of humans and AI systems in physics education. While humans excel at physical intuition, experimental design, contextual understanding and embodied exploration of phenomena, GenAI demonstrates strengths in pattern recognition, rapid computation, information synthesis and generating multiple perspectives on complex problems. Thus, at its core, my third vision proposes an epistemic division of labour—a partnership where human teachers and learners focus on developing physical intuition through embodied experiences, designing experiments and making meaning of physical phenomena, while GenAI partners contribute computational power, pattern recognition and the ability to generate diverse explanatory models through genuine discursive interactions.

Conceptualising GenAI as a collaboration partner can be understood through a broader embodied cognition perspective, which recognises that human cognition regularly extends beyond the boundaries of the individual brain to incorporate environmental resources and external supports (Clark and Chalmers 1998). This extended view of cognition—where thinking processes are distributed across brain, body and environment—offers a productive framework for conceptualising human-AI partnerships in physics education. As Annie Murphy Paul (2021) observes, 'Our brains, it might be said, are like magpies, fashioning their finished products from the materials around them, weaving the bits and pieces they find into their trains of thought… the kinds of materials available to 'think with' affect the nature and quality of the thought that can be produced'.

From this embodied perspective, GenAI becomes part of our cognitive ecology—not a replacement for human thinking but an extension of it, offering new 'materials to think with' that can enhance our understanding of physics. This extends the embodied approaches discussed in Vision I by acknowledging that our thinking processes naturally incorporate both physical experiences and cognitive tools that expand our capabilities.

For teachers, this third vision transforms their role to that of orchestrators of complex learning ecosystems that include both human and AI participants. Teachers become guides who help students navigate this new collaborative landscape, teaching them to critically evaluate AI-generated content, leverage AI capabilities to enhance their learning, and develop the embodied and epistemic meta-cognitive skills needed to work effectively with these non-human partners. Furthermore, teachers themselves can form collaborative relationships with GenAI, co-designing learning experiences, analysing student work and reflecting on pedagogical approaches in ways that enhance their professional practice.

Ultimately, this vision of collaboration with GenAI in physics education is not about replacing human elements with technological ones, but about creating new possibilities for teaching and learning that neither humans nor machines could achieve alone. In other words, my final vision embraces the transformative potential of AI while preserving and enhancing the embodied, exploratory and social dimensions of learning that make physics education meaningful in the first place.

4.3 The Example: A Research Agenda for Human-AI Collaboration in Physics Education

Rather than presenting a completed case study, this section outlines an emerging research agenda for exploring human-AI collaboration in physics education. The rapid evolution of GenAI technologies means that empirical research often lags behind technological developments, creating an opportunity for collaborative knowledge building among teachers, researchers and even AI systems themselves.

To assist such research and knowledge building, I have formulated three research questions that can guide future investigations into the dynamics of human-AI collaboration in physics education:

- How do the distinctive capacities of humans and GenAI systems interact during collaborative learning processes, and how does this interaction influence physics learning specifically and scientific literacy practices more generally?
- What are the characteristic features of collaborative learning episodes involving GenAI in physics education?
- Which design principles can underpin the development of instructional resources tailored to support physics learning specifically and scientific literacy more generally in conjunction with GenAI collaboration?

These questions are accompanied by hypotheses that articulate the potential for synergistic relationships between human learners and AI systems:

- First, GenAI systems can serve as non-human collaboration partners in the physics learning process, engaging in interactions conducive to developing novel insights. Rather than merely providing information, GenAI can participate in dialogic exchange, challenging students' thinking and offering alternative perspectives that promote deeper conceptual understanding.
- Second, a complementary relationship will likely emerge between students and GenAI, where each contributes different strengths to the learning process. Students bring critical thinking, embodied experiences with physical phenomena and contextual understanding, while GenAI contributes computational power, pattern recognition capabilities and the ability to generate multiple representations of physical concepts. This division of epistemic labour creates learning opportunities that neither humans nor AI could achieve independently.
- Third, instructional approaches that strategically leverage both human and AI capabilities will enhance conceptual understanding in physics and, ultimately, scientific literacy, particularly in areas such as conducting scientific inquiry, articulating scientific explanations and interpreting data and scientific evidence.

This research agenda invites physics teachers to become co-investigators in exploring the potential of human-AI collaboration. By documenting their experiences with GenAI in the classroom, reflecting on the changing dynamics of learning and experimenting with different collaborative structures, teachers can contribute valuable insights to our collective understanding of this emerging educational frontier. Physics education researchers, meanwhile, can develop methodological approaches for studying these new forms of collaboration, design conceptual frameworks that bridge embodied and algorithmic 'disembodied' perspectives on learning, and investigate how different pedagogical approaches influence the quality of human-AI interactions.

Indeed, this vision of human-AI collaboration is particularly well-positioned to benefit from existing theoretical and methodological tools in physics and science education research. A critical insight in this regard is the foundational role of language in both science learning and GenAI systems. While students construct scientific understanding through discursive practices, multimodal representations and disciplinary language use, GenAI is fundamentally powered by large language models that process and generate text through statistical relationships among words. Given this shared linguistic foundation, research in languages and literacies in science education offers valuable ideas for investigating and designing the instructional use of GenAI (Kersting et al. 2025; Tang 2024). For instance, established frameworks for analysing classroom discourse, multimodal meaning-making and disciplinary literacy can be adapted to study how students and teachers engage with AI-generated content. Similarly, methods for investigating representational practices in physics—such as how learners interpret graphs, equations and diagrams—can illuminate how AI-generated representations mediate understanding. By leveraging these existing research traditions while developing new approaches specific to human-AI interaction, researchers can systematically investigate how collaborative learning with GenAI shapes physics understanding, influences disciplinary practices and transforms educational interactions.

By embracing a research stance that views both teachers and AI systems as potential collaborators rather than merely subjects or tools, we can collectively navigate the transformative potential of GenAI in physics education. This research agenda thus exemplifies the broader vision of collaboration with AI—it demonstrates how the physics education community can study human-AI collaboration and embody collaborative principles in the research process itself.

5 Conclusion

Against the backdrop of rapid societal transformations, this chapter has presented three collaborative visions for the future of physics education, each addressing a distinct yet interconnected challenge: collaborating across disciplines to broaden student success and access, partnering research physicists with teachers to integrate contemporary scientific advancements and reconceptualising physics education through human-AI collaboration. These visions share a common thread—they position physics education research at productive intersections, leveraging our field's epistemological flexibility to bridge diverse domains. Moreover, these visions acknowledge teachers as essential partners in educational transformation, recognising that meaningful change requires sustained professional learning opportunities that empower educators to navigate instructional, curricular and technological challenges.

5.1 *Implications for Physics Education Research and Teacher Professional Learning*

Turning now to implications, we see that the three collaborative visions offer several promising avenues for advancing PER and teacher professional learning. First, the visions suggest how interdisciplinary approaches and methodologies can address persistent challenges in physics education, suggesting that future research might increasingly adopt interdisciplinary and mixed-methods approaches that combine insights from cognitive science, educational psychology and technology studies, among many others. Second, the visions highlight the need for new analytical frameworks capable of examining collaborative learning processes across human and non-human actors, particularly as AI integration becomes more prevalent in educational settings. Vision III particularly has implications for how physics educators design learning environments, assess student understanding, and conceptualise the nature of physics learning itself. Future research might explore how human-AI collaboration affects the development of scientific literacy and the changing nature of epistemic agency. Third, the visions collectively suggest a shift towards more collaborative, inquiry-based models of teacher professional development. Rather than traditional top-down approaches, these models acknowledge the professional expertise of teachers and position them as co-researchers and knowledge creators, working alongside disciplinary experts and educational researchers. This approach aligns with converging research in teacher education and professional learning that has identified agency, reflection and collaborative knowledge construction as key factors contributing to successful professional learning (Darling-Hammond et al. 2017; Farmer 2024; Sims et al. 2021).

The examples of SENSES, KASTEM and our emerging research agenda for human-AI collaboration illustrate how these visions can be grounded in practical implementation. In presenting these examples, I hope to have demonstrated how the physics education community can harness our cross-disciplinary superpower to create more inclusive, relevant and forward-looking educational experiences.

As we continue to embrace changes together, these collaborative approaches offer promising pathways for physics education to remain relevant and responsive in transformative times. By pooling insights across disciplines, professions and technologies, we can collectively advance physics education that equips learners with the knowledge, will and skills to navigate the complexities of our world—and transform scientific understanding into meaningful action towards positive change.

References

Archer L, Dewitt J, Osborne J, Dillon J, Willis B, Wong B (2010) 'Doing' science versus 'being' a scientist: examining 10/11-year-old schoolchildren's constructions of science through the lens of identity. Sci Educ 94(4):617–639. https://doi.org/10.1002/sce.20399

Avraamidou L (2020) "I am a young immigrant woman doing physics and on top of that I am Muslim": identities, intersections, and negotiations. J Res Sci Teach 57(3):311–341. https://doi.org/10.1002/tea.21593

Balta N, Eryilmaz A, Oliveira AW (2022) Increasing the presence of Einsteinian physics in high school: the impact of a professional development program on teacher knowledge and practice. Teach Dev 26(2):166–188. https://doi.org/10.1080/13664530.2021.2019102

Bitzenbauer P (2023) ChatGPT in physics education: a pilot study on easy-to-implement activities. Contemp Educ Technol 15(3). https://doi.org/10.30935/cedtech/13176

Blair D, Lonshakova A, Santoso J, Kaur T, Adams K, Boublil S, Treagust D, Wood D, Ju L, Stalley J, Scott S (2024) Collapsing school physics enrolments: micro-credential courses for upskilling teachers in Einsteinian physics. In: Proceedings of The 30th Australian conference on science and mathematics education

Blandford RD, Thorne KS (2020) Post-pandemic science and education. Am J Phys 88(7):518–520. https://doi.org/10.1119/10.0001390

Bøe MV, Henriksen EK, Lyons T, Schreiner C (2011) Participation in science and technology: young people's achievement-related choices in late-modern societies. Stud Sci Educ 47(1):37–72. https://doi.org/10.1080/03057267.2011.549621

Bøgild P (2024) Med kroppen ind i fysikken—Kongruens betydning for embodiment i fysikundervisning på STX [Bringing the body into physics—the importance of congruence for embodiment in physics teaching at TGE Higher General Examination Programme (stx)], University of Copenhagen. https://www.ind.ku.dk/publikationer/studenterserien/studenterserien-publikationer/120-pelle-boegild/120-Pelle_B_gild-2024.pdf

Bouchée T, De Putter-Smits L, Thurlings M, Pepin B (2022) Towards a better understanding of conceptual difficulties in introductory quantum physics courses. Stud Sci Educ 58(2):183–202. https://doi.org/10.1080/03057267.2021.1963579

Bybee RW, Taylor J, Gardner A, Van Scotter P, Powell JC, Westbrook A, Landes N (2006) The BSCS 5E instructional model: origin and effectiveness. BSCS, p 65

Cheung K, Long Y, Liu Q, Chan H (2024) Unpacking epistemic insights of artificial intelligence (AI) in science education: a systematic review. Sci Educ. (WOS:001183987000001). https://doi.org/10.1007/s11191-024-00511-5

Clark A, Chalmers D (1998) The extended mind. Analysis 58(1):7–19

Cooper G, Berry A (2020) Demographic predictors of senior secondary participation in biology, physics, chemistry and earth/space sciences: students' access to cultural, social and science capital. Int J Sci Educ 42(1):151–166. https://doi.org/10.1080/09500693.2019.1708510

Cress U, Kimmerle J (2023) Co-constructing knowledge with generative AI tools: reflections from a CSCL perspective. Int J Comput-Support Collab Learn 18(4):607–614. https://doi.org/10.1007/s11412-023-09409-w

Dare EA, Roehrig GH (2016) 'If I had to do it, then I would': understanding early middle school students' perceptions of physics and physics-related careers by gender. Phys Rev Phys Educ Res 12(2):1–11. https://doi.org/10.1103/PhysRevPhysEducRes.12.020117

Darling-Hammond L, Hyler M, Gardner M (2017) Effective teacher professional development. Learning Policy Institute. https://doi.org/10.54300/122.311

Dellermann D, Ebel P, Söllner M, Leimeister JM (2019) Hybrid intelligence. Bus Inf Syst Eng 61(5):637–643. https://doi.org/10.1007/s12599-019-00595-2

Di Paolo LD, White B, Guénin-Carlut A, Constant A, Clark A (2024) Active inference goes to school: the importance of active learning in the age of large language models. Philos Trans Royal Soc B: Biol Sci 379(1911):20230148. https://doi.org/10.1098/rstb.2023.0148

Docktor JL, Mestre JP (2014) Synthesis of discipline-based education research in physics. Phys Rev Spec Top Phys Educ Res 10(2):020119. https://doi.org/10.1103/PhysRevSTPER.10.020119

Duit R, Gropengießer H, Kattmann U, Komorek M, Parchmann I (2012) The model of educational reconstruction—a framework for improving teaching and learning science. In: Jorde D, Dillon J (eds) Science education research and practice in Europe: retrospective and prospective. Sense Publishers, pp 13–37

Erduran S, Levrini O (2024) The impact of artificial intelligence on scientific practices: an emergent area of research for science education. Int J Sci Educ 46(18):1982–1989. https://doi.org/10.1080/09500693.2024.2306604

Erduran S (2023) AI is transforming how science is done. Science education must reflect this change. Science 382(6677). Scopus. https://doi.org/10.1126/SCIENCE.ADM9788

European Commission (2015) Strengthening teaching in Europe: new evidence from teachers compiled by Eurydice and CRELL, June 2015. https://nesetweb.eu/en/news/new-evidence-compiled-by-eurydice-and-crell-on-strengthening-teaching-in-europe/

Farmer S (2024) The alignment of policy and practice for the career-long professional learning of teachers in Scotland. Strathclyde Institute of Education

Gifford JD, Finkelstein ND (2020) Categorical framework for mathematical sense making in physics. Phys Rev Phys Educ Res 16(2):020121. https://doi.org/10.1103/PhysRevPhysEducRes.16.020121

Hattie J (2009) Visible learning: a synthesis of over 800 meta-analysis relating to achievement. Routledge

Häussler P, Hoffmann L (2002) An intervention study to enhance girls' interest, self-concept, and achievement in physics classes. J Res Sci Teach 39(9):870–888. https://doi.org/10.1002/tea.10048

International Commission on the Futures of Education (2023) Guidance for generative AI in education and research. UNESCO

Jeppsson F, Kersting M, Haglund J, Danielsson K (2025) Dual lenses on science classroom interaction: a multimodal and embodied analysis of learning newton's laws in a primary physics classroom. Sci Educ. https://doi.org/10.1007/s11191-025-00630-7

Kaldaras L, Haudek KC (2022) Validation of automated scoring for learning progression-aligned next generation science standards performance assessments. Front Educ 7. Scopus. https://doi.org/10.3389/feduc.2022.968289

Kapon S, Laherto A, Levrini O (2018) Disciplinary authenticity and personal relevance in school science. Sci Educ 102(5):1077–1106. https://doi.org/10.1002/sce.21458

Kaur T, Blair D, Choudhary RK, Dua YS, Foppoli A, Treagust D, Zadnik M (2020) Gender response to Einsteinian physics interventions in school. Phys Educ 55(3). https://doi.org/10.1088/1361-6552/ab764d

Kaur T, Kersting M, Adams K, Blair D, Treagust D, Lonshakova A, Boublil S, Santoso J, Ju L, Zadnik M, Wood D, Horne E, McGoran D, Scott S, Venville G (2024a) Developing and implementing an Einsteinian science curriculum from years 3–10: B. Teacher upskilling: response to training and teacher's classroom experience. Phys Educ 59(6):065014. https://doi.org/10.1088/1361-6552/ad6dcb

Kaur T, Kersting M, Blair D, Adams K, Treagust D, Santoso J, Lonshakova A, Boublil S, Zadnik M, Ju L, Wood D, Horne E, McGoran D (2024b) Developing and implementing an Einsteinian science curriculum from years 3–10: A. Concepts, rationale and learning outcomes. Phys Educ 59(6):065008

Kelly AM (2016) Social cognitive perspective of gender disparities in undergraduate physics. Phys Rev Phys Educ Res 12(2):020116. https://doi.org/10.1103/PhysRevPhysEducRes.12.020116

Kersting M, Haglund J, Steier R (2021a) A growing body of knowledge: on four different senses of embodiment in science education. Sci Educ. https://doi.org/10.1007/s11191-021-00232-z

Kersting M, Danielsson K, Mortimer EF, Olander C, Siry C, Tang K (2025) From founding voices to future visions: languages and literacies in science education. Int J Sci Educ 47(14):1923–1940. https://doi.org/10.1080/09500693.2024.2377424

Kersting M, Schrocker G, Papantoniou S (2021b) 'I loved exploring a new dimension of reality'—a case study of middle-school girls encountering Einsteinian physics in the classroom. Int J Sci Educ, 1–21. https://doi.org/10.1080/09500693.2021.1950943

Kersting M, Amin TG, Euler E, Gregorcic B, Haglund J, Hardahl LK, Steier R (2023) What is the role of the body in science education? A conversation between traditions. Sci Educ. https://doi.org/10.1007/s11191-023-00434-7

Kersting M, Blair D, Sandrelli S, Sherson J, Woithe J (2024) Making an IMPRESSion: mapping out future directions in modern physics education. Phys Educ 59{1):015501

Klette K, Blikstad-Balas M, Roe A (2017) Linking Instruction and Student Achievement. A research design for a new generation of classroom studies. Acta Didactica Norge 11(3):19. https://doi.org/10.5617/adno.4729

Kortemeyer G, Babayeva M, Polverini G, Gregorcic B, Widenhorn R (2025) Multilingual performance of a multimodal artificial intelligence system on multisubject physics concept inventories. arXiv. https://doi.org/10.48550/arXiv.2501.06143

Kortemeyer G (2023) Could an artificial-intelligence agent pass an introductory physics course. Phys Rev Phys Educ Res 19(1). Scopus. https://doi.org/10.1103/PhysRevPhysEducRes.19.010132

Lin-Siegler X, Ahn JN, Chen J, Fang FFA, Luna-Lucero M (2016) Even Einstein struggled: effects of learning about great scientists' struggles on high school students' motivation to learn science. J Educ Psychol 108(3):314–328. https://doi.org/10.1037/edu0000092

McDermott LC (2001) Oersted medal lecture 2001: "Physics Education Research—The Key to Student Learning". Am J Phys 69(11):1127–1137. https://doi.org/10.1119/1.1389280

McLoughlin, Grimes P, Lovatt J (2024) Embracing an inquiry stance in physics teacher professional learning. In: Kaczmarek M, Sokołowska D (eds) Proceedings—book of extended abstracts: 4th world conference in physics education, pp 424–425

Mujtaba T, Reiss MJ (2013) What sort of girl wants to study physics after the age of 16? findings from a large-scale UK survey. Int J Sci Educ 35(17):2979–2998. https://doi.org/10.1080/09500693.2012.681076

Nissen JM, Shemwell JT (2016) Gender, experience, and self-efficacy in introductory physics. Phys Rev Phys Educ Res 12(2):020105. https://doi.org/10.1103/PhysRevPhysEducRes.12.020105

OECD (2023) PISA 2022 results (Volume I): the state of learning and equity in education. OECD. https://doi.org/10.1787/53f23881-en

OECD (2005) Teachers matter: attracting, developing and retaining effective teachers, education and training policy. OECD Publishing

OECD (2019) Learning Compass 2030. https://www.oecd.org/education/2030-project/teaching-and-learning/learning/learning-compass-2030/OECD_Learning_Compass_2030_Concept_Note_Series.pdf

Paul AM (2021) The extended mind: the power of thinking outside the brain. Houghton Mifflin Harcourt

Polverini G, Gregorcic B (2024a) Evaluating vision-capable chatbots in interpreting kinematics graphs: a comparative study of free and subscription-based models. Front Educ 9:1452414. https://doi.org/10.3389/feduc.2024.1452414

Polverini G, Gregorcic B (2024b) Performance of ChatGPT on the test of understanding graphs in kinematics. Phys Rev Phys Educ Res 20(1):010109. https://doi.org/10.1103/PhysRevPhysEducRes.20.010109

Prefontaine B, Mullen C, Güven JJ, Rispler C, Rethman C, Bergin SD, Hinko K, Fracchiolla C (2021) Informal physics programs as communities of practice: how can programs support university students' identities? Phys Rev Phys Educ Res 17(2). https://doi.org/10.1103/physrevphyseducres.17.020134

Reid N, Skryabina EA (2003) Gender and physics. Int J Sci Educ 25(4):509–536. https://doi.org/10.1080/0950069022000017270

Ross K, Galaudage S, Clark T, Lowson N, Battisti A, Adam H, Ross AK, Sweaney N (2023) Invisible women: gender representation in high school science courses across Australia. Aust J Educ 67(3):231–252. https://doi.org/10.1177/00049441231197245

Sadler TD (2009) Situated learning in science education: socio-scientific issues as contexts for practice. Stud Sci Educ 45(1):1–42. https://doi.org/10.1080/03057260802681839

Sands D (2014) Concepts and conceptual understanding: what are we talking about? New Dir 10(1):7–11. https://doi.org/10.11120/ndir.2014.00030

Schön D (1983) The reflective practitioner. Basic Books

Serio M, Piccione A, Marino T, Marocchi D, Alessandra Massa A, Rinaudo M, Luca Ruggiero M, Scalzo G (2025) School—University collaboration to train teachers on new topics and new tools in physics education. J Phys: Conf Ser 2950(1):012048. https://doi.org/10.1088/1742-6596/2950/1/012048

Shapiro L (2014) The Routledge handbook of embodied cognition. Routledge

Shapiro L, Spaulding S (2021) Embodied cognition. In: Zalta EN (ed) The stanford encyclopedia of philosophy (Winter 2021). Metaphysics Research Lab, Stanford University. https://plato.stanford.edu/archives/win2021/entries/embodied-cognition/

Sims S, Fletcher-Wood H, O'Mara A, Cottingham S, Stansfield C, Herwegen JV, Anders J (2021) What are the characteristics of effective teacher professional development? A systematic review and meta-analysis. Education Endowment Foundation

Solomon F, Champion D, Steele M, Wright T (2022) Embodied physics: utilizing dance resources for learning and engagement in STEM. J Learn Sci 31(1):73–106. https://doi.org/10.1080/10508406.2021.2023543

Speirs JC, Stetzer MR, Lindsey BA, Kryjevskaia M (2021) Exploring and supporting student reasoning in physics by leveraging dual-process theories of reasoning and decision making. Phys Rev Phys Educ Res 17(2). https://doi.org/10.1103/physrevphyseducres.17.020137

Tang K-S, Cooper G (2024) The role of materiality in an era of generative artificial intelligence. Sci Educ. https://doi.org/10.1007/s11191-024-00508-0

Tang K (2024) Informing research on generative artificial intelligence from a language and literacy perspective: a meta-synthesis of studies in science education. Sci Educ, sce.21875. https://doi.org/10.1002/sce.21875

Uhden O, Karam R, Pietrocola M, Pospiech G (2012) Modelling mathematical reasoning in physics education. Sci Educ 21(4):485–506

Wilson K, Low D, Verdon M, Verdon A (2016) Differences in gender performance on competitive physics selection tests. Phys Rev Phys Educ Res 12(2):020111. https://doi.org/10.1103/PhysRevPhysEducRes.12.020111

Zhai X, He P, Krajcik J (2022) Applying machine learning to automatically assess scientific models. J Res Sci Teach 59(10):1765–1794. Scopus. https://doi.org/10.1002/tea.21773

Insights into Practicing Teachers' Pedagogical Content Knowledge for Teaching Physics

Stephen Gammell and Eilish McLoughlin

Abstract Pedagogical content knowledge (PCK) is widely recognised as an important component of successful teaching. Definitions of PCK vary across studies and details of how, when and to what extent PCK develop are contested. This study presents the influence of a weeklong summer school on the development of teachers' PCK for teaching physics. During the summer school, teachers completed a range of online workshops, which addressed a range of topics relevant to the teaching of physics including active learning, teaching for inclusion, assessment, making connections, use of digital technologies and using education research to inform practice. A core element of the programme was to support teachers collaborating as part of a professional learning community to develop a unit of learning suitable for second-level physics students. Findings from a qualitative analysis of these units are discussed, including insights into the influence of this professional learning experience on teachers' PCK for teaching physics and recommendations for providing further support to address teachers' needs and learning.

Keywords Pedagogical content knowledge · Physics teaching · In-service teachers

1 Introduction

The past 50 years have seen accelerated change in global economic patterns, including a decrease in the proportion of the workforce of western countries employed in manufacturing, a phenomenon often referred to as deindustrialisation (Rowthorn and Ramaswamy 1997). In response, many countries, including Ireland, have placed an

S. Gammell (✉) · E. McLoughlin
School of Physical Sciences & Centre for Advancement of STEM Teaching and Learning, Dublin City University, Dublin, Ireland
e-mail: stephen.gammell@dcu.ie

E. McLoughlin
e-mail: eilish.mcloughlin@dcu.ie

D. Sokołowska and E. McLoughlin (eds.), *Embracing Changes Together in Physics Education*, Challenges in Physics Education,
https://doi.org/10.1007/978-3-032-27385-7_19

emphasis on developing a highly skilled, scientifically literate and innovative population with an internationally recognised programme of STEM education central to this Government of Ireland Department of Education (2023). This aim is compromised, however, by another global phenomenon, that of teacher supply (Darling-Hammond et al. 2017), with many countries, including Ireland, experiencing an acute shortfall in personnel qualified to teach certain subjects, including those historically identified as falling under the STEM banner (O'Doherty and Harford 2018).

1.1 Pedagogical Content Knowledge

In his seminal paper on teacher knowledge, Shulman (1986) challenged the consensus on how teaching is evaluated and the understanding of teaching more generally, arguing that a focus on process and product ignored the complex nature of teaching. As part of his argument, Shulman introduced a new type of knowledge—PCK—which he defined as 'the particular form of content knowledge that embodies the aspects of content most germane to its teachability' and which includes the ways of representing and formulating the subject to make it understandable to others, and an understanding of what makes the learning of specific topics hard. One corollary of this definition is that PCK is the knowledge which separates a physics teacher from a physicist. Another is that PCK is subject and topic specific. Subsequently, Shulman (1987) identified seven categories of knowledge important for teaching while subsequent publications have varyingly extended, contracted and rearranged these categories. In general, teacher knowledge is described as comprising three main categories: content knowledge, pedagogical content knowledge and pedagogical knowledge. Etkina (2010) describes the structure of physics teacher knowledge in physics in the discipline of physics, as presented in Table 1.

Development of each of the three components of knowledge is a key aim of pre-service teacher education and presents several challenges. Pre-service physics teachers generally develop their content knowledge through undergraduate modules designed and delivered by physics academics. In many cases these modules are

Table 1 The structure of physics teacher knowledge

Pedagogical content knowledge Orientation towards teaching Knowledge of physics curriculum Knowledge of student ideas Knowledge of effective instructional strategies Knowledge of assessment methods	Content knowledge Knowledge of physics concepts, relationships among them and methods of developing new knowledge
	Pedagogical knowledge Knowledge of brain development Knowledge of cognitive science Knowledge of collaborative learning Knowledge of classroom management and school laws

delivered simultaneously to a variety of cohorts of students following different study tracks and do not focus on development of students' PCK. Furthermore, it is unlikely that these modules model teaching approaches consistent with contemporary education reform, and given it is widely recognised that teachers often teach in the way that they experienced a subject for themselves, these further challenge the development of their PCK. Pre-service teachers (PSTs) generally develop their pedagogical knowledge through modules designed and delivered by academics from a university education department. While such modules may promote pedagogical approaches in line with contemporary education reform, they are often delivered to cohorts of PSTs studying a variety of subject specialisms. Because of the highly specific nature of PCK (Etkina 2010), this limits the opportunity to develop PSTs' PCK for teaching physics. Given the recognised importance of developing teachers' PCK, PST education programmes need to consider how this can be best achieved in each subject area. Understanding how best to support teachers to develop their PCK is further complicated by a lack of consensus on when, how and to what extent PCK is developed. Van Driel et al. (1998) state that all scholars agree that PCK is developed through classroom practice and that newly qualified teachers (NQTs) have little or no PCK. However, Grossman et al. (2005) argue that some aspects of PCK, including understanding students' ideas in the subject and how to build on students' existing knowledge, can be developed during pre-service education. Magnusson et al. (1999) identify five components of PCK which can begin to develop in PST education (see the intersecting area of Etkina's scheme in Table 1). While the boundaries and the interaction between the three categories of knowledge are also contested, there is consensus that well-developed content knowledge is a prerequisite for developing PCK and that where deep conceptual understanding is lacking, PSTs tend to focus on teaching facts and algorithms (Cess-Newsome 1999).

1.2 Professional Diploma in Teaching Physics

In 2020, the Department of Education in Ireland set out to address the shortage of qualified physics teachers at post-primary and to reduce the number of teachers teaching physics out of field, by establishing the Professional Diploma in Teaching Physics (PDITP) (McLoughlin 2025). The PDITP is a level 8, two-year, part-time programme designed to upskill in-service, post-primary teachers to meet the requirements for teaching physics to Leaving Certificate level in Ireland (end of upper second level). The programme comprises 75 ECTS, with the aim of developing teachers' physics content knowledge, their pedagogical content knowledge for teaching physics and their reflective practice. The programme is delivered in a blended fashion, comprising a mix of online, face-to-face, synchronous and asynchronous modes. Enrolment is free to qualified second-level teachers through funding provided by the Department of Education.

Opportunities for developing teachers' PCK for teaching physics are built into the design of the PDITP programme (McLoughlin 2025). Teachers complete two physics pedagogy modules (10 ECTS) via 2 week-long summer schools at the end of first year, followed by a year-long inquiry into their own practice during second year (5 ECTS). The first summer school focuses on physics education research and different approaches to teaching physics, while the second summer school focuses on enquiry-based learning, reflective practice and practitioner inquiry. During the first summer of school, teachers engaged in several online workshops which address a range of topics relevant to the teaching of physics including active learning, teaching for inclusion, assessment strategies, connecting physics and STEM learning, use of digital technologies, and using education research to inform practice. All discussions are situated within the practice of teaching physics in the classroom. Facilitators include experienced second-level physics teachers and university physics educators.

Teachers are facilitated to form professional learning communities (PLCs) of 2–3 members and work within these to develop a unit of learning on a specific physics topic. It is expected that teachers incorporate their learning from the various workshops to the design of the unit of learning. The PLCs offer teachers the opportunity to debate and discuss the process of teaching physics. Given that some teachers enrolled in the PDITP are teaching physics out of field, while others are not, there are frequent opportunities for peer learning. Each PLC prepares and records a 20-min presentation, providing an overview of their unit of learning. This recording forms the basis of teachers' grading for the module. Participants in the PDITP programme represent a unique cohort due to the purpose of the programme—that is to upskill practising second-level teachers to become qualified to teach physics. As a result, the group comprises experienced teachers, most of whom aren't experienced in teaching physics (Note: Some teachers have previously taught physics out of field). Given the consensus around PCK being subject specific, but the lack of agreement on when teachers develop PCK and when it's most opportune to work on the development of PCK, analysis of this group's PCK after they engage in a pedagogical module offers an opportunity to gain insights into the nature of PCK development. The following research question will be addressed in this study: What is the nature of practising teachers' pedagogical content knowledge for teaching physics?

2 Data Collection and Analysis

A core aspect of the summer school was to support the 45 participating teachers to collaborate in 15 PLC groups to develop a unit of learning on a physics topic of their choosing. Each group submitted a 20-min video presentation where they provided an overview of the design of their unit and explained the rationale for the activities and approaches, they incorporated in their unit. Teachers were provided with the assessment criteria used to grade their presentations, shown in Table 2.

Qualitative content analysis (QCA) (Hsieh and Shannon 2005) was used for analysing the data from these videos based on: the ontological points of view that meaning is not present in data, rather it is constructed as data is interrogated; that meaning is complex, context-dependent and not immediately obvious, and that the

Table 2 Assessment criteria for physics unit of learning

Criteria	Mark [100]
Topic chosen and rationale	5
Overview of content and approach (what will be taught and why), links to curriculum and theory informing approach to teaching the unit	15
Demonstration of active learning pedagogical strategies for expected difficulties (common misconceptions/issues and how they will be dealt with)	10
Demonstration of active learning pedagogical strategies for creative teaching	10
Demonstration of active learning pedagogical strategies for assessment for and of learning with a focus on variety and with examples	10
Curriculum design—incorporation of key skills, literacy and numeracy, and integration with other STEM subjects	10
Socio-scientific issues and making the topic alive for the students, connections to industry and careers, STS	5
Digital learning resources and strategies	10
Practical work/experiments, including non-mandatory practical work	5
References to the literature/sources	5

importance of aspects of meaning is not necessarily attributable to the frequency of those aspects in the data (Kracauer 1952); and the needs for a method which is systematic, valid and reliable (Neuendorf 2002). QCA was carried out by following the eight-step method prescribed by Schreier (2012) and using the software package Nvivo. Initially, only transcriptions of the audio from the video presentations were selected for analysis; however, while constructing the coding frame, it was noted that many statements made by teachers referred to additional materials, and that these statements were often ambiguous without reference to these. As a result, the analysis materials were expanded to include all files submitted by each PLC, alongside audio transcriptions of the video presentations.

A coding frame was constructed using a hybrid approach, with a subset of submissions ($n = 5$) analysed using an initial coding frame built from the learning outcomes of the module. During this process, to ensure validity (Kripendorff 2018), the coding frame was extended and reorganised based on the content of the data. Reliability of the coding frame was confirmed after re-coding a subset of the materials after a 2-week break to ensure that there was agreement between the outcomes of both passes. During this process, five dimensions (or main categories) were identified in the data: rationale, focus of learning, teaching strategies, assessment and use of digital technologies. This paper reports on the analysis of one dimension only—focus of learning—defined as what the teacher described as the purpose of a portion of class time. The focus-of-learning dimension was divided into three additional category levels using a deductive approach. Within the first level, a 'miscellaneous' category was added into which any data which was deemed not relevant to this dimension could be placed. This was done in keeping with the principle of exhaustiveness and as another check of validity (Schreier 2012). Once a final coding frame was established, a codebook was developed, explaining each of the different categories via a description, example and decision rule. The entire dataset was then segmented into units of coding using a thematic approach (Schreier 2012) before the coding frame was applied to code the entire dataset. A subset of the dataset ($n = 5$) was re-coded after a period of 1 week as a reliability check.

3 Findings and Discussion

Results of the QCA are presented in Table 3, showing the number of cases and instances of each category. The entries at each level are organised in decreasing order, based on the number of instances. While the frequency of a given theme does not indicate importance (Schreier 2012), it provides a general sense of the topics discussed by teachers.

3.1 Content Knowledge

Teachers' units of learning exhibited a strong focus on the development of content knowledge with almost half of the units of codes (699/1603) falling into this category. Within this, several sub-categories were identified. The sub-category of connections was strongly represented (15/244) and was further categorised. The most numerous categories were real life, for example: 'The starter activity for this will be what happens when a bullet is shot out of a gun' which seemed to be motivated by a desire to make physics more tangible, to highlight the usefulness of physics and to be able to recall examples of physics concepts in action.

Connecting to prior knowledge was also heavily referenced, for example: 'the students will be asked to draw a mind map whose central question will be "What do you know about energy?"'. This focus was motivated by a need to start learning from where students are at. Much of connecting to prior knowledge focused on the recall of facts, as opposed to conceptual understanding, and there was little mention of connecting to prior skills learned. Connections across the curriculum and cross-curricularly were discussed in most units of learning, with links to mathematics most highlighted. Discourse on these connections tended not to develop beyond identification of where such connections exist and there was no mention of how such connections could enhance students' learning.

Most units included references to physics careers, for example: 'we can see these in real-life applications, such as in engineering as a career and things like Formula One design cars, testing out cars in general, shoe design for athletes'. This focus was motivated by a desire to show the value of physics but where it was included it failed to extended beyond simply stating related careers. In spite of being highlighted in the summer school, there was limited connection to history and philosophy of

Table 3 Results of coding of the dataset. Level 1 categories are represented by emboldened text and Level 2 categories by ●. The numbers beside each category represent, in order, the number of units of learning in which a given category is present and the number of instances of a given category

Content knowledge (15/699)	
• Connections (15/244)	• History and philosophy of science (4/6)
• Real life (15/102)	• Future study (3/4)
• Prior knowledge (15/69)	• Description (15/161)
• Cross-curricular (12/28)	• Conceptual understanding (14/98)
• Within the curriculum (9/16)	• Misconceptions (15/79)
• Careers (6/9)	• Formal definition (13/56)
• Prior skills (3/7)	• Units (11/35)
Practical work (15/266)	
• Complete experiment (9/40)	• Analysing data (8/10)
• Connect to theory (10/29)	• Following instructions (3/9)
• Observing (10/26)	• Recording data (7/8)
• Drawing conclusions (10/19)	• Hypothesising (5/6)
• Designing (8/16)	• Working safely (2/4)
• Predicting (5/16)	• Writing up (3/3)
• Controlling variables (5/11)	• Making (2/2)
Key skills (15/244)	
• Personally effective (14/50)	• Communicating (12/32)
• Working with others (10/44)	• Numeracy (9/26)
• Literacy (13/41)	• Information processing (6/11)
• Critical and creating thinking (10/33)	
Attitudes (15/192)	
• Accessible to all (15/76)	• Practical work (4/9)
• Hook (12/44)	• Working together (3/5)
• Student-led (10/20)	• Different voices (1/3)
• Real life (9/17)	• Recognising learning (1/2)
Mathematics (15/176)	
• Calculating (15/92)	• Deriving formulae (4/8)
• Graphing (6/22)	• Building formulae (3/3)
• Mathematical tools (8/19)	• Positive attitude (2/2)
• Manipulating formulae (4/9)	
Exam prep (12/26)	

science and, where there was, this focused on the life and work of famous physicists, landmark technological developments and popular stories from the media. Based on these findings, we recommend that time is allocated to support teachers to develop approaches to prior-knowledge activation which target more than the recall of facts (Osman and Hannafin 1994) and that opportunities are provided for teachers to understand how to use connections to real life and careers more effectively to enhance students' learning (Monk and Osborne 1997).

The *Description* sub-category refers to descriptions by teachers of the content areas they planned to focus on in a lesson or unit. This included description from the teacher's point of view and also description with regard to making students aware of what they are doing, for example: '*Today we are going to be learning about light. As part of this we will discuss light as a form of energy and light as a type of wave. We will recap our current knowledge of light and we will explore a number of properties of light, including reflection and refraction*'. While such discourse appears superficial on one level, it shows that teachers have knowledge of the curriculum, have thought about the trajectory of learning and understand that sharing the intention of learning with students can aid learning (Wiliam 2003).

Most cases included reference to formal definitions, with an ability to remember and recall definitions accurately, seen as an important aspect of learning physics. In most instances where formal definitions are referenced, they are presented at the start of a sequence of learning and followed by an exploration of what they mean, for example: '*The approach we're going to take when we study acceleration is we're going to go through the definitions and real-life examples of acceleration*'. In a small number of cases, teachers described exploring concepts first and then using definitions as a summary/formulising of students' own findings, for example: '*So once they've gone through the worksheet and with a bit of questioning from the teacher, hopefully they'll have come up with Newton's first law themselves. But we'll just go through it, make sure it's really clear what the exact definition is*'.

Units were also discussed by most groups and were described as one of the first things to discuss with students. Most discourse on units centred on knowing/recalling SI units and the skill of converting from one unit to another, for example: '*So we're going to start with units and converting units, kind of emphasizing SI units as well. So the learning outcomes we came up with for this particular topic was to kind of highlight to students the importance of SI units and being able to recall what the SI units are for common measurements*'. Some groups discussed developing the skill of dimensional analysis; '*they have to perform the calculations themselves, and we will finally then perform a dimensional analysis whereby students have to come up with the unit for speed and the unit for acceleration themselves*'. Formal definitions and units were not discussed during the summer school; however, teachers' focus on these is likely due to the physics curriculum and physics textbooks highlighting both in their description of each content area and exam papers including questions which test for these. This focus on facts and information may also be evidence of a lack of PCK as previously described by Cess-Newsome (1999).

While conceptual understanding is a goal of physics teaching and learning and was a theme of the summer school, there was little mention of it in the units of learning. There were, however, several tasks described which had the potential to support students developing conceptual understanding. These included applying physics to understand a real-life situation, exploring how change in one quantity affects another, writing causal explanations for different phenomena, predicting what will happen and why and utilising models to understand non-visible phenomena, for example: '*Got a nice AFL exercise which is a PHET simulation. Where you have a balloon charged off a jumper and the location in between the jumper and the wall. The idea is I'll split the class and ask them to actually stand with their feet—to vote for where the balloon is going to go. Is it going to go to the jumper or to the wall? Kind of poke at their understanding and maybe try and get a discussion between the two sides of the room as to why something could go to the jumper, or why something could go to the wall*'. Such approaches are consistent with research-based recommendations which suggest that opportunities for students to reason qualitatively (Van Heuvelen 1991) and to explain their thinking (McDermott 1990a, b) are important for developing conceptual understanding; however, in many cases, it is unclear if conceptual understanding was the purpose of these activities or if they were simply nice activities for students to engage in. As a result, we recommend that there is a greater focus on the importance of conceptual understanding in physics during the summer school and that time is given for teachers to engage with such activities and to reflect on them.

Related to conceptual understanding, all groups referenced student misconceptions, focusing on ways to identify misconceptions which students are likely to present with, for example: '*These are the misconceptions that we have from our own understanding and from the CFA, Harvard, 2022 some of these misconceptions are that a force is needed to keep an object in motion, or that a moving object stops when its force is actually used up and only one force is acting on an object if it is accelerating, or similarly things at rest don't have any forces on them*'. The majority of groups described dealing with misconceptions at the start of a sequence of learning as opposed to misconceptions being something to be cognisant of throughout learning. A minority of groups discussed how to deal with student misconceptions once identified, and one group identified students' own inquiry as a potential contributor to misconceptions: '*there's a danger with a sort of bottom-up approach that a lot of these misconceptions can be reinforced and misunderstandings can be reinforced. So we have to be very mindful of that*'. Given that knowledge about misconceptions does not guarantee high diagnostic competence and that the choice of pedagogy to remediate misconceptions is imperative to success (Timothy et al. 2023), we recommend that, along with discussing the impact of misconceptions on learning, teachers are supported to develop their capacity for identifying and remediating misconceptions (Stofflet 1999).

3.2 Practical Work

A significant focus of practical work centred on completion of mandatory experiments, where a known result/relationship is verified, for example: '*As part of the mandatory practical for this unit of learning, students will carry out the experiment to verify Boyle's law*'. Teachers also discussed the use of experiments for developing students' practical skills with a greater emphasis on skills associated with carrying out a pre-prescribed experiment (following instructions, data analysis, drawing conclusions and writing up), for example: '*when we're doing the mandatory experiments, we will be giving them worksheets with prompt questions to guide them through the experiments*' compared to skills associated with the earlier stages of the scientific method (questioning, hypothesising, designing), for example: '*Now we'll begin to look at designing the experiment. Focusing on the purpose of an experiment, and how its purpose is achieved*'. Teachers also highlighted the use of practical work, in the form of demonstrations to promote understanding of physics concepts. In many such cases, teachers referenced the importance of actively engaging students with the demonstration, for example: '*Gold Leaf Electroscope demonstration...we will begin by putting the students into groups and trying to predict what they think will happen... and more importantly why*'. In several units, teachers discussed the use of simulations for carrying out investigations, with reasons for doing so including reproducibility, access to equipment, safety, ability to complete at home, preparation for carrying out the physical experiment and access to more-advanced experiments. While the overriding theme here was on completion of mandatory practical work (in preparation for exams), it is encouraging to see many groups describe using practical work to promote understanding and actively engaging students with demonstrations through prediction, hypothesis, debate and drawing conclusions. Given the limited reference to inquiry, we recommend that additional focus is given to teachers understanding the positive impact of inquiry on students' conceptual understanding, process skills and attitudes (Blanchard et al. 2010).

3.3 Attitudes

All groups discussed ways to develop a positive attitude towards the study of physics. In making physics accessible to all, teachers described using differentiated practice, scaffolding, supporting different learning styles, formative assessment, catering for the language needs of students, catering for under-represented groups, focus on a growth mindset and the use of different representations. Examples of this include: '*the teacher will be observing their work to offer scaffolding to students who may need it*' and '*An issue we would like to focus on during this unit is the gender imbalances in the classroom. With this in mind, we're going to focus on some group and peer work*'. Most groups discussed the use of physics hooks to engage students, including the use of demonstrations, videos, images, historical events, stories, practical work and surprising results, for example: '*We felt it would be beneficial to include a story-telling hook ... Joule himself had a fascinating backstory, ... He famously brought thermometers on his honeymoon*' and '*our hook for this lesson will be a short video clip of a hot metal ball when dropped into cold water*'. Teachers also highlighted the importance of linking to real-life to motivate and engage students, for example: '*Real life examples... keeps their interest going ... so for athletes, so you could look at Usain Bolt and his acceleration, his velocity, different surfaces you can bring in friction*'. Most units included approaches which promote a sense of ownership around physics, for example: '*We think that' it's really important for students to have a kind of feeling that physics is for them, and physics is about them trying to work things out for themselves, and that they are the centre of the classroom. And it's not just the teacher giving them information and giving them all the things that they need*'. Based on their units of learning, teachers appear to be aware of the importance of motivating students and developing positive attitudes towards physics, with several common strategies utilised including hooks, differentiated learning and collaboration. We recommend that the programme continues to provide an opportunity for teachers to discuss their beliefs and approaches towards nurturing positive attitudes in physics and by supplementing these with recommendations from literature.

3.4 Mathematics

The role of mathematics in teaching and learning physics was highlighted in all units of learning. Teachers highlighted how mathematics presents difficulty for many physics students, from both a skill and attitude point of view, for example: '*We will use a scaffolding activity ... which will aim to encourage all the different types of learners in the room as maths can be a topic where students may not feel very comfortable with. Hoping as well to encourage them to challenge their mathematical abilities*'. This is in keeping with the findings of Pietrocola (2008). Discussion of mathematical formulae focused on calculation, with the ability to correctly identify the meaning of different symbols, correctly substituting into formulae and manipulating formulae seen as priorities, for example: '*we're going to spend the first lesson really focusing on calculations*' and '*we're first going to get students to get the formula for acceleration in terms of initial and final velocity. We will then get students to rearrange this formula so that they can get a v in terms of u, a & t*'. There was some discussion of deriving formulae, with an emphasis on correctly following a prescribed procedure, for example: '*the last part then dealing with linear motion is the derivation of the kinematic equations...in order to do that, they have to follow the following three steps*'. There was limited discussion on understanding the structure of formulae or the skill of developing formulae based on an understanding of physics principles, for example: '*We'll be using directed questioning to try and build up the idea of something loses heat and something gains heat, and students will be coming up with a formula that we can use*'. Given that calculation of correct numerical answers does not necessarily imply a corresponding level of conceptual understanding (McDermott 1990a, b) and that students need to be taught how to read physics equations (Romer 1993), we recommend that teachers are given the opportunity to examine their beliefs around equations/formulae in physics and to develop strategies for developing students' understanding of how equations relate to a given physics concept. Graphs were discussed by a minority of groups with emphasis placed on correctly drawing graphs, for example: '*As graph drawing is often difficult for some students, we feel it is hugely beneficial to spend time discussing the success criteria for graphs... how to scale them correctly and making sure axes labels are used*' and interpreting graphs, for example: '*Dealing with the different aspects of a velocity–time graph and what each part of the graph represents*' and '*to further enhance the understanding ... we would discuss what would happen to the shape of the graph if the current was not squared compared to when it is*'.

4 Conclusion

The QCA of teachers' units of learning provides a number of insights into teachers' PCK which will now be discussed in terms of four of Etkina's five elements of PCK shown in Table 1. The fifth element (knowledge of assessment methods) was not included as part of the original QCA.

4.1 Orientation Towards Teaching

Magnusson et al. (1999) defined nine distinct orientations towards teaching science. The predominant orientations evidenced in this analysis are didactic and activity driven, with an emphasis on teachers telling and instructing students to develop their content knowledge. Students were regularly involved in activities but there was little discussion of the purpose of such activities. There is some evidence for orientations towards conceptual change but, in many cases, it is unclear if activities were included for that purpose. There is also some evidence of orientation towards process, with a focus on following instructions, practical set-up, data collection and analysis but a lesser focus on skills such as hypothesising, predicting, experimental design and interpretation. There is limited evidence of orientations towards discovery, inquiry or project-based learning. The gaps in teachers' knowledge are not surprising given the teachers' background, in terms of learning and teaching physics. These findings highlight the importance of developing teachers' curricular knowledge so they can explain and make connections between physics concepts with their students.

4.2 Knowledge of Physics Curriculum

All presentations referred to the expected learning outcomes described in the physics syllabus and created learning paths which were appropriately sequenced; however, it is not clear if these sequences were based on a clear rationale or simply the order they are presented in the syllabus. A majority of activity descriptions focused on what was being taught (declarative PCK), with lesser discourse around how (procedural PCK) and less still on why (conditional PCK) (Kirschner et al. 2016). There was also limited evidence of teachers connecting concepts from within the physics curriculum or across related curricula, and where presentations did so, discourse didn't extend beyond simply stating that such connections exist. This is indicative of teachers' lack of a deep knowledge of the physics curriculum.

4.3 Knowledge of Students' Ideas

All units emphasised the importance of understanding students' knowledge of a particular topic before exploring new concepts. Some examples of strategies described to elicit students' prior knowledge included peer discussion, mind mapping, quizzes and worksheets. Discourse tended to be quite general and where it was more specific, it tended to focus on the recall of basic knowledge (e.g. definitions, units and examples of principles). Related to this, misconceptions that may arise were discussed in all presentations. In some cases, specific misconceptions around a given topic were highlighted; however, there was no discussion of how to diagnose these misconceptions or to remediate them. The importance of connecting new knowledge to existing knowledge is recognised across education (Simonsmeier et al. 2022). Strategies used to assess students' existing knowledge are applicable across multiple subjects, and the lack of specificity here is indicative of teachers having limited knowledge of students' physics ideas and teachers actively trying to take their PCK from other subjects or their general PK and apply it to their teaching of physics.

4.4 Knowledge of Effective Instructional Strategies

A wide variety of instructional strategies were evidenced across all units of learning, and these strategies were used appropriately. For example, in assessing prior knowledge of a given topic, some groups asked students to work collaboratively to construct a mind map. Such an approach focuses students on recall of knowledge and how elements of knowledge are connected. In another situation, where students were exploring static electricity using a PHET simulation, the teacher promoted engagement by getting students to predict the impact of changes made to the simulation. This demonstrates an understanding of the effective use of demonstrations and digital technologies. The wide variety of instructional strategies and their appropriate use may be evidence of teachers transferring elements of their PCK from a different subject area or their PK more generally, to their teaching of physics, as described in Van Driel (2023). Findings from the QCA of teachers' units of learning show that the teachers are aware of several important elements of PCK for teaching physics, but that their PCK for teaching physics is limited. Teachers utilise and draw on their general PK or PCK from other subject areas to support them in teaching physics.

References

Blanchard MR, Southerland SA, Osborne JW et al (2010) Is inquiry possible in light of accountability? A quantitative comparison of the relative effectiveness of guided inquiry and verification laboratory instruction. Sci Educ 94(4):577–616. https://doi.org/10.1002/sce.20390

Cess-Newsome J (1999) Secondary teachers' knowledge and beliefs about subject matter and their impact on instruction. In: Examining pedagogical content knowledge: the construct and its implications for science education. Springer, Netherlands, pp. 51–94. https://doi.org/10.1007/0-306-47217-1_3
Darling-Hammond L, Hyler EE, Gardner M (2017) Effective teacher professional development. Learning Policy Institute, Palo Alto, CA. https://doi.org/10.54300/122.311
Etkina E (2010) Pedagogical content knowledge and preparation of high school physics teachers. Phys Rev Spec Top—Phys Educ Res 6(2):020110. https://doi.org/10.1103/physrevstper.6.020110
Government of Ireland, Department of Education (2023) STEM education. Implementation Plan to 2026. https://assets.gov.ie/static/documents/stem-education-implementation-plan-to-2026.pdf.
Grossman P, Schoenfeld A, Lee C (2005) Teaching subject matter. In: Darling-Hammond L, Bransford J (eds) Preparing teachers for a changing world. Jossey-Bass, San Francisco, CA, pp 201231. https://doi.org/10.5860/choice.43-1083
Hsieh HF, Shannon SE (2005) Three approaches to qualitative content analysis. Qual Health Res 15(9):1277–1288. https://doi.org/10.1177/1049732305276687
Kirschner S, Borowski A, Fischer HE et al (2016) Developing and evaluating a paper-and-pencil test to assess components of physics teachers' pedagogical content knowledge. Int J Sci Educ 38(8):1343–1372. https://doi.org/10.1080/09500693.2016.1190479
Kracauer S (1952). The challenge of qualitative content analysis. Public Opinion Quart, 631–642. https://doi.org/10.1086/266427
Krippendorff K (2018) Content analysis: an introduction to its methodology. Sage Publications. https://doi.org/10.4135/9781071878781
Magnusson S, Krajcik J, Borko H (1999) Nature, sources, and development of pedagogical content knowledge for science teaching. In: Gess-Newsome J, Lederman NG (eds) Examining pedagogical content knowledge: the construct and its implications for science education. Springer Netherlands, Dordrecht, pp 95–132. https://doi.org/10.1007/0-306-47217-1_4
McDermott LC (1990a) A perspective on teacher preparation in physics and other sciences: the need for special science courses for teachers. Am J Phys 58(8):734–742. https://doi.org/10.1119/1.16395
McDermott LC (1990b) Millikan Lecture 1990: What we teach and what is learned—closing the gap. Am J Phys 59(4):301–315.https://doi.org/10.1119/1.16539
McLoughlin E (2025) Remaking physics teacher professional learning: a national upskilling programme for "out-of-field" physics teachers. In: Jeskova Z, Hanc J (eds) Advancing physics education—innovations from early learning to higher education. Springer. https://doi.org/10.1007/978-3-031-94207-5_3
Monk M, Osborne J (1997) Placing the history and philosophy of science on the curriculum: a model for the development of pedagogy. Sci Educ 81(4):405–424. https://doi.org/10.1002/(sici)1098-237x(199707)81:4%3c405::aid-sce3%3e3.0.co;2-g
Neuendorf KA (2002). Defining content analysis. In: Content analysis guidebook, pp 10–27. https://doi.org/10.4135/9781071873045
O'Doherty T, Harford J (2018) Teacher recruitment: reflections from Ireland on the current crisis in teacher supply. Eur J Teach Educ 41(5):654–669. https://doi.org/10.1080/02619768.2018.1532994
Osman ME, Hannafin MJ (1994) Effects of advance questioning and prior knowledge on science learning. J Educ Res 88(1):5–13. https://doi.org/10.1080/00220671.1994.9944829
Pietrocola M (2008). Mathematics as structural language of physical thought. In: Connecting research in physics education with teacher education, vol 2. https://doi.org/10.1088/0031-9120/33/4/021
Romer RH (1993) Reading the equations and confronting the phenomena—the delights and dilemmas of physics teaching. Am J Phys 61(2):128–142. https://doi.org/10.1119/1.17327
Rowthorn R, Ramaswamy R (1997) Deindustrialization: its causes and implications, vol 10. International Monetary Fund, Washington, DC. https://doi.org/10.5089/9781557756435.051

Schreier M (2012) Qualitative content analysis in practice. Jacobs University Bremen. https://doi.org/10.4135/9781529682571
Shulman LS (1986) Those who understand: knowledge growth in teaching. Educ Res 15(2):4–14. https://doi.org/10.3102/0013189x015002004
Shulman LS (1987). Knowledge and teaching: foundations of the new reform. Harv Educ Rev 57(1). https://doi.org/10.17763/haer.57.1.j463w79r56455411
Simonsmeier BA, Flaig M, Deiglmayr A et al (2022) Domain-specific prior knowledge and learning: a meta-analysis. Educ Psychol 57(1):31–54. https://doi.org/10.1080/00461520.2021.1939700
Stofflet RT (1999) Putting constructivist teaching into practice in undergraduate introductory science. Electron J Res Sci Math Educ 1(2)
Timothy V, Watzka B, Stadler M et al (2023) Fostering preservice teachers' diagnostic competence in identifying students' misconceptions in physics. Int J Sci Math Educ 21(5):1685–1702. https://doi.org/10.1007/s10763-022-10311-4
Van Driel JH, Verloop N, De Vos W (1998) Developing science teachers' pedagogical content knowledge. J Res Sci Teach: Off J Natl Assoc Res Sci Teach 35(6):673–695. https://doi.org/10.1002/(sici)1098-2736(199808)35:6%3c673::aid-tea5%3e3.0.co;2-j
van Driel JH, Hume A, Berry A (2023) Research on science teacher knowledge and its development. In: Handbook of research on science education. Routledge, pp 1123–1161. https://doi.org/10.4324/9780367855758-41
Van Heuvelen A (1991) Learning to think like a physicist: a review of research-based instructional strategies. Am J Phys 59(10):891–897. https://doi.org/10.1119/1.16667
Wiliam D (2003). Formative assessment in mathematics. In: Aspects of teaching secondary mathematics. Routledge, pp 289–300. https://doi.org/10.4324/9780203165874-25

Schreier M (2012) Qualitative content analysis in practice. Sage Publications, [illegible]. https://doi.org/10.4135/9781529682571

Shulman LS (1986) Those who understand: knowledge growth in teaching. Educ Res 15(2):4–14. https://doi.org/10.3102/0013189X015002004

[illegible]

Shaping Academic Physics Teacher Education Together: How to Promote Interdisciplinary Collaboration?

Sebastian Schellhammer, Frank Beier, Rolf Puderbach, and Gesche Pospiech

Abstract Teacher education, and academic teacher education in particular, requires collaboration between a multitude of disciplines and institutions in order to be effective. However, different professions, biographies, concepts regarding teacher education, and missing resources lead to a rather fragmented reality, which makes a coherent, competence-based curriculum difficult. The connection between the courses provided by scientific content disciplines, subject-matter didactics, and educational sciences is regularly made by the students themselves reducing student motivation and learning progress. In our contribution, we present recent advances at Technische Universität Dresden, Germany, at different levels of establishing collaborations in academic teacher education. The module "Applications in physics and their didactics" is discussed as a best practice entangling concepts from physics education and insights into current physics research to create a holistic learning environment for advanced teacher students. The high level of student satisfaction demonstrates the opportunities of synergetic teacher education. Structural support for strengthening such collaborations is discussed in terms of a multiplier program as a bottom-up approach and a cross-faculty project supervised by the local center for teacher education as a top-down strategy. The experiences made and lessons learned provide inspiration and guidance for institutions to promote interdisciplinary collaborations in their academic teacher education programs.

S. Schellhammer (✉)
Institute of Applied Physics, Technische Universität Dresden, Dresden, Germany
e-mail: sebastian.schellhammer@tu-dresden.de

F. Beier
Institute of Education, Technische Universität Dresden, Dresden, Germany

R. Puderbach
Center for Teacher Education and Educational Research, Technische Universität Dresden, Dresden, Germany

G. Pospiech
Chair of Physics Education, Technische Universität Dresden, Dresden, Germany

D. Sokołowska and E. McLoughlin (eds.), *Embracing Changes Together in Physics Education*, Challenges in Physics Education,
https://doi.org/10.1007/978-3-032-27385-7_20

Keywords Teacher education · Multidisciplinarity · Higher education · Network · Cooperation · Pre-service teacher

1 Introduction

Teacher education essentially revolves around three fundamental questions: What is it that in-service teachers need to know and be capable of? How should teacher education be designed to effectively foster these competencies and connected productive habits (Etkina et al. 2017)? How are the answers to these questions evolving in time (Yun 2020)? Despite the significant differences among various teacher education systems and their cultural contexts (Blömeke 2006), it seems that no teacher education program can avoid addressing these questions in some form.

As teacher education has increasingly shifted to universities, several challenges have emerged regarding the professionalization of pre-service teachers. In some regions, such as Japan, teacher education is limited to an add-on master's degree or practical training, meaning it lacks independent academic status. In other countries, like Germany, teacher education has become an integral part of academic culture at universities, aiming to produce a distinct body of knowledge for teachers. However, unlike other professions, such as law or medicine, there is no single core discipline and no distinct organizational unit that provides the academic foundation for the teaching profession in the same cohesive manner. Accordingly, different views on teacher education and required competencies collide within the curricula (Ligozat et al. 2015; Hudson and Zgaga 2017; Werler and Tahirsylaj 2020). As presented in Fig. 1 for the case of Technische Universität Dresden (TUD), Germany, usually stakeholders from scientific content disciplines, subject-matter didactics, and educational sciences as well as organizational units, such as the Center for Teacher Education and Educational Research, shape the academic teacher education program. In addition, different external stakeholders shape boundary conditions and can be fruitful partners, if collaborations are efficiently fostered (Schellhammer et al. 2020). Collaboration between internal stakeholders occurs rather poorly, although being considered beneficial for the promotion of student learning (Hanke et al. 2021; Meier et al. 2018). Students perceive a weak horizontal coherence of educational sciences, subject-matter didactics, and scientific content disciplines (Mehlmann and Bikner-Ahsbahs 2018; Schwalbe et al. 2022). Such a fragmented responsibility and low level of cooperation pose a risk to the efficiency of academic teacher education programs (Hellmann et al. 2019), also considering that schools with pronounced collaborations among teachers promote student learning more successfully (Scheerens and Bosker 1997).

In our contribution, we present experiences and lessons learned over the last 10 years at TUD for strengthening interdisciplinary collaboration in teacher education—from a best practice in teaching to strategic bottom-up and top-down approaches which create a fruitful environment for collaboration.

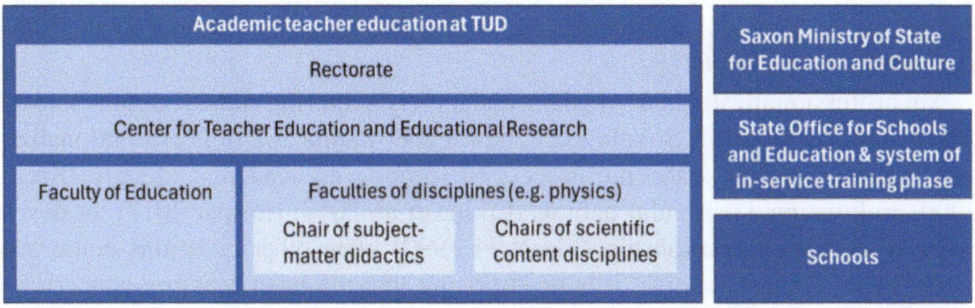

Fig. 1 Simplified illustration of the academic teacher education system at TUD including involved internal and external stakeholders. For simplicity, this contribution focuses on interdisciplinary collaboration between academic stakeholders

2 Fragmented Academic Teacher Education

Academic teacher education programs face the challenge of incorporating a wide array of disciplines, each with its own unique knowledge culture. This "multi-paradigm" structure poses significant obstacles to creating a coherent teacher education curriculum (Hudson and Zgaga 2017; Hellmann et al. 2019). Often, it is left to pre-service teachers to make sense of these fragmented and partially even opposing knowledge cultures. As Colin Cramer argues (Cramer et al. 2023), teacher students must develop a sense of "meta-reflexivity" to navigate the high degree of uncertainty inherent in the teaching profession. This requires the ability to shift perspectives and to redefine problems in regard to teaching and other educational practices (Sokolowska 2024).

"In a meta-reflexive manner, it is necessary to take different perspectives on the same object, to understand the foundations of individual perspectives, as well as to be able to comprehend the mode of the respective observation regarding potentials and limitations and to interpret it in relation to the other perspectives. This seems especially relevant in teacher education because it is characterized by a plurality of (epistemological) theories and empirical approaches due to the numerous scientific (sub-)disciplines involved" (Cramer et al. 2023, p 472).

The concept of meta-reflexivity challenges the notion of a straightforward separation of labor in teacher education, where specific disciplines are responsible for the cumulative development of competencies and personalities. Following Shulman's framework (Shulman 1986), many approaches in teacher education distinguish between various types of correlated knowledge: content knowledge, curricular knowledge, pedagogical knowledge, and pedagogical content knowledge. This division might suggest that different scientific disciplines should take responsibility for cultivating pre-service teachers' professional knowledge. However, the question remains whether such education programs truly enable students to grasp the underlying logic of these diverse scientific paradigms and to synthesize this knowledge during their professionalization process. Especially subject-matter courses are considered less relevant by students, e.g., because of their abstract nature at the academic level and a missing link to teaching practice (Cramer et al. 2009; Hefendehl-Hebecker 2013).

Yet professionalization has to go beyond merely acquiring and developing knowledge (Etkina et al. 2017; Sokolowska 2024; Fabel-Lamla 2024). If professionalization is understood as a specific process of forming an academic identity (Kreitz 2000), cultivating a particular habitus (Etkina et al. 2017; Helsper 2018), or developing a world-self relationship (Marotzki 1990), then other questions come into focus. From this perspective, it is unsurprising that the most pressing concern for pre-service teachers regarding academic content in teacher education programs is related to their professional identity. Instead of purely academic questions of knowledge, practical concerns dominate: What does the academic knowledge discussed at university mean for me as a future teacher? Is the knowledge provided in academic programs applicable to the professional realities of teaching? However, those questions are regarded as problematic in the context of academic teacher education (Wenzl et al. 2018), because they might get in conflict with the scientific questions raised by scientific disciplines regarding epistemic problems.

3 Synergistic Teacher Education at Course Level: "Applications in Physics and Their Didactics" as a Role Model

With numerous scientific disciplines being involved in academic teacher education, the students generally need to connect and make sense of the plurality of content, competences, perspectives, and potential role models within the courses themselves regarding their future careers as teachers (Larsson et al. 2020). Accordingly, course designs that synergistically combine physics didactics and physics as a science as well as the respective educators may foster student motivation and competence development. Despite strong efforts over the last decade (Hanke et al. 2021; Mehlmann and Bikner-Ahsbahs 2018; Sokolowska 2024; Barzel et al. 2016; Diehr 2018), best practices are still widely missing because the disciplines involved in teacher education are separated by academic structures as well as different scientific cultures and biographies within the disciplines.

However, the module "Applications in physics and their didactics" at TUD formally forces a collaboration between the Chair of Physics Education and the Institute of Applied Physics. Each of the partners contributes half of the module with a session of 90 min per week and a combined assessment of the students' performance. The full potential of a vivid collaboration between the two partners has been realized with a redesign of the educational concept in 2023, leading to excellent student feedback and winning the university's teaching award. The corresponding course structure is schematically depicted in Fig. 2. The course is attended by advanced pre-service teachers for secondary education. Accordingly, during the sessions advanced topics in lesson design are treated, such as interdisciplinary teaching, out-of-school

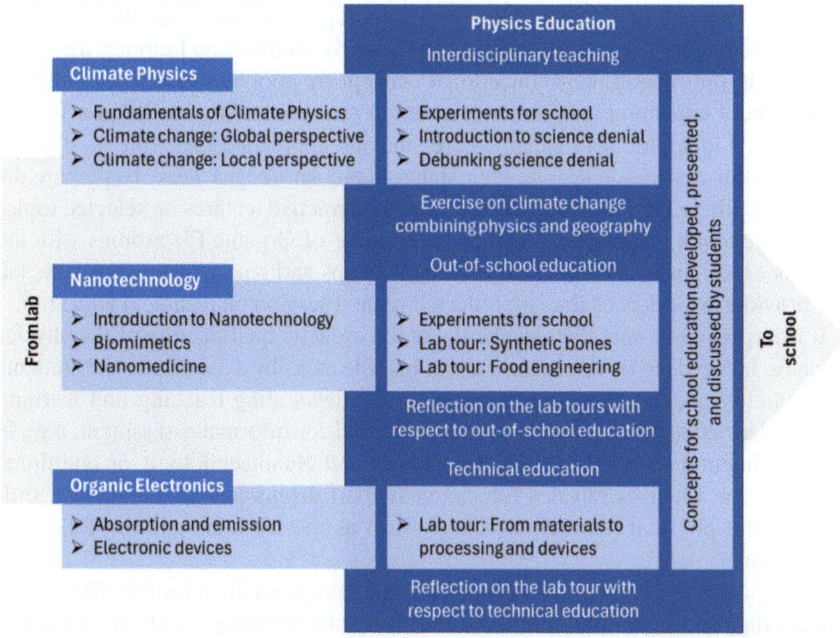

Fig. 2 Schematic illustration of the didactic concept within the course "Applications in physics and their didactics" demonstrating how physics sessions and physics education interact with each other within the interdisciplinary main topics Climate Physics, Nanotechnology, and Organic Electronics

education, and perspectives on career-oriented education with a focus on aspects of technical education. These concepts are mostly introduced and discussed in traditional, partly interactive, lectures and practiced in hands-on seminars in the following session.

The physical part of the module has been redesigned in order to address these concepts and provide authentic scenarios for discussion in the seminars. The three topics covered were chosen to be interdisciplinary to provide a basis for discussion in interdisciplinary teaching. In addition, they provide exemplary cases for general interdisciplinary principles that are rarely addressed within the curriculum: debunking science denial is linked to Climate Physics, risk assessment of new technologies to Nanotechnology, and the opportunities of computational studies and artificial intelligence to Organic Electronics.

In addition, to get authentic insights into research environments, the students take part in three 3-h guided tours of scientific laboratories, which are linked to the educational aspects. Related to the topic of Nanotechnology, they create and characterize materials for bone substitutes at the Institute of Materials Science and reflect on their experience with respect to out-of-school education by performing a potential analysis for the lab and their learning experiences. This includes an evaluation of criteria such as relevance to the real world, scientific relevance, primary experiences (smells, sounds, etc.), and interaction with experts. Second, they visit the labs at the Institute of Natural Materials Technology and experience their program designed for secondary schools. Afterward, they give feedback and support the scientific staff in optimizing their educational concept by applying their knowledge on out-of-school education and career orientation at school. This results in a fruitful exchange of ideas for both sides, where students appreciate their role as didactic experts in this discussion despite their status as pre-service teachers. To prepare the students for these lab tours, they attend related interactive lectures on selected topics of Nanotechnology. The same applies to the topic of Organic Electronics with the students experiencing different experimental setups and research projects. This lab tour provides experiences that are reflected on in a session on technical education.

It is important to note that despite the broad didactic qualification of the physics educator, he needs to be transparent about his role as a physicist. Either the students are explicitly addressed as ongoing experts for developing teaching and learning in secondary education, and peer-feedback is used for informal assessment, e.g., in the experimental sessions on Climate Physics and Nanotechnology, or additional staff from the Chair of Physics Education support discussions that address a shift from current physical research to school, such as the lab tours and the follow-up discussions.

The students appreciate that the course indeed prepares them for their later work as secondary teachers. They highlight that it is, unfortunately, rarely the case that physics courses explicitly address their needs as pre-service teachers, for example, because they are usually a minority among regular physics students. Still, there are some lessons learned during the two implementations we had so far. In the first year, the students should acquire skills in reading and reflecting on original scientific publications. However, a considerable number of them showed strong reluctance to work with literature in English. In the second implementation, this learning goal was canceled and mostly visual and interactive materials based on scientific results were used. This was greatly appreciated by the students, as they were able to incorporate these materials directly into their teaching materials. While we started with Nanotechnology in the first year, we changed the order of topics in the second year to the one shown in Fig. 2. The discussion of Climate Physics is underrepresented in the current curriculum despite its high relevance for pre-service teachers, leading to high student motivation and high participation rates. This helped to promote students' confidence in the benefits of this relatively unusual course for their learning and secured above-average participation rates along the entire semester, although Nanotechnology and Organic Electronics felt far away from usual secondary school physics classes. In fact, several students highlighted in the final feedback on the course concept that they appreciated being able to tackle these topics in the course because it opened their eyes regarding their hidden relevance in everyday life.

4 Structural Support to Promote Synergistic Academic Teacher Education: Bottom-Up and Top-Down Approaches

Aside from a few best practices, how can one structurally overcome the strong separation of the disciplines involved and create a culture of synergistic academic teacher education? In the following subsections, we present two potential approaches: the strengthening of the collegial network of educators involved via a multiplier program as a bottom-up strategy and the creation of a scaffold for collaboration through a cross-faculty project at the Center for Teacher Education and Educational Research.

4.1 Establishing a Collegial Culture: Multiplier Program and Monthly Get-Together

In the system of academic education, where time and resources are scarce and careers are regularly defined by research output rather than the quality of teaching (Teichler 2017), ambitious educators are usually driven by an exceptional intrinsic motivation. This, along with an already well-established relationship of trust between the educators, has led to the redesign of the module "Applications in physics and their didactics." Between 2016 and 2019, a multiplier program for such committed educators was established by the Center for Continuing Education at TUD to empower them, provide advanced educational competences, and create an interdisciplinary network (Weller and Schellhammer 2019). Its fundamental objectives and activities, which are a combination of individual support and group activities, are presented in Fig. 3. Here, we reflect on the initialization of such a multiplier program and those elements that bring educators from different disciplines together and create a breeding ground for potential further synergistic teacher education projects, i.e., the monthly get-together allowing a regular collegial exchange of experiences as well as the annual symposia celebrating successful best practices and commitment to teaching in general.

In 2016, the multiplier program at TUD was created to strengthen educators in STEM-fields. Its mission goes far beyond regular task forces, e.g., for improving digitalization in teaching and learning, as it also aims for a promotion of individual careers trying to outbalance the pronounced academic reward system for strong research output (Teichler 2017). For a successful start of the program, approx. 20 suitable educators were identified based on teaching awards, third-party funding for teaching projects, suggestions from student representatives, and personal contacts. Care was taken to ensure that the starting group had a balanced mix of familiarity and difference with multiple educators per faculty and a heterogeneous composition in terms of academic positions from students to postdocs and professors. Indeed, it was most challenging to attract professors to such a program. Already before the kickoff workshop that put special emphasis on networking, team building, and the development of

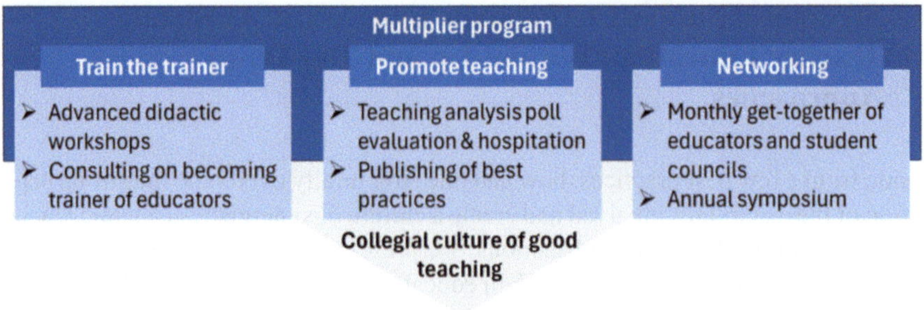

Fig. 3 Schematic illustration of the concept of the multiplier program for educators at TUD

common goals, all candidates were contacted personally and interviewed regarding their teaching activities and personal requirements for the program. Although the program can be considered a bottom-up approach as it elevated the activities of educators, it was not fully self-driven but depended on the support of the Center for Continuing Education staff, who organized meetings, proactively initiated collaborations, and provided additional support. However, the network formed outlasted third-party funding in 2019, sustainably strengthening a culture for collaborations and appreciations in teaching and learning.

One key element of the multiplier program was the monthly get-together, which is still offered at 2-month intervals. Within its 90 min, a specific topic is discussed based on a keynote speech by an educator. To underline its low-threshold nature, coffee and cake are regularly provided—or a BBQ and beverages in the summer. Already because of its pleasant atmosphere, a core group of approx. Ten people is attending the get-togethers relatively independent of the topic. In addition, depending on the topic, 10 to 20 more people are attending. It needs to be noted that already with more than 20 people it becomes challenging to maintain the non-binding nature of the meeting. The same applies to hybrid meetings to enable the participation of other universities. The topics are usually suggested by the participants. Especially in the beginning of the get-together, it was challenging to convince educators to give insights into their teaching because they did not think their work was sufficient. Consultation with the educators and, if necessary, feedback in advance of the presentation were regularly required to create a safe space for the keynote speech. These presentations not only discuss best practices but also create a space to jointly find solutions to didactic challenges. Accordingly, it creates a room for collaboration—sometimes, for example, educators also sit in on the educator's course and give feedback afterward independent of the discipline or faculty. While the get-together can be considered a first step toward an even more pronounced professionalization in teaching, the annual symposium created an even more professional stage with best practice presentations in a university-wide conference-like environment. After funding, an attempt was made to continue the get-together as a monthly self-organized room for discussion without a set topic, but this failed. With time being a scarce resource, educators need to recognize a clear benefit for their teaching. Although this appears partially in contrast with the idea of a low-threshold format, such as the get-together, it turned out that a coordination of the activities and advertisement with clear topics are crucial for success.

Despite the positive experiences we made, e.g., in attracting educators, who usually do not attend didactic programs, several challenges remain. Based on the funding received to establish the multiplier program, it only addressed educators from STEM-fields. Attempts have been made to extend it to all faculties. However, while established members appreciated the program because it was specific to STEM-fields and avoided such activities, the pull for new members was not enough to compensate for the loss of participants. Bottom-up programs rely on the motivation of the people they address. This means that aside from potential benefits, such as an exchange of experiences and insights into didactic concepts, it needs to be low threshold and create a pleasant atmosphere. Discussing teaching and learning with educators beyond STEM-fields was unfortunately considered outside of their personal comfort zone and less transferable to their own teaching activities. The same applied to activities in English because of limited linguistic confidence when discussing teaching and learning. It yet remains to be seen, if such a program can efficiently strengthen collaborations within the teacher education system, but the connecting objective to improve it might work as a similar trigger as the optimization of courses in STEM-fields. This can already be observed regularly at the monthly get-together, where educators from scientific content disciplines and from subject-matter didactics exchange ideas and experiences about their courses, as well as in task forces, e.g., on digitalization and training of in-service teachers, hosted by the Center for Teacher Education and Educational Research.

4.2 Creating a Scaffold for Interdisciplinary Teacher Education within the Project TUD-Sylber

As bottom-up approaches especially rely on the commitment of educators and can efficiently promote their activities and visibility, they should only be considered one component toward a synergistic teacher education. Structural support and additional resources by top-down strategies are beneficial to provide an ideal environment for educators (Hellmann et al. 2019) and can sustainably establish improvements beyond committed individuals (Barzel et al. 2016; Schellhammer and Kainer 2018). Initiating and intensifying the cooperation between the stakeholders involved in teacher education courses at TUD were the central aim of the development project "Synergetic University" (TUD-Sylber). It was conducted as part of the Qualitätsoffensive Lehrerbildung, a joint program the German Federal Government and the Länder have implemented between 2016 and 2023 to promote an improvement of German teacher education (BMBF 2023). Professorships from seven faculties (Faculty of Education, Faculty of Philosophy, Faculty of Language, Literature and Cultural Studies, Faculty of Psychology, Faculty of Physics, Faculty of Computer Science, and Faculty

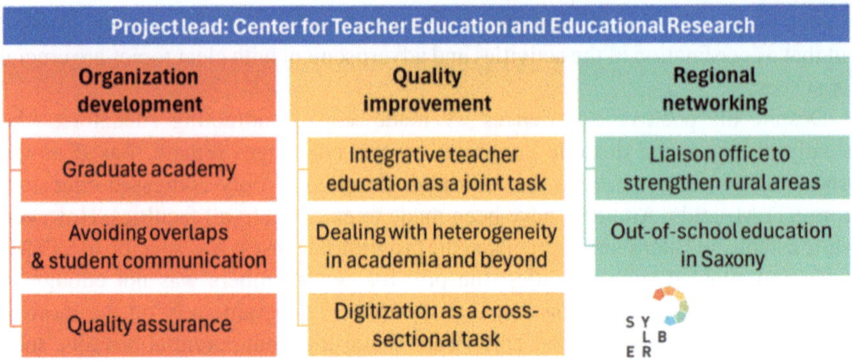

Fig. 4 Project structure of TUD-Sylber[2] between 2019 and 2023. In particular, the activities under the central pillar for quality improvement aimed at overcoming integral challenges by bringing together stakeholders from different disciplines and partially even beyond the academic teacher education system

of Environmental Sciences) and the Center for Teacher Education and Educational Research were involved in several interdisciplinary sub-projects of TUD-Sylber as schematically depicted in Fig. 4.

The core element of the project was the establishment of cross-faculty exchange and working structures to overcome fragmentation. At the heart of this strategy is the Center for Teacher Education and Educational Research as an overarching structure that coordinates the numerous stakeholders involved in the academic teacher training program at the TUD. It is not a question of centralizing the responsibilities for teacher education, as the responsibility for the content and format of teacher education courses remains with the TUD faculties, whose autonomy will not be curtailed. The Center for Teacher Education supports the faculties and provides services in areas such as study organization, administration, evaluation, and monitoring as well as networking with non-university partners. It creates discourse spaces for interfaculty collaboration and initiates joint activities. The acquisition and coordination of the project TUD-Sylber, which brings in relevant third-party funds for all participants, helped strengthen the role of the center as a recognized overarching structure of TUD teacher education.

Activities of TUD-Sylber provided the opportunity for educators or professors from different subjects, subject-matter didactics, and educational sciences to meet in person for the first time. Working together on overarching teacher education topics helped to identify common interests and concerns and build trust over the course of the project. The joint activities ranged from evaluation studies on teacher education (Schwalbe et al. 2022) to measures to improve the organization of studies and the establishment of a joint structure to promote young academics. The project participants jointly developed innovative teaching formats in which educational sciences, subject-matter didactics, and subjects worked together. The teaching formats were dedicated to current cross-cutting topics such as inclusion, dealing with heterogeneity, digitalization, and out-of-school education and significantly enriched teacher training courses at TUD.

One example of the interdisciplinary activities in TUD-Sylber is onboarding measures for new TUD lecturers: Often, lecturers of the subjects are not aware that prospective teachers are among their students. They are not familiar with the structure of teacher education programs and the special requirements of this group of students and are therefore unable to respond appropriately to their needs. In order to support lecturers for their role in teacher education, an interdisciplinary project group designed information materials and events for newly appointed professors and teaching staff.

Another successful example of cooperation between several subject-matter didactics is a project to promote interdisciplinary out-of-school education, which is included in the school curriculum but not sufficiently covered by teacher education as it is traditionally structured along disciplinary lines. Starting from the development of a common conceptual and theoretical framework, colleagues from the chairs of physics, chemistry, German studies, and geography didactics carried out joint seminars and developed teaching concepts that have been piloted in cooperation with local schools, museums, and other educational organizations.

Experience in the TUD-Sylber project has shown that additional financial resources can be used to initiate collaborations that were not previously possible. Although there is a fundamental willingness to cooperate among many teacher education stakeholders, limited personnel and time resources often prevent implementation. In addition to the availability of financial resources, it was the central coordination of the funding application and the implementation of the project by the Center for Teacher Education and Educational Research that made it easier for each of the many professorships to participate in the joint project. The first step on the path to interdisciplinary cooperation is to find a common language and to discuss, if not agree on, central concepts and theoretical principles, which is a challenge even for representatives of educational science and different subject-matter didactics.

After the end of the project funding, not all joint activities were continued to the same extent. What remains is the widespread awareness of both the necessity and potential of interdisciplinary collaboration in teacher education and the common concern of strengthening the interlinking of scientific content disciplines, subject-matter didactics, and educational sciences and increasing the sense of coherence among pre-service teachers.

5 Conclusions

To overcome the fragmentation in academic teacher education and establish a coherent competence-based curriculum, interdisciplinary collaborations between the different stakeholders from educational sciences, subject-matter didactics, the sciences themselves, and organizational units are required. Although there is widespread agreement on this, the reality is dominated by division of labor at best. Lacking time and personnel resources as well as disciplinary boundaries or the separation between organizational units are only a few reasons for this situation.

In our contribution we presented several experiences made at TUD that illustrate the challenges but also the opportunities by establishing interdisciplinary collaborations. The synergistic combination of activities from physics and physics education in the module "Applications in physics and their didactics" supports students in making sense of the knowledge and acquired competencies, which leads to a high level of student satisfaction. The multiplier program provides a bottom-up strategy for cultivating such interdisciplinary collaborations by creating an appreciative culture for designing teaching and learning and forging bonds of trust between educators beyond disciplinary borders. While this approach mostly promotes individuals, structural improvements can be made based on top-down approaches—as presented here in the case of the third-party-funded project TUD-Sylber. The strengthening of the local center for teacher education as an overarching structure for academic teacher education has already promoted collaboration because coordinating staff became available. The distribution of third-party funds strongly motivated those involved to work together, e.g., in the introduction of an onboarding program for new lecturers.

Despite the success achieved, most of the activities discussed were based on individuals or project funding, which means that the achievements are jeopardized by the high staff turnover in academic teacher training. This applies particularly to best practices in teaching and learning, as a synergistic teaching approach is more resource intensive, as well as bottom-up approaches that are aimed at promoting individuals rather than structural change. In such cases, an overarching structure for academic teacher education, i.e., the Center for Teacher Education and Educational Research at TUD, can help to promote continuous success, for example by supporting handovers in teaching (Schellhammer and Kainer 2018).

The fragmentation of academic teacher education remains a challenge, but the experiences presented enabled valuable progress toward interdisciplinary collaboration and a coherent curriculum at TUD.

References

Barzel B, Eichler A, Holzäpfel L, Leuders T, Maaß K, Wittmann G (2016) Vernetzte Kompetenzen statt trägen Wissens—Ein Studienmodell zur konsequenten Vernetzung von Fachwissenschaft, Fachdidaktik und Schulpraxis. In: Hoppenbrock A, Biehler R, Hochmuth R, Rück H-G (eds) Lehren und Lernen von Mathematik in der Studieneingangsphase. Springer Spektrum, Wiesbaden, pp 33–50. https://doi.org/10.1007/978-3-658-10261-6_3

Blömeke S (2006) Struktur der Lehrerausbildung im internationalen Vergleich. Ergebnisse einer Untersuchung zu acht Ländern. Zeitschrift für Pädagogik 52:393–416. https://doi.org/10.25656/01:4466

BMBF, Bundesministerium für Bildung und Forschung (2023) "Qualitätsoffensive Lehrerbildung". A look back, a look ahead. Berlin.

Cramer C, Horn K-P, Schweitzer F (2009) Zur Bedeutsamkeit von Ausbildungskomponenten des Lehramtsstudiums im Urteil von Erstsemestern. Erste Ergebnisse der Studie Entwicklung

"Lehramtsstudierender im Kontext institutioneller Rahmenbedingungen" (ELKiR). Zeitschrift für Pädagogik 55:761–780. https://doi.org/10.25656/01:4274
Cramer C, Brown C, Aldridge D (2023) Meta-reflexivity and teacher professionalism: facilitating multiparadigmatic teacher education to achieve a future-proof profession. J Teach Educ 74:467–480. https://doi.org/10.1177/00224871231162295
Diehr B (2018) Wissenschaftliche Englischlehrerbildung—Eine Herausforderung für Fachwissenschaft und Fachdidaktik. In: Diehr B (ed) Universitäre Englischlehrerbildung: Wege zu mehr Kohärenz im Studium und Korrespondenz mit der Praxis. Peter Lang, Berlin, pp 75–102
Etkina E, Gregorcic B, Vokos S (2017) Organizing physics teacher professional education around productive habit development: a way to meet reform challenges. Phys Rev Phys Educ Res 13:010107. https://doi.org/10.1103/PhysRevPhysEducRes.13.010107
Fabel-Lamla M (2024) Professionalisierung von Lehrkräften aus Perspektive des (berufs-) biografischen Ansatzes. Forschungsbefunde und Konsequenzen für die Lehrer*innenbildung. In: Kratz M, Dlugosch A, Heß M (eds) Biographisches Erzählen in der Hochschulbildung. Impulse für pädagogische Professionalisierungsprozesse, Psychosozialverlag, Gießen, pp 97–126. https://doi.org/10.23668/psycharchives.16112
Hanke E, Hehner S, Bikner-Ahsbahs A (2021) Reducing fragmentation in university pre-service teacher education: conditions and strategies. Educ Des Res 5:1–32. https://doi.org/10.15460/eder.5.2.1613
Hefendehl-Hebecker L (2013) Doppelte Diskontinuität oder die Chance der Brückenschläge. In: Ableitinger C, Kramer J, Prediger S (eds) Zur doppelten Diskontinuität in der Gymnasiallehrerbildung. Springer, Wiesbaden, pp 1–15. https://doi.org/10.1007/978-3-658-01360-8_1
Hellmann K, Kreutz J, Schwichow M, Zaki K (2019) Kohärenz in der Lehrerbildung. Theorien, Modelle und empirische Befunde Springer VS, Wiesbaden
Helsper W (2018) Lehrerhabitus. In: Paseka A, Keller-Schneider M, Combe A (eds) Ungewissheit als Herausforderung für pädagogisches Handeln. Springer VS, Wiesbaden, pp 105–140
Hudson B, Zgaga P (2017) History, context and overview: Implications for teacher education policy, practice and future research. In: Hudson B (ed) Overcoming fragmentation in teacher education policy and practice. University Press, Cambridge, pp 1–25
Kreitz R (2000) Vom biographischen Sinn des Studierens. Die Herausbildung fachlicher Identität im Studium der Biologie. Leske+Budrich, Opladen
Larsson J, Airey J, Danielsson AT, Lundqvist EA (2020) Fragmented training environment: discourse models in the talk of physics teacher educators. Res Sci Educ 50:2559–2585. https://doi.org/10.1007/s11165-018-9793-9
Ligozat F, Amade-Escot C, Östman L (2015) Beyond subject specific approaches of teaching and learning: comparative didactics. Interchange 46:313–321. https://doi.org/10.1007/s10780-015-9260-8
Marotzki W (1990) Entwurf einer strukturalen Bildungstheorie. Biographietheoretische Auslegung von Bildungsprozessen in hochkomplexen Gesellschaften. Deutscher Studien Verlag, Weinheim
Mehlmann N, Bikner-Ahsbahs A (2018) Spotlights Lehre—Ein Ansatz zur Vernetzung von Fachwissenschaft und Fachdidaktik an der Universität Bremen. In: Glowinski I, Borowski A, Gillen J, Schanze S, von Meien J (eds) Kohärenz in der universitären Lehrerbildung. Vernetzung von Fachwissenschaft, Fachdidaktik und Bildungswissenschaften, Universitätsverlag, Potsdam, pp 77–102
Meier M, Ziepprecht K, Mayer J (2018) Lehrerausbildung in vernetzten Lernumgebungen. Waxmann, Münster
Scheerens J, Bosker RJ (1997) The foundations of educational effectiveness. Pergamon, Oxford
Schellhammer KS, Kainer F (2018) Nachhaltige Verstetigung von Lehr-Lern-Konzepten: Wie Lehrveranstaltungen von ihren Entwickler_innen losgelöst werden können, HDS J:49–56. https://nbn-resolving.org/urn:nbn:de:bsz:15-qucosa2-332299

Schellhammer S, Puderbach R, Gehrmann A (2020) Das Verbindungsbüro Lehrerbildung der TU Dresden: Synergetisch Lehrerbildung denken, regional Verantwortung übernehmen. Netzwerke: Magazin zu gelingender Bildungsnetzwerkarbeit der Zentren für Lehrer*innenbildung und Schools of Education 76–84. https://doi.org/https://doi.org/10.24355/dbbs.084-202010221241-0

Schwalbe A, Puderbach R, Schmechtig N, Gehrmann A (2022) In: Lehramtsstudierendenbefragung 2021 (ed) Die Studiensituation im Lehramt. ZSLB, Dresden

Shulman L (1986) Those who understand: knowledge growth in teaching. Educ Res 15:4–14. https://doi.org/10.3102/0013189X015002004

Sokolowska D (2024) Inquiry for meaningful learning. In: Faletic S, Pavlin J (eds) Teaching and learning physics effectively in challenging times, challenges in physics education. Springer, Cham, pp 3–21. https://doi.org/10.1007/978-3-031-72541-8_1

Teichler U (2017) Teaching versus research: an endangered balance? In: Machado-Taylor M, Soares V, Teichler U (eds) Challenges and options: the academic profession in Europe. Springer, Cham, pp 11–28. https://doi.org/10.1007/978-3-319-45844-1_2

Weller A, Schellhammer KS (2019) Förderung der Lehrkultur durch Multiplikator-Innenprogramme: Zwischen Empowerment, Vernetzung und kollegialer Beratung. In: Gotzen S, Heuchemer S, van Treeck T (eds) Hochschuldidaktik forscht zur Kultur des Ermöglichens: Profilbildung und Wertefragen in der Hochschulentwicklung II. TH Köln, Cologne, pp 75–85

Wenzl T, Wernet A, Kollmer I (2018) Praxisparolen. Dekonstruktionen zum Praxiswunsch von Lehramtsstudierenden, Springer VS, Wiesbaden

Werler TC, Tahirsylaj A (2020) Differences in teacher education programmes and their outcomes across Didaktik and curriculum traditions. Eur J Teach Educ 45:154–172. https://doi.org/10.1080/02619768.2020.1827388

Yun E (2020) Review of trends in physics education research using topic modeling. J Balt Sci Educ 19:388–400. https://doi.org/10.33225/jbse/20.19.388

Development of Digital Teaching Scenarios on Advanced STEM Topics for Teacher Education

Dimitris Stavrou, Athanasia Kokolaki, Lucy Avraamidou, Chara Bitsaki, Eleni Botzaki, Dirk Brockmann-Behnsen, Gunnar Friege, Stephen Gammell, Eilish McLoughlin, Emily Michailidi, and Argyris Nipyrakis

Abstract This study presents the development of digital teaching scenarios on advanced STEM topics for implementation with pre-service STEM teachers (PSTs). The context of the study is the Erasmus + project 'STEM Digitalis' which aimed at the development of blended learning environments for teaching PSTs advanced

D. Stavrou (✉) · A. Kokolaki · E. Botzaki · E. Michailidi · A. Nipyrakis
Department of Primary Education, University of Crete, Rethymno, Greece
e-mail: dstavrou@uoc.gr

A. Kokolaki
e-mail: akokolaki@edc.uoc.gr

E. Botzaki
e-mail: e.botzaki@edc.uoc.gr

E. Michailidi
e-mail: e.michailidi@uoc.gr

A. Nipyrakis
e-mail: agnipyrakis@uoc.gr

L. Avraamidou · C. Bitsaki
Centre for Learning and Teaching, Faculty of Science and Engineering, University of Groningen, Groningen, the Netherlands
e-mail: l.avraamidou@rug.nl

C. Bitsaki
e-mail: ch.bitsaki@edc.uoc.gr

D. Brockmann-Behnsen · G. Friege
Institute for Mathematics and Physics Education, Leibniz University Hannover, Hannover, Germany
e-mail: brockmann-behnsen@idmp.uni-hannover.de

G. Friege
e-mail: friege@idmp.uni-hannover.de

S. Gammell · E. McLoughlin
Centre for the Advancement of STEM Teaching and Learning &, School of Physical Sciences, Dublin City University, Dublin, Ireland

D. Sokołowska and E. McLoughlin (eds.), *Embracing Changes Together in Physics Education*, Challenges in Physics Education,
https://doi.org/10.1007/978-3-032-27385-7_21

STEM topics such as climate change, interferometry, renewable energy sources, water quality and sustainability, through the integration of digital technologies like serious games, augmented and virtual reality application, interactive experiments, etc. The study examines the influence of the blended implementation of the digital scenarios on PSTs' engagement, inquiry skills and professional learning, and offers insights into the design of blended learning environments for STEM education.

Keywords STEM education · Digital technologies · Blended learning environments · Pre-service STEM teachers' education

1 Introduction

The COVID-19 pandemic forced institutions worldwide to quickly change instruction from face-to-face to distance or blended learning environments. In most universities, educators and students were not prepared for this rapid transition to online environments and lacked relevant knowledge and skills on using distance or blended learning approaches and methodologies. Further challenges arose in STEM education, which uses hands-on activities to engage students in STEM practices to develop not only their content knowledge but also their inquiry and practical skills, such as data collection and processing, use of scientific equipment, etc. The closure of universities over a long period has raised several critical questions for policy makers, regarding the readiness of higher education institutions for effectively dealing with future global or local crises, and their preparedness to switch to accessible, inclusive and engaging education online, particularly for STEM courses (Dillon and Avraamidou 2020).

In this context, there has been increased focus over recent years related to developing digital resources for use in online and blended learning environments for tertiary STEM education, especially in cases where face-to-face teaching is limited or not possible. From a pedagogical point of view, the use of online or blended digital resources is reported to have a positive impact on teaching, learning and assessment strategies (Cook and Steinert 2013). Various pedagogical and instructional strategies and tools such as gamification, interactive models, virtual environments, computer simulations and animations, etc. have been implemented. Studies report on increased student engagement and meaningful knowledge construction through enriched learning experiences involving collaborative and multimedia learning environments that promote feedback and reflection both synchronously and asynchronously (Eichler and Peeples 2013).

e-mail: stephen.gammell@dcu.ie

E. McLoughlin
e-mail: eilish.mcloughlin@dcu.ie

It is critical that pre-service STEM teacher education, both primary and secondary, should prepare teachers for future uncertainties in local and global education, and support them to develop relevant STEM knowledge and skills as part of their initial teacher education. PSTs need to be supported in designing and developing authentic STEM learning environments through critical reflection on their own teaching practices and their beliefs about STEM education, the integration of digital technologies in STEM learning and the use of online and/or blended learning contexts (Butler and Leahy 2021).

This study reports on the development of digital learning scenarios which were implemented with primary and secondary PSTs and discusses the influence of this approach and resources on the teachers' professional learning. The research question addressed in this study is '*How does the development and implementation of digital learning scenarios impact the professional learning of primary and secondary pre-service STEM teachers?*'.

2 Theoretical Background

PST education that embeds digital technologies and digital learning within STEM contexts can support the development of PSTs' competence and confidence in using digital tools, as well as their understanding of how to effectively integrate these tools in their own classroom practices (Nipyrakis and Stavrou 2022). These core considerations for PST education are discussed in the following sections.

2.1 STEM in Pre-Service Teacher Education

Recent trends have emphasized the importance of the integration of STEM disciplines (science, technology, engineering and mathematics) in education as well as in the professional learning of pre- and in-service STEM teachers (Nipyrakis and Stavrou 2022; Hurley et al. 2021; Nipyrakis et al. 2024). While there are diverse perspectives about the meaning of STEM, STEM education is seen as an approach which interconnects STEM disciplines, often emphasizing the teaching of science and mathematics through inquiry, use of (digital) technology, engineering design and twenty-first century skills (Li et al. 2020; Radloff and Guzey 2016). STEM concepts are combined with authentic real-world contexts that make connections between school, community, work and the global enterprise (Radloff and Guzey 2016; Luo et al. 2017), and facilitate learners in solving real-world problems by connecting content and practices of various STEM disciplines (Ryu et al. 2019).

PST education aims to develop teachers' STEM content knowledge, STEM conceptualization as well as their pedagogical content knowledge (PCK) for STEM teaching (Radloff and Guzey 2016; Ryu et al. 2019). However, preparing teachers to adopt an integrated approach to STEM education brings additional challenges, such as teachers' lack of disciplinary and interdisciplinary knowledge, lack of STEM teaching experience, as well as the structure of the current curricula which reflect monodisciplinary approaches (Nipyrakis et al. 2024; Radloff and Guzey 2016; Ryu et al. 2019). To address such barriers, teacher education programmes should aim to develop PSTs' knowledge across and between STEM disciplines, provide PSTs with experiences of STEM integration as learners, foster collaboration between practicing teachers and PSTs, and promote the use of authentic problem-based and real-world contexts in STEM Education (Hurley et al. 2021; Radloff and Guzey 2016; Luo et al. 2017; Ryu et al. 2019; Nipyrakis et al. 2023).

2.2 *Digital Technologies in STEM Education*

The use of digital technologies is inherent in STEM education. Research reveals the possibilities and affordances of digital tools in teaching and learning processes, particularly in science education and by extension in STEM education (Skinner and Preece 2003), contributing to formulating dynamic-, interactive- and inquiry-based learning environments (Sokoloff et al. 2007). Specifically, using digital technologies (modelling tools, sensor software, etc.) has the potential to effectively support teachers' efforts in adopting inquiry-based learning in their classrooms and lead to students developing deeper conceptual understanding of scientific ideas, as well as encouraging students to engage in STEM roles and careers (Nipyrakis and Stavrou 2022; Iliaki et al. 2019). Examples of digital tools used in STEM education, such as interactive videos, mobile applications, dataloggers, serious games, virtual and augmented reality applications, etc., promote increased student engagement and interest, support the use of inquiry and problem-based learning approaches as well as the development of students' interpersonal skills (e.g., communication, collaboration) (Nipyrakis and Stavrou 2022; Iliaki et al. 2019; Zourmpakis et al., 2023).

2.3 *Online and Blended Learning Contexts and Methodologies*

Contemporary learning contexts include a range of face-to-face- and technology-mediated instruction types and a diverse range of methodologies for teaching and learning in these contexts. PSTs may experience a variety of contexts in their initial teacher education, to prepare them for teaching in these variety of contexts (Luo et al. 2017). Learning contexts expand on a continuum from face-to-face instruction, where learners gather with the instructor in person, to wholly technology-based learning, as in online and distance learning (Margulieux et al. 2016). Mixed instruction types use a combination of face-to-face and instruction via technology, and include hybrid, blended and combination modalities (Margulieux et al. 2016). However, there is a lack of clarity regarding the use of these terms and the features of such learning contexts (Weiß and Friege 2021). Additionally, the terms synchronous and asynchronous are often used when referring to online and blended courses. Synchronous activities, such as online meetings, enable real-time interactions between learners and teachers while asynchronous activities, such as video-recorded lectures, may be carried out at a time of choosing, not in real time (Heilporn et al. 2021).

To create effective online or blended learning contexts, with both synchronous and asynchronous aspects, requires consideration of the social and engagement aspects of the activities, such as communication, interaction, reflection and feedback (Farrell et al. 2019). Such considerations are important for fostering inclusive online learning environments. Educators require relevant knowledge and expertise for utilizing digital technologies in a way that encourages learners' engagement and interaction, both with the educator and with each other. Professional learning opportunities for educators, resources, and the development of infrastructure are essential for effective design and implementation of online and blended instruction (Heilporn et al. 2021; Farrell et al. 2019).

3 Methodology

The study was carried out in the context of the Erasmus + project 'STEM Digitalis' (https://stemdigitalis-project.eu/), involving a collaboration across five European Universities from Estonia, Germany, Greece, Ireland and the Netherlands. The focus of the project was on the design and development of online and blended learning environments for teaching primary and secondary PSTs' advanced STEM topics. The project aims to digitally equip PSTs not only to gain confidence in using digital tools (e.g. educational robotics, data loggers, VR and AR, etc.) but also to be able to effectively integrate such tools in STEM teaching and learning. Initially, several digital resources and materials were designed and developed to engage PSTs with advanced STEM topics, to facilitate PSTs' conceptualization regarding STEM learning and to provide authentic learning experiences for supporting PSTs to develop their PCK for STEM teaching. Subsequently, the digital resources were organized in five STEM scenarios: 1. Climate change; 2. Interferometry; 3. Energy at home; 4. Ocean batteries and energy farms; and 5. Water quality and sustainability. Each scenario was presented over several units with each unit comprising several learning activities. These activities were designed around an instructional framework with five dimensions, focusing on mode, social, media, timing and approach (Fig. 1). The framework was informed by the experiences of the partners involved in the STEM Digitalis project and international literature on integrated STEM (Radloff and Guzey 2016) and blended learning (Margulieux et al. 2016).

The activities within each unit were designed such that educators could select subsets of the activities and design a suitable learning path for their programme's context. Learning paths were visualized using the LePlanner programme and

Fig. 1 Five-dimensional instructional framework for STEM Digitalis scenarios

published on the STEM Digitalis Pressbooks platform: (https://web.htk.tlu.ee/stem/stem2/).

The content of each scenario was initially tested during STEM Digitalis summer school held in July 2022, involving 25 PSTs and 15 educators from the five participating countries. The structure of the five scenarios was refined based on the feedback received from PSTs and educators during the summer school. The five scenarios were implemented using a blended learning approach with PSTs in local educational contexts by each of the project partners during January to September 2023. The data gathered from local implementation of these scenarios supported partners in developing guidelines and recommendations for PST education using blended and online learning environments. The methodology of STEM Digitalis project is depicted in Fig. 2.

This study describes three of the digital scenarios developed in the context of the STEM Digitalis project, as indicative examples of the different types of digital tools were used in this project, namely: climate change, interferometry and ocean batteries and energy farms. The influence of STEM Digital scenarios on PSTs' professional learning was examined using data collected from the blended implementation of these scenarios with 75 PSTs (15 from each country) during January 2023 to September 2023. Data was collected through worksheets completed by PSTs during online or face-to-face seminars. A mixed type of questionnaire was distributed to PSTs after each seminar to collect their views on the use of digital tools in STEM education, as well as their feedback on the structure of the blended learning experience. In addition, each of the five project partners prepared a case study report to describe the implementation and influence of the STEM Digitalis approach and digital scenarios

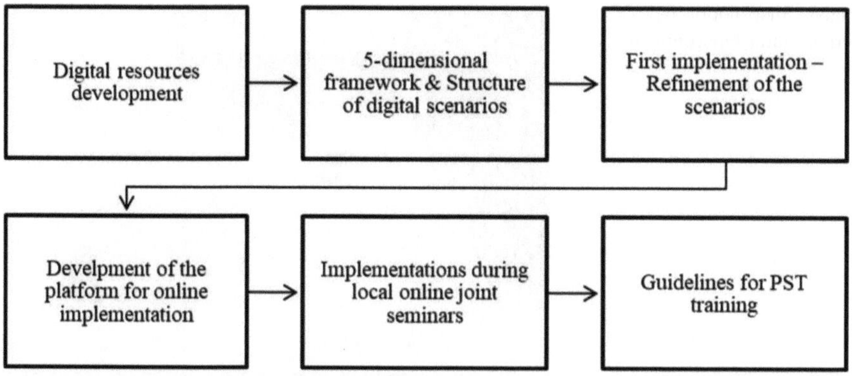

Fig. 2 Methodology of STEM Digitalis project

in their local context. In this study, we focus on the qualitative findings, primarily drawn from the thematic analysis of the five partner case study reports, to identify common experiences and differences in local implementations of the digital scenarios. Finally, the findings were used to inform the development of guidelines for the design and implementation of digital teaching material on advanced STEM topics for PST education.

4 STEM Digitalis Teaching Scenarios

4.1 Digital Scenario on Climate Change

The digital teaching scenario on climate change is structured into three units. In particular, the first unit addresses the fundamental causes of climate change, the second explores the human-induced carbon dioxide emissions as well as corresponding mitigation strategies prioritizing the use of renewable energy sources while the third engages participants in the examination of a socio-scientific issue linked to the European Union's climate neutrality agenda.

A diverse set of digital technologies supports the scenario including Unity 3D, mobile applications, H5P and online communication and collaboration platforms. Within the first unit, a Unity 3D-based serious game (Fig. 3) enables participants' access to datasets on global temperature and atmospheric carbon dioxide covering the years 1980–2010 and supports the identification of trends in global temperature and carbon dioxide concentration rise. The first unit is also complemented by an interactive experiment with embedded hotspots providing explanatory content, multiple-choice questions and links to external online dashboards for sharing interpretations (Fig. 4). The interactive experiment allows participants to further investigate the correlation between the two variables under examination.

Fig. 3 Snapshot of the serious game environment

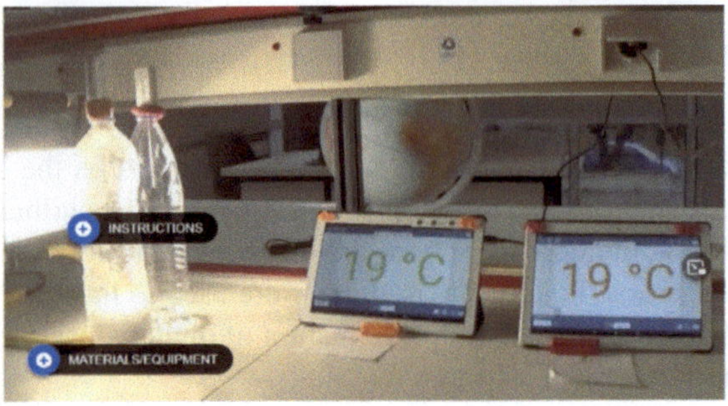

Fig. 4 Snapshot of the interactive video experimental environment

The second unit incorporates interactive maps (Fig. 5) to visualize national and sectoral carbon dioxide emissions as well as a mobile application (Fig. 6) that facilitates the calculation of individual carbon footprint. The third unit focuses on the debate of a socio-scientific issue regarding climate neutrality by considering the viewpoints of different stakeholders. The activity is mediated by an augmented reality 'treasure hunt', whereby participants access stakeholder-specific information through image scanning and collaboratively solve tasks to trace the profile of the stakeholders involved in the debate.

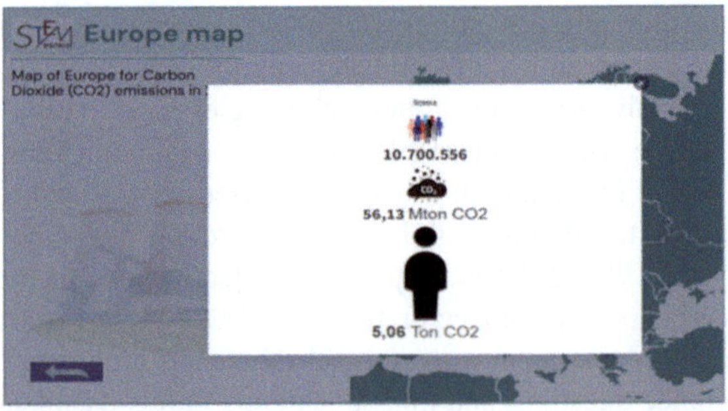

Fig. 5 Snapshot of the interactive map

Fig. 6 Snapshot of the mobile application

4.2 Digital Scenario on Interferometry

The digital scenario of interferometry covers fundamental aspects of the wave theory of light, the superposition of wave trains and the phenomenon of interference. Different types of interferometers and their application are presented while special attention is also given to the detection of gravitational waves. The digital teaching scenario of interference consists of three units through which PSTs familiarize with (i) the fundamentals of the wave theory of light and interference, (ii) the functionality of different types of interferometers and measurement basics of interferometry and (iii) the use of interferometers for the detection of gravitational waves. Among the developed digital tools for variable use in higher education were interactive screen experiments on various types of interferometers like the Michelson interferometer (List 2023) or the Sagnac interferometer (Maas 2023) and a VR application which gave PSTs the opportunity to carry out experiments on different types of interferometers in the virtual environment and to analyse the obtained experimental data there (Bublitz 2023).

Interactive screen experiments which are photographic representations of real experiments are used in this activity. All technically and didactically relevant actions for the experiment are ideally recorded in advance and displayed as required. Technically, interactive screen experiments are realized by embedding the photos in visual programming environments. Interactive screen experiments show real experiments that can be manipulated on the computer and can be carried out individually by all PSTs at the same time. Interactive screen experiments thus combine the advantages of simulations and real experiments, the latter of which are often not available in class sets and sometimes not at all. Four different interactive screen experiments were developed through which PSTs can determine the wavelength of lasers as well as the refractive index of air, the thickness of a glass plate and the coefficient of linear thermal expansion of aluminium. By clicking on the arrows embedded, the position of one of the interferometer mirrors can be changed, whereupon the interference pattern changes accordingly (Fig. 7).

The innovative VR laboratory named 'Interferometer building' included experiments that can be carried out on different types of interferometers. One room of the

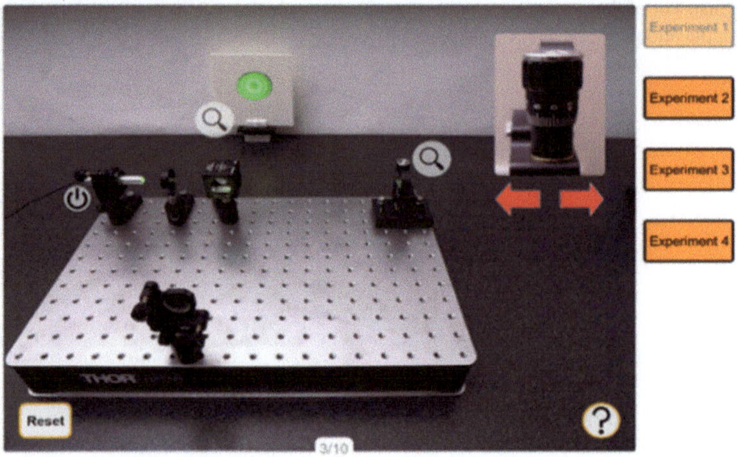

Fig. 7 Michelson interferometer Interactive screen

virtual laboratory hosts a Michelson interferometer. This device can first be used to determine the wavelength of the laser and, in a further experiment, the coefficient of thermal expansion of aluminium. The virtual laboratory also comprises a Mach–Zehnder interferometer and a gravitational wave detector (like the planned LISA space detector). The aim for PSTs is initially to experiment individually and independently with VR interferometers and then to record and analyse measured values as realistically as possible. The application also provides feedback on the quality of the obtained values while also addressing safety aspects when handling lasers.

4.3 Digital Scenario on Ocean Batteries and Energy Farms

The digital scenario on ocean batteries and energy farms aims to engage students in engineering practices by exploring, modifying, testing, revising and reflecting on virtual simulations of an energy farm including Ocean batteries. Specifically, the participants navigate their avatars in an online collaborative platform (gather.town) which was customized accordingly to meet the educational goals. The participants engage in activities related to Ocean batteries and energy farms, interact with objects, virtual assistants and peers, and answer questions to fulfil tasks and get the passwords for progressing to the next stage, following an escape-room mode.

In the first unit, the participants are introduced to the first stages of the online platform which have the form of ships and rivers; they identify their peers and workspaces and go through a historical overview of energy sources through interaction with embedded text-based questions. Subsequently, they get problematized concerning the real-world problems of increasing efficiency in energy production and environmental pollution by coming across facts and international reports. They are also introduced to the rationale and operation principles of wind turbines, wave energy harvesting and ocean batteries. By succeeding in answering the questions that virtual AR agents pose, they progress to the next unit. In the second unit, the participants engage with interactive Matlab-based simulations of ocean batteries and wave energy converters that visualize their operation and calculate the efficiency of the designed system, as shown in Fig. 8. They carry out variable testing on several design variables (e.g. geometrical dimensions and structure, the efficiency of the pump/turbine, height and frequency of the waves, etc.) and collect data regarding the system's efficiency. AR agents provide assistance and/or evaluate their inferences on the optimal design in order to allow them to proceed.

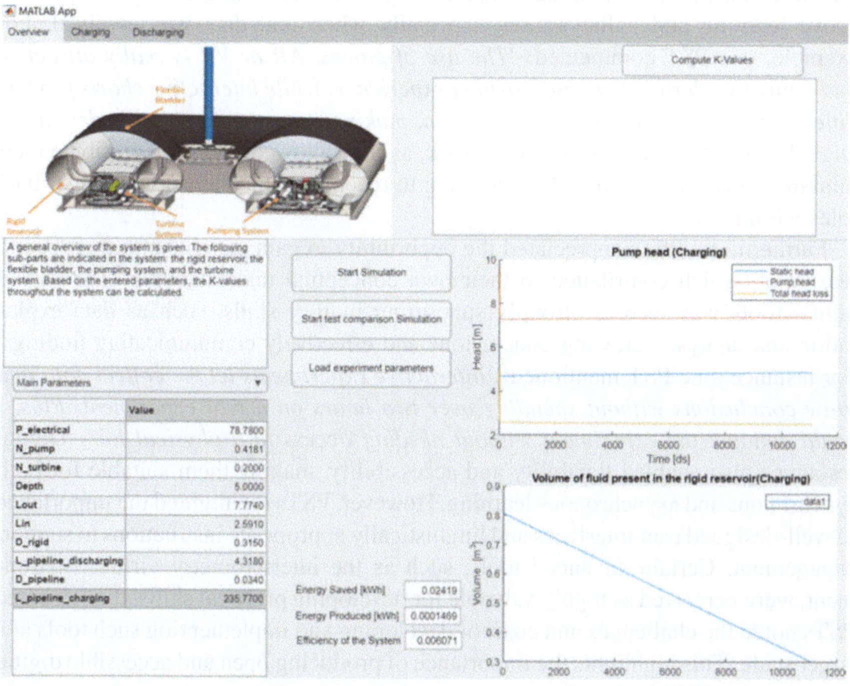

Fig. 8 Snapshot of the ocean battery simulation

5 Findings

5.1 Use of Digital Technologies in STEM Teaching

Based on the analysis of the five case study reports and the open-ended responses, several key themes emerged regarding PSTs' experiences with the digital scenarios. PSTs reported increased interest, engagement and motivation when engaging with a variety of digital tools and resources, such as educational games, mobile and AR applications, and collaborative communication platforms. Scenarios that combined multiple tools and diverse pedagogical approaches were perceived as particularly effective in supporting learning. PSTs consistently noted that these tools supported active learning and collaboration, especially when real data was involved. For example, one PST commented: *'The use of games, AR or VR is really attractive and funny by offering a unique learning experience, while interactive charts provide different ways of real data representation, making it easy to use and understand'.* Digital tools also encouraged group work, as they allowed PSTs to contribute their thinking individually before collaborating to reach shared conclusions or distribute roles within a task.

Furthermore, PSTs appreciated the opportunity to gain insights into their peers' reasoning, which contributed to their own conceptual understanding. The use of digital tools was seen as strongly supporting inquiry skills, such as data exploration and analysis, drawing conclusions and effectively communicating findings. For instance, one PST mentioned: *'Interactive experiments let me collect data and draw conclusions without spending over two hours on a real experiment. Plus, I could conduct the experiment without needing access to a physical lab'.* Digital resources also enabled flexibility and accessibility, making them suitable for both synchronous and asynchronous learning. However, PSTs highlighted the importance of well-designed user interfaces and linguistically appropriate instructions to support engagement. Certain advanced tools, such as the interferometry virtual environment, were perceived as highly valuable for developing practical skills, though some PSTs noted the challenges and costs of developing and implementing such tools at a larger scale. This highlights the importance of producing open and accessible digital resources for STEM education.

In summary, the analysis underscores the importance of offering PSTs opportunities to engage with sophisticated digital tools, which have high motivational potential and afford immersive learning experiences not easily achieved through other media.

5.2 Design and Development of Blended Learning Environments for STEM Teaching

Another emerging theme from the data relates to the design of blended learning environments. PSTs emphasized the need for clear goals and objectives at the start of each course—whether online, face-to-face or blended. One PST noted: *'I would find it easier to follow the blended format if the overall goals of the scenario were clearly defined before starting'*. PSTs found that asynchronous digital environments supported deeper content understanding and allowed self-paced learning, especially for those who found the synchronous pace less suitable for their individual needs. In particular, many PSTs appreciated the flexibility of asynchronous sections to complete tasks in their own time, to explore questions independently and to have access to their peers' responses and opinions. However, the lack of interaction and feedback during asynchronous sessions was noted as a limitation. For example, one PST mentioned: *'I found it very convenient to choose when and how to complete assignments and tasks, while also having time to search for my questions. However, I couldn't share my views with my peers'*. Conversely, face-to-face sessions were valued for fostering teamwork and peer interaction. These findings suggest that effective blended learning designs should balance autonomy and flexibility with opportunities for feedback and collaboration.

6 Conclusions

The findings of this study highlight key considerations for developing blended learning environments, and how face-to-face and online sessions are combined. In this study, it was more effective for the PSTs to complete easier activities asynchronously and facilitate activities that demanded higher order thinking during face-to-face sessions where PST learning could be supported by their peers and their instructors. Use of a blended format enabled flexible and independent learning, but asynchronous sessions need to be enriched with technologies that promote interaction and feedback. For successful implementation of blended learning approaches in STEM education, clear learning goals and alignment between face-to-face and online sessions are critical. This study highlights the importance of using a wide range of digital tools, e.g., interactive/virtual experiments, podcasts and silent videos in PST education. The integration of digital tools and blended learning environments into STEM teacher education has the dynamics to enhance PSTs' engagement, inquiry skills and content knowledge (Nipyrakis and Stavrou 2022; Iliaki et al. 2019). Tools such as interactive experiments, VR applications and gamification platforms can be used to support PSTs' conceptual understanding and foster collaboration (Heilporn et al. 2021; Farrell et al. 2019).

Acknowledgements The research work is supported by 'STEM DIGITALIS' project which was funded by the Erasmus+ Programme of the European Union. Grant agreement n. 2020-1-EL01-KA226-HE-094691.

References

Bublitz M (2023) Digitale Experimente in einer Virtual Reality Lernumgebung am Beispiel von Interferometern. Unpublished master's thesis, Hochschule Hannover

Butler D, Leahy M (2021) Developing preservice teachers' understanding of computational thinking: a constructionist approach. Br J Educ Technol 52:1060–1077. https://doi.org/10.1111/bjet.13090

Cook DA, Steinert Y (2013) Online learning for faculty development: a review of the literature. Med Teach 35:930–937. https://doi.org/10.3109/0142159X.2013.827328

Dillon J, Avraamidou L (2020) Towards a viable response to COVID-19 from the science education community. J Activist Sci Technol Educ 11:1–6 https://doi.org/10.33137/jaste.v11i2.34531

Eichler JF, Peeples J (2013) Online homework put to the test: a report on the impact of two online learning systems on student performance in general chemistry. J Chem Educ 90:1137–1143. https://doi.org/10.1021/ed3006264

Farrell O, Brunton J, Costello E, Donlon E, Trevaskis S, Eccles S, Ní Shé C (2019) An investigation of effective online teaching: a needs analysis of online educators and online students. Dublin: #Openteach Project. https://doi.org/10.13140/RG.2.2.19462.32324

Heilporn G, Lakhal S, Bélisle M (2021) An examination of teachers' strategies to foster student engagement in blended learning in higher education. Int J Educ Technol High Educ 18:1–25. https://doi.org/10.1186/s41239-021-00260-3

Hurley M, Butler D, McLoughlin E (2021) Immersive STEM learning experiences to shape shared futures. Dublin City University, Dublin. https://doi.org/10.5281/zenodo.5589758

Iliaki G, Velentzas A, Michailidi E, Stavrou D (2019) Exploring the music: a teaching-learning sequence about sound in authentic settings. Res Sci Technol Educ 37:218–238. https://doi.org/10.1080/02635143.2018.1526170

Li Y, Wang K, Xiao Y, Froyd JE (2020) Research and trends in STEM education: a systematic review of journal publications. Int J STEM Educ 7:1–16. https://doi.org/10.1186/s40594-020-00207-6

List K (2023) Entwicklung medialer Elemente zum Thema Interferometrie. Unpublished master's thesis, Leibniz University Hannover, Physics Education Group

Luo T, Murray A, Crompton H (2017) Designing authentic learning activities to train pre-service teachers about teaching online. Int Rev Res Open Distrib Learn 18(7). https://doi.org/10.19173/irrodl.v18i7.3037

Maas H (2023) Digitale Medien zum Thema Interferometrie: Entwicklung von Materialien für blended learning Szenarien. Unpublished master's thesis, Leibniz University Hannover

Margulieux LE, McCracken WM, Catrambone R (2016) A taxonomy to define courses that mix face-to-face and online learning. Educ Res Rev 19:104–118. https://doi.org/10.1016/j.edurev.2016.07.001

Nipyrakis A, Stavrou D, Avraamidou L (2023) Designing technology-enhanced science experiments in elementary teacher preparation: the role of learning communities. Res Sci Technol Educ 42:889–911. https://doi.org/10.1080/02635143.2023.2202386

Nipyrakis A, Stavrou D, Avraamidou L (2024) In-service teachers' views about STEM integration: a case study. Res Integr STEM Educ 2:85–119. https://doi.org/10.1163/27726673-bja00021

Nipyrakis A, Stavrou D (2022) Integration of ICT in science education laboratories by primary student teachers. In: Papadakis S, Kalogiannakis M (eds) STEM, robotics, mobile apps in early childhood and primary education. Springer, Singapore. https://doi.org/10.1007/978-981-19-0568-1_4

Radloff J, Guzey S (2016) Investigating pre-service STEM teacher conceptions of STEM education. J Sci Educ Technol 25:759–774. https://doi.org/10.1007/s10956-016-9633-5

Ryu M, Mentzer N, Knobloch N (2019) Preservice teachers' experiences of STEM integration: challenges and implications for integrated STEM teacher preparation. Int J Technol des Educ 29:493–512. https://doi.org/10.1007/s10798-018-9440-9

Skinner NC, Preece PF (2003) The use of information and communications technology to support the teaching of science in primary schools. Int J Sci Educ 25:205–219. https://doi.org/10.1080/09500690210126757

Sokoloff DR, Laws PW, Thornton RK (2007) Real-time physics: active learning labs transforming the introductory laboratory. Eur J Phys 28:S83–S94. https://doi.org/10.1088/0143-0807/28/3/S08

Weiß LF, Friege G (2021) The flipped classroom: media hype or empirically based effectiveness? Probl Educ 21st Century 79:312–329. https://www.ceeol.com/search/article-detail?id=941745

Zourmpakis A-I, Kalogiannakis M, Papadakis S (2023) Adaptive gamification in science education: an analysis of the impact of implementation and adapted game elements on students' motivation. Computers 12:143. https://doi.org/10.3390/computers12070143

Greek Science Teachers' Views about the Use of Educational Simulations in their Practice

Chrysanthi Gidari-Gounaridou and Ioannis Lefkos

Abstract The integration of digital media into the educational process has become a focal point of discourse at both international and national levels, with particular emphasis on the utilisation of simulations in science education. This study examines the attitudes of secondary school teachers towards the integration of simulations in science teaching, employing a mixed-methods approach and an explanatory research design. The findings reveal that positive attitudes towards simulations are significantly influenced by the educational level (middle school versus high school), while supplementary training in Information and Communication Technologies (ICT) plays a crucial role. Specifically, the absence of advanced ICT training contributes to the development of negative attitudes towards the use of simulations. Additionally, limitations in teaching time and the lack of adequate technical infrastructure further hinder the widespread adoption of digital tools in the classroom. The research findings are discussed in the context of informing the development of effective educational policies.

Keywords Simulations · Science teachers · Attitudes · ICT · Science education

1 Introduction

In recent years, the Ministry of Education in Greece has invested in training teachers to integrate ICT into the educational process (B-Level ICT Teacher Training). The integration and use of simulations in science classrooms are part of this strategy. Additionally, the educational policy seems to be promoting the use of simulations, as the Ministry's instructions for teaching science suggest their use as supporting

C. Gidari-Gounaridou · I. Lefkos (✉)
Laboratory of Informatics and Robotics in Education and Society, University of Macedonia, Thessaloniki, Greece
e-mail: lefkos@uom.edu.gr

D. Sokołowska and E. McLoughlin (eds.), *Embracing Changes Together in Physics Education*, Challenges in Physics Education,
https://doi.org/10.1007/978-3-032-27385-7_22

material to enhance students' understanding of concepts and phenomena. Moreover, during the pandemic and emergency remote teaching, even more teachers were involved in using simulations (Rodgríguez et al. 2021).

The global educational community recognises the potential of simulation-based education and its benefits for teachers and students (Ouahi et al. 2022). Simulations can improve students' conceptual understanding by visualising abstract scientific concepts (Wieman et al. 2010; Psillos 2023), develop their experimental skills (Lefkos et al. 2011) and increase their interest in the course and positive attitudes towards science (Perkins et al. 2015).

However, despite the positive results of research on the benefits of integrating simulations into teaching at a practical level, their use in the science classroom is more complex. Some of the obstacles frequently reported are the lack of school equipment, teachers' unfamiliarity with the technology and their lack of education and training in specific tools, the limited teaching material/resources and the time needed to organise a lesson (Jean Justice and Ritzhaupt 2015; Bo et al. 2018).

Consequently, there appears to be a gap between the research data, educational policy and actual educational practice. Furthermore, the situation does not seem stable, as new factors could emerge. Therefore, it is intriguing to explore the current situation.

2 Research Background

Information, communication, and technology (ICT) play a pivotal role in supporting and enhancing the teaching and learning process. The integration of technology in education has become increasingly important, particularly in science education, where abstract concepts and complex phenomena often require dynamic visualisations to foster understanding. Effective teaching of science subjects necessitates the use of well-designed and differentiated instructional tools, such as multimedia, throughout the teaching and learning process (Ndihokubwayo et al. 2020; Papadakis and Kalogiannakis 2020). One tool of particular significance in the context of experiments is simulations. Simulations serve as a powerful complement to real-world experiments, offering valuable support in the exploration and understanding of scientific concepts (Stinken-Rösner 2020) or the development of experimental skills (Lefkos et al. 2011).

Simulations are computer-generated models or real-world (or even fictional or theoretical) processes or systems, allowing students to explore, experiment and observe phenomena in a controlled and interactive virtual environment. They provide a platform for students to manipulate variables and observe outcomes, making them particularly useful in science education (Bell and Smetana 2008). Computer simulations, such as PhET, are increasingly accessible in science education and have become an essential component of science teaching and learning. These simulations offer the potential to complement traditional instructional methods and enhance the overall learning experience. In science education, simulations have been shown to improve students' conceptual understanding and engagement with the material. By providing dynamic, visual representations of scientific concepts, simulations bridge the gap between theoretical knowledge and real-world applications, fostering deeper learning (De Jong and Van Joolingen 1998). Studies have demonstrated that simulations support students in developing a more comprehensive understanding of scientific principles, especially in subjects such as Physics, Chemistry and Biology, where abstract concepts can be difficult to grasp (Finkelstein et al. 2005). Furthermore, the interactive nature of simulations enables students to experiment with variables in ways that would not be feasible in traditional laboratory settings due to time, cost or safety constraints (Smetana and Bell 2012) but has also been shown to be equally effective in students' positive attitudes and learning (Papalazarou et al. 2024).

Additionally, simulations allow for the differentiation of instruction to meet diverse learning needs. By adapting the complexity and level of interactivity, simulations cater to a wide range of learning styles and abilities, offering personalised learning experiences that can enhance both student motivation and performance (Papadakis and Kalogiannakis 2020), enabling students to engage in inquiry-based learning, where they can hypothesise, test and reflect on their findings, following scientific methodologies in a virtual environment (Smetana and Bell 2012).

The growing accessibility and availability of simulations, such as PhET simulations, have further solidified their place in science classrooms worldwide. These simulations are freely available and cover a wide range of scientific topics, making them invaluable resources for educators aiming to enrich traditional teaching methods (Bell and Smetana 2008). Simulations enhance traditional instruction, fostering a more interactive and engaging learning environment that encourages students' active participation and critical thinking (Papadakis and Kalogiannakis 2020).

While there is strong evidence supporting the positive impact of simulations on student learning, there are also challenges related to their integration into the classroom. Educators must possess the necessary technological skills to effectively utilise simulations and integrate them into their lesson plans. Moreover, research indicates that while simulations can be powerful tools for learning, they must be carefully designed and aligned with curriculum objectives to ensure their effectiveness (De Jong and Van Joolingen 1998). Therefore, continuous professional development for teachers and research into best practices for simulation-based instruction are crucial to maximise the potential of these digital tools in science education. In conclusion, simulations offer significant potential for enhancing science education by making abstract scientific concepts more accessible and engaging for students. However, integrating them into the classroom requires thoughtful planning, teacher preparation and ongoing research to ensure their effective use in fostering meaningful learning experiences.

This paper explores science teachers' attitudes and views towards using simulations in teaching and learning. It aims to identify key factors that may influence the implementation of simulations in their educational practice. Specifically, the research is guided by the following questions:

(RQ1) What factors shape teachers' attitudes and intention to use simulations in science education?

(RQ2) How are their attitudes and intention to use simulations influenced by their demographics?

(RQ3) What are the reasons for teachers' willingness (or not) to continue using simulations in science teaching?

3 Research Methodology

In order to identify the key factors influencing the attitudes of secondary education teachers who teach Science regarding the use of simulations, a mixed-methods research approach was employed because attitudes are a multifactorial variable (Kay 1993). To enhance understanding of the research problem, the selected research design was explanatory (Creswell 2012), a mixed-methods design involving data collection in two phases: quantitative and qualitative in the first and qualitative in the second. In the first phase, data were gathered using an electronic questionnaire, the results of which are analysed in this paper. The qualitative data collection, through semi-structured interviews with teachers, is planned for a future phase of the research and is not presented here.

A research instrument using Google Forms was developed to collect data regarding the factors influencing science teachers' attitudes towards simulations. This instrument was based on similar earlier work (Mahdi et al. 2018), with appropriate modifications to accommodate our specific focus on simulations and secondary education science. The questionnaire primarily consists of closed-ended questions, including Likert-type scaled items and binary Yes/No questions. Additionally, it includes several open-ended questions that allow for detailed textual inputs.

The sample comprised secondary education science teachers and the factors examined concerning their attitudes towards simulations were organised into specific domains (Table 1), as found in similar earlier work (Kostaki and Kalogiannakis 2019). The sampling method utilised was convenience sampling, as participation in the questionnair e was voluntary, following postings through social media platforms.

4 Results

This study involved 139 secondary education teachers from the Physics, Chemistry, Life and Earth Sciences disciplines. Most of the participants have reported using interactive simulations in their classrooms.

In the following sections, initially, the demographics of the sample are presented, then the results from the qualitative analysis (Likert scale statements) and lastly, the results from the qualitative analysis (open-ended questions).

Table 1 Factors influencing the formation of teachers' attitudes towards simulations

Category	Investigated factors
Demographic	Gender, age in years, education, teachers' specialisation, years of experience, training in ICT, school location, teaching grade
Technological	Available technological resources of the school, knowledge of the existence of simulations
Technical characteristics of simulations	Reusability, interoperability, adaptability, accessibility, granularity, interactivity, discoverability, manageability
Educational	Student engagement, interactivity, activation of interest, suitability of the simulation
Pedagogical	Compatibility with the curriculum, understanding of concepts

4.1 Demographics

Participants' ages range from 24 to 69 years, while years of service span from 1 to 35 years, highlighting the diversity in age and professional experience among the participants. The mean age of the participants is 45.79 years, while the average year of service is 12.73 years. The median age, which represents the central value of the distribution, is 45 years, while the median year of service is 11.5 years. This suggests that half of the participants are under 45 years of age and have less than 11.5 years of professional experience. The standard deviation for age is 10.144 years, indicating some variability in the participants' ages. For years of service, the standard deviation is 9.567 years, also reflecting variation in the participants' work experience.

Moreover, regarding the descriptive measures, the large majority of participants, 132 individuals or 95%, is aware of the existence of simulations such as PhET Colorado or similar Greek repositories. In contrast, only seven individuals, representing 5.0% of the sample, reported being unaware of such simulations. Overall, the results indicate that awareness of simulations is highly prevalent among educators, with 95% being knowledgeable about their existence.

Regarding the technological equipment in the schools of the participants, 66.8% have internet access in at least a sufficient number of classrooms, while 48.9% have computers and projectors in each classroom. On the other hand, 46.8% of schools do not have interactive whiteboards in any classrooms.

4.2 Teachers' Attitudes Towards Simulations: A Quantitative Analysis

The analysis of educators' attitudes towards simulations was based on 30 statements from the questionnaire and presented in the following sections.

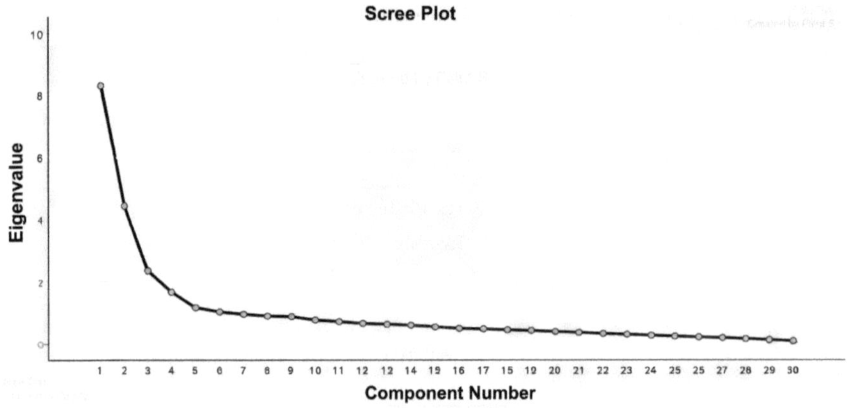

Fig. 1 Eigenvalue plot of factor analysis

4.2.1 What Are the Factors that Influence Educators' Attitudes Towards Simulations?

To identify the key factors underlying their attitudes, a factor analysis was performed using SPSS (version 20.0). Principal axis factoring was applied with Varimax rotation and Kaiser normalisation to improve factor interpretability. Bartlett's test of sphericity was significant ($\chi^2 = 2122.684$, $p < 0.001$), confirming adequate correlation among variables for factor analysis. The Kaiser–Meyer–Olkin (KMO) Index was 0.822, exceeding the acceptable threshold of 0.6, indicating strong intercorrelation among the variables. Figure 1 presents the scree plot, which illustrates the distribution of eigenvalues for each factor. It is evident that the most significant factors are located among the first four, as a sharp decline in eigenvalues is observed at this point.

Table 2 presents the categorisation of educators' statements based on the four factors, each assigned a suitable label. Specifically, the first factor, titled 'Positive Attitude towards Simulations', accounts for 18.70% of the total variance; the second factor, 'Negative Attitude towards Simulations', explains 17.73%; the third factor, 'Interest in Simulation Application', explains 11.27%; and the fourth factor, 'Perception of Complexity', accounts for 8.19% of the total variance. Together, these four factors explain 55.89% of the total variance.

4.2.2 How Are the Attitudes of Secondary Education Science Teachers Influenced by their Demographics?

To examine the effect of gender, age, education, work experience, professional development, school location, teaching grade and technological equipment on the four

Table 2 Categorisation of teachers' attitudes towards simulations and naming of factors

Factors	Example statements from the factors
Positive attitude towards simulations	Simulations do not scare me at all I can handle difficulties that may arise with simulations I feel confident in learning how to handle simulations
Negative attitude towards simulations	I am hostile towards simulations I hesitate to use simulations for fear of appearing clumsy Hearing colleagues mention the use of simulations makes me anxious
Interest in simulation application	I enjoy working with the possibility of utilising new technologies in teaching I like the idea of using simulations in teaching Simulations are particularly useful in teaching
Perception of complexity	One must be young to learn how to use simulations One must have special knowledge to make use of these tools Their usage is intended for individuals with specialised knowledge

factors, separate analysis of covariance (ANCOVA) tests were conducted for each of these variables.

Concerning the factor 'Positive Attitude towards Simulations', the analysis revealed a marginal effect of teaching grades on educators' attitudes. In the overall model, the p-value for this variable was 0.056, slightly above the 0.05 significance threshold. This implies that teaching grades do not have a statistically significant effect at the 5% level. However, in the analysis of the 'Grade' variable, the 'Secondary School' category exhibited a t-test value of 2.071, with a coefficient $B = 0.415$ and $p\text{-value} = 0.041$, indicating a positive contribution to educators' attitudes. The R^2 value was 0.207, accounting for 20.7% of the total variance, while the adjusted R^2 stood at 0.001, suggesting limited explanatory power of the independent variables in the model.

Regarding the factor 'Negative Attitude towards Simulations', the ANCOVA analysis revealed that the level of ICT training statistically influences educators' negative attitudes. Specifically, the 'A1 Level Training' category presented a t-test value of 2.153, with a coefficient $B = 0.606$ and $p\text{-value} = 0.034$ (< 0.05). This indicates that the variable increases negative attitudes towards simulations by an average of 0.606 units compared to the 'B2 Level Training' category, set as the reference. The R^2 value was calculated as 0.216, while the adjusted R^2 was 0.012, suggesting that the model explains only 21.6% of the total variance in the dependent variable through the independent variables.

For the factor 'Interest in Simulation Application', the analysis showed that educators' interest is influenced by the availability of interactive whiteboards in schools

($F = 3.222$, $df = 4$, p-value = 0.015). Specifically, the variable 'Interactive whiteboards available in most classrooms' had a t-test value of 1.883, with a coefficient $B = 0.861$ and p-value = 0.063, indicating a marginal statistical significance. This suggests a tendency for this category to positively influence the factor, though it does not strictly meet the p-value <0.05 threshold. Additionally, the analysis revealed that the variable 'Internet/WiFi available in a few classrooms' significantly affects the factor ($B = -0.512$, $t = -2.001$, p-value = 0.048). This category reduces interest in simulations by an average of 0.512 units, compared to 'Internet/WiFi available in all classrooms'. Other independent variables did not show significant effects, as their p-values exceeded 0.05. Finally, the R^2 coefficient was 0.215, indicating that the model explains 21.5% of the total variance in 'Interest in Simulations', while the adjusted R^2 of 0.011 suggests limited explanatory power of the independent variables.

Regarding the factor 'Perception of Complexity', the analysis indicates that the model explains 21.7% of the total variance ($R^2 = 0.217$), while the adjusted R^2 is relatively low at 0.013. Although no independent variables show statistically significant effects at the $\alpha = 0.05$ significance level, the variable 'PhD holder' presented a t-test value of 2.222, with a coefficient $B = 0.991$ and p-value = 0.028. This suggests a statistically significant positive effect on the factor, indicating that educators with a PhD tend to perceive the use of simulations as 0.991 units more complex compared to those holding a second degree and PhD (the reference category).

Finally, the other variables did not demonstrate statistically significant effects, as their corresponding p-values exceeded the 0.05 threshold.

4.2.3 What Factors Shape the Intention to Use Simulations in the Teaching of Science Education?

The analysis of teachers' attitudes concerning the intention to use simulations in science education (SE) was based on 17 statements, specifically designed to capture their perceptions, opinions and emotions regarding simulations. To identify the key factors influencing their attitudes, factor analysis was employed. The analysis was conducted using the principal axis factoring method, with Varimax rotation and Kaiser normalisation applied for optimal factor interpretability.

The Bartlett's test of sphericity was statistically significant ($\chi^2 = 1750.436$, $p < 0.001$), indicating that the variables exhibit adequate correlations for the application of factor analysis. Furthermore, the Kaiser–Meyer–Olkin (KMO) measure of sampling adequacy yielded a value of 0.895, significantly exceeding the acceptable threshold of 0.6. This high value suggests a strong intercorrelation among the variables, making factor analysis suitable for the current study. From the scree plot (Fig. 2), it is observed that the most significant factors are identified within the first two.

The categorisation of the teachers' statements based on the two identified factors, with each factor appropriately labelled, is presented in Table 3. Specifically, the first factor, labelled 'Educational Benefits of Simulations', accounts for 36.67% of the total variance, while the second factor, 'Ease of Use and Integration of Simulations

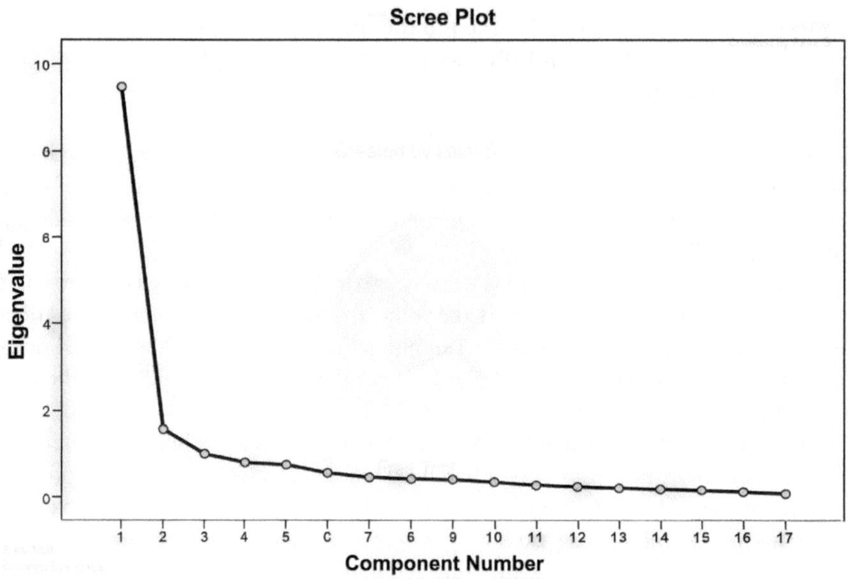

Fig. 2 Eigenvalue plot of factor analysis

in Teaching', explains 28.19% of the total variance. Collectively, these two factors account for 64.86% of the overall variance.

Table 3 Categorisation of teachers' attitudes towards the intent to use simulations in the teaching of science and naming of factors

Factors	Example statements from the factors
Educational benefits of simulations	They assist me in teaching They are well-designed and help in understanding the concepts they address They actively engage students in the learning process They allow interaction among students
Ease of use and integration of simulations in teaching	I can integrate them into my teaching without any problems The required equipment is available at the school I can easily adapt them to my teaching needs and objectives I can easily and freely use any simulation that meets my needs The information provided with them is clear, and I can always identify the most appropriate one

4.2.4 How Are the Factors Related to the Intention to Use Simulations in Science Education Influenced by the Demographics of the Sample?

In terms of educational background, teachers who held a second degree (University + Master's) had a significantly lower perception of the didactic advantages of simulations (B = −0.942, p = 0.04), when compared to those holding a second degree (University + PhD).

Regarding age-related perceptions, teachers who strongly disagreed or were undecided about the statement 'You have to be young to learn how to use simulations' rated the didactic advantages of simulations significantly lower by 8.172 and 7.924 units, respectively (p = 0.038 and p = 0.046). This model explained 95.8% of the total variability in the dependent variable (adjusted $R^2 = 0.646$).

Concerning technological infrastructure, teachers in schools where computers and projectors were available in only a few classrooms reported a lower perception of the ease of use and integration of simulations (B = −0.649, p = 0.01), compared to those in schools where these tools were widely available.

Lastly, teachers' comfort with using technology significantly influenced their perceptions of the ease of use and integration of simulations. Those who disagreed, were undecided, or agreed with the statement 'I feel very comfortable using such tools' reported lower ease of use perceptions by 2.725, 2.022 and 1.180 units, respectively (all p-values <0.05). This model explained 96.5% of the total variability in perceptions of ease of use and integration (adjusted $R^2 = 0.705$).

These findings highlight the importance of teachers' educational backgrounds and comfort with technology in shaping their attitudes towards integrating simulations in science teaching.

4.3 Teachers' Attitudes Towards Simulations: A Qualitative Analysis

A qualitative analysis was conducted on the open-ended questions of the questionnaire. The analysis focused on the responses of educators regarding the use or desire to use simulations in the sciences. It encompasses positive, negative and neutral views, along with the motivations or barriers identified by the educators.

First, the responses were coded, and the emerging thematic categories from the answers were analysed. Codes were assigned to each category, with subcategories included where necessary. Additionally, each educator's responses were classified within the corresponding categories (Table 4). As the open-ended questions were not mandatory, only a fraction of the participants completed them.

Table 4 Summary of the coding table of responses to the question: 'In summary, I would/would not continue using simulations for Science because..'

Topic	Code	Responses
Perceptions of the educational value of simulations	A1/A2	Teachers 1, 2, 7, 9, 10, 17, 18, 19, 30, 36, 38, 49, 51, 53, 56, 66, 74, 80, 84, 98, 114, 138, 139 (A1)
		Teachers 43, 78, 125 (A2)
Problems and limitations in the use of simulations	B1	Teachers 43, 61, 69, 107, 117, 139
Benefits for the learning environment	C1	Teachers 9, 19, 30, 49, 51, 53, 66, 74, 80, 84, 135
Time-saving and easy access	D1	Teachers 21, 56, 80, 81, 103
Contradictions in teaching methods	E1	Teachers 78, 125
Assessment of the quality and usefulness of simulations	F1	Teachers 62, 139

4.3.1 Willingness to Continue Using Simulations in Science Teaching

Teachers' willingness to continue using simulations (or not) and their reasoning were documented in the first open-ended question.

A. Perceptions Regarding the Educational Value of Simulations.

Simulations are valued for their capacity to enhance comprehension and facilitate the visualisation of physical concepts that may otherwise be difficult for students to grasp through traditional teaching methods. Educators generally recognise the potential of simulations to make abstract phenomena more accessible and to promote a deeper understanding of scientific principles.

Teachers who hold positive perceptions of simulations view them as valuable tools that can significantly improve student comprehension. These educators believe simulations enrich the teaching process by providing students with opportunities for active engagement. For example, Teacher 9 notes, 'Simulations help us understand physical concepts better, especially when we cannot experiment directly in the classroom'. Similarly, Teacher 19 highlights, 'They are very useful for visualising phenomena that are difficult for students to follow using traditional methods'.

In contrast, some teachers express reservations about the application of simulations in the educational process. Those with negative perceptions often cite technical, pedagogical or personal concerns. Teacher 78 asserts, 'No matter how well-designed the simulation is, it cannot replace experimental teaching'. Teacher 125 echoes this sentiment, stating, 'I would not continue using them because they do not promote experiential teaching, which has better learning outcomes'. These teachers argue that simulations, while useful, cannot substitute the hands-on, immersive learning experiences that live experiments offer.

B. Challenges and Limitations in the Use of Simulations.

One significant barrier to the widespread use of simulations in schools is the lack of appropriate infrastructure and equipment, such as computers, software and access to reliable technology. Many teachers face limitations in integrating simulations into their lessons due to these constraints. Teacher 61 explains, ‘We don’t always have the ability to use simulations since our school does not have the necessary equipment’. Moreover, the preference for live experiments or the need for tangible materials often supersedes the advantages of simulations. Teacher 107 notes, ‘Simulations are useful, but they are limited by the need for materials and real experiments for more authentic learning’.

C. Benefits for the Learning Environment. Despite these challenges, simulations are appreciated for their ability to engage students and make the learning process more dynamic. Teachers observe that simulations help create excitement and foster a sense of enthusiasm among students, making them more active participants in their education. According to Teacher 49, ‘Students are more excited about simulations and actively participate in the learning process’. Furthermore, simulations offer students the opportunity to practice concepts repeatedly, without the constraints of time or material availability. Teacher 66 highlights, ‘With simulations, students can repeat the activities as many times as they want, which helps them better understand the concepts’.

D. Time Savings and Easy Access. Another advantage of simulations is the time they save. Teachers can complete multiple examples in a short period, allowing them to cover more material within a limited timeframe. Additionally, simulations are easily accessible at any time, without requiring specialised materials or equipment. Teacher 80 notes, ‘With simulations, we can show students many experiments in just a few minutes without requiring extensive preparation’. Teacher 103 further emphasises the practicality of simulations, stating, ‘Using simulations does not require special equipment, making them easy to use in every lesson’.

E. Contradictions in Teaching Methods. Despite the clear benefits, many educators continue to prefer physical hands-on experiments as the most effective method of teaching, especially in the context of Physical Sciences. Teachers frequently express the need for a balance between simulations and traditional experimental methods. Teacher 78 argues, ‘Simulations are good, but they cannot replace the real experience of a live experiment’. Similarly, Teacher 125 advocates for a balanced approach, saying, ‘There must be a balance between simulations and live experiments for students to have a complete experience’. These educators maintain that hands-on experiments provide a level of engagement and understanding that simulations cannot replicate.

F. Evaluating the Quality and Usefulness of Simulations. Simulations are generally valued for their quality and effectiveness in the classroom, yet teachers acknowledge that for simulations to be truly effective, they must be adapted to meet the specific needs of their curriculum. Teachers report that while simulations are beneficial, they require proper integration into the teaching process to be fully useful. Teacher 62 remarks, ‘I believe simulations are useful, but they need to be adapted to meet the needs of our teaching’. Teacher 139 further stresses the importance of evaluation, stating, ‘There needs to be greater evaluation of simulations to ensure they are fully suitable for the classroom’.

G. Reasons for Not Using Simulations in Teaching Sciences. Several reasons emerged for teachers' reluctance to adopt simulations in their teaching of sciences. One primary reason is the preference for experiential learning, particularly through physical experiments. Educators who favour this approach emphasise the importance of hands-on, direct interaction with scientific phenomena as essential to student understanding. Teacher 90 asserts, 'For a student to understand and love Chemistry, they must hold a test tube, see a gas being produced, smell it, and even burn their fingers from the exothermic reaction... Chemistry is a real laboratory, not a virtual one'. Teacher 118 similarly reflects, 'Theory becomes more understandable to students, but the fact that they cannot use the equipment makes them feel uncomfortable'.

Another reason for the non-use of simulations is teachers' difficulty or lack of familiarity with technology. Some educators report feeling insecure or uncertain about using simulation tools, which contributes to their reluctance to incorporate them into their lessons. Teacher 48 comments, 'I have difficulty with technology', while Teacher 130 expresses hesitation, stating, 'I struggle and am hesitant. I'm retiring soon'. These teachers cite the challenges posed by technology as a significant barrier to the integration of simulations into their classrooms.

4.3.2 Teachers' Familiarity with Simulations in Science Teaching

Teachers' familiarity with using simulations (or not) and their reasoning were documented in the second open-ended question. Despite the challenges, there is a growing interest among teachers in learning and adopting simulation tools. Some educators are eager to explore new technologies, believing that their proper use can lead to significant advancements in teaching and learning. Teacher 27 expresses enthusiasm for new tools, saying, 'I would like to because the correct use of technology can bring leaps forward for humanity'. Teacher 37 similarly notes, 'I would like to get to know these tools because I believe they enhance teaching, help students with photographic memory, and sometimes make the lesson more interesting'. However, other teachers remain uncertain or hesitant about adopting these tools. These educators may express reluctance due to personal challenges with technology or a lack of confidence in their ability to incorporate simulations effectively. Teacher 42 offers a neutral response, saying, 'Neither yes nor no', while Teacher 71 directly expresses reluctance: 'I wouldn't want to'.

5 Discussion and Conclusions

The results of the study suggest that the science curriculum in high schools should be revised, as the pressure of the syllabus and time constraints leave little room for experimentation and innovation in teaching, enhancing the development of negative attitudes towards simulations. Moreover, it was somewhat unexpected that several teachers still employ a contrasting, rather than complementary (Papalazarou et al. 2024), approach regarding the use of interactive simulations versus traditional hands-on experiments.

Furthermore, technological infrastructure and the availability of interactive whiteboards have a positive influence on teachers' willingness to use simulations. Specifically, the presence of interactive whiteboards in more classrooms and stable internet connectivity enhances teachers' positive attitudes towards using digital teaching tools. The accessibility of appropriate technological equipment in schools facilitates the use of digital media in teaching (Ouahi et al. 2022). The availability of interactive whiteboards, in particular, has been recognised as critical for creating interactive and active learning environments (Mahdi et al. 2018). A curriculum reform planned for implementation in 2025 across subjects in Greek schools aims to develop textbooks and, simultaneously, linked digital objects, such as physics simulations, thereby fostering their wider adoption by teachers (Demkanin and Sands 2023).

Nevertheless, there is a clear need for continuous and specialised professional development courses in ICT training to overcome the resistance teachers have towards using digital tools and simulations, as well as promoting a complementary approach alongside the traditional physical lab. Targeted training at the B level provides the necessary pedagogical foundation for teachers to realise the benefits of simulations and develop a positive attitude towards their use in teaching. These conclusions align with other studies that emphasise the significance of training, evidence-based technological resources tied to the curriculum and classroom dynamics for the successful adoption of simulations in science education (Hsu et al. 2018; Hu et al. 2021; Ouahi et al. 2022; Demkanin and Sands 2023).

Although our research is rooted in the national context, we argue that these findings may be relevant worldwide, as they highlight a broader lack of understanding of the affordances of interactive simulations.

6 Limitations

This study provides an initial overview of the application of simulations in science education, from which implications for future development, practice, and research have been derived. However, two limitations must be acknowledged: the sample size and the potential bias introduced by the voluntary nature of teacher participation. The relatively small sample size may limit the generalisability of the findings. Additionally, it can be inferred that teachers who already utilise simulations in their teaching may be more motivated to participate in the study, which could lead to an overrepresentation of such teachers. As a result, the data may not fully capture the perspectives of those who do not use simulations, potentially skewing the interpretation of the findings.

References

Bell RL, Smetana LK (2008) Using computer simulations to enhance science teaching and learning. In: Bell RL, Gess-Newsome J, Luft J (eds) Technology in the Secondary Science Classroom. NSTA Press, pp 23–32

Bo WV, Fulmer GW, Lee CKE, Chen VDT (2018) How do secondary science teachers perceive the use of interactive simulations? The affordance in Singapore context. J Sci Educ Technol 27:550–565. https://doi.org/10.1007/s10956-018-9744-2

Creswell JW (2012) Educational research: planning, conducting, and evaluating quantitative and qualitative research, 4th edn. Pearson Education

De Jong T, Van Joolingen WR (1998) Scientific discovery learning with computer simulations of conceptual domains. Rev Educ Res 68:179–201. https://doi.org/10.3102/00346543068002179

Demkanin P, Sands D (2023) The development of experimental skills, the role of digital technologies and multimedia in physics teacher education. In: Challenges in Physics Education, pp 65–76

Finkelstein ND, Adams WK, Keller CJ et al (2005) When learning about the real world is better done virtually: a study of substituting computer simulations for laboratory equipment. Phys Rev Spec Top Phys Educ Res 1:1–8. https://doi.org/10.1103/PhysRevSTPER.1.010103

Hsu H-Y, Wang S-K, Coster D (2018) New literacy implementation. In: K-12 STEM education, IGI global, pp 526–546

Hu D, Yuan B, Luo J, Wang M (2021) A review of empirical research on ICT applications in teacher professional development and teaching practice. Knowl Manag E-Learning An Int J 13:1–20. https://doi.org/10.34105/j.kmel.2021.13.001

Jean Justice L, Ritzhaupt AD (2015) Identifying the barriers to games and simulations in education. J Educ Technol Syst 44:86–125. https://doi.org/10.1177/0047239515588161

Kay RH (1993) An exploration of theoretical and practical foundations for assessing attitudes toward computers: the computer attitude measure (CAM). Comput Human Behav 9:371–386. https://doi.org/10.1016/0747-5632(93)90029-R

Kostaki SM, Kalogiannakis M (2019) How do primary teachers deal with the digital learning objects of Photodentro for teaching natural sciences? (in Greek). Open Educ J Open Distance Educ Educ Technol 15:160–183. https://doi.org/10.12681/jode.20888

Lefkos I, Psillos D, Hatzikraniotis E (2011) Designing experiments on thermal interactions by secondary-school students in a simulated laboratory environment. Res Sci Technol Educ 29:189–204. https://doi.org/10.1080/02635143.2010.533266

Mahdi K, Laafou M, Janati-Idrissi R (2018) The impact of continuous distance training on teachers of physics in computer simulation software. J Educ Teach Trainers 9:27–41

Ndihokubwayo K, Uwamahoro J, Ndayambaje I (2020) Effectiveness of PhET simulations and YouTube videos to improve the learning of optics in Rwandan secondary schools effectiveness of PhET simulations and YouTube videos to improve the learning of optics in Rwandan secondary. 7295. https://doi.org/10.1080/18117295.2020.1818042

Ouahi MB, Lamri D, Hassouni T, Al Ibrahmi EM (2022) Science teachers' views on the use and effectiveness of interactive simulations in science teaching and learning. Int J Instr 15:277–292. https://doi.org/10.29333/iji.2022.15116a

Papadakis S, Kalogiannakis M (2020) A research synthesis of the real value of self-proclaimed Mobile educational applications for young children. https://services.igi-global.com/resolvedoi/resolve.aspx?doi=10.4018/978-1-7998-1486-3.ch001. IGI Global Scientific Publishing, pp 1–19

Papalazarou N, Lefkos I, Fachantidis N (2024) The effect of physical and virtual inquiry-based experiments on students' attitudes and learning. J Sci Educ Technol 33:349–364. https://doi.org/10.1007/s10956-023-10088-3

Perkins KK, Moore E, Chasteen SV (2015) Examining the use of PhET interactive simulations in US college and high school classrooms. In: 2014 physics education research conference proceedings. American Association of Physics Teachers, pp 207–210

Psillos D (2023) The role and impact of virtual laboratories in physics teaching and learning: a synthesis of literature. In: The international handbook of physics education research: teaching physics. AIP Publishing LLCMelville, New York

Rodgríguez CL, Mula-Falcón J, Segovia JD, Cruz-González C (2021) The effects of Covid-19 on science education: a thematic review of international research. J ofTurkish Sci Educ 18:26–45. https://doi.org/10.36681/tused.2021.70

Smetana LK, Bell RL (2012) Computer simulations to support science instruction and learning: a critical review of the literature. Int J Sci Educ 34:1337–1370. https://doi.org/10.1080/09500693.2011.605182

Stinken-Rösner L (2020) Simulations in science education–status Quo. Prog Sci Educ 3:26–34. https://doi.org/10.25321/PRISE.2020.996

Wieman CE, Adams WK, Loeblein P, Perkins KK (2010) Teaching physics using PhET simulations. Phys Teach 48:225–227. https://doi.org/10.1119/1.3361987

MIX
Papier aus verantwortungsvollen Quellen
Paper from responsible sources
FSC® C105338

If you have any concerns about our products,
you can contact us on
ProductSafety@springernature.com

In case Publisher is established outside the EU,
the EU authorized representative is:
Springer Nature Customer Service Center GmbH
Europaplatz 3, 69115 Heidelberg, Germany

Printed by Libri Plureos GmbH
in Hamburg, Germany